# Digital Communication for Agricultural and Rural Development

This volume presents insights on the challenges of digital communication and participation in agricultural and rural development. The COVID-19 pandemic has revealed that digital technology and mediated participation is more important and essential in managing ongoing communication for development projects than ever before. However, it has also underscored the various challenges and gaps in knowledge with digital participatory practices, including the further exclusion of marginalized groups and those with limited access to digital technology. The book considers how the concept of participation has been transformed by the realities of the pandemic, reflecting on essential principles and practical considerations of communication for development and social change, particularly in the context of global agriculture and food security, the well-being of rural communities, and evolving environmental challenges, such as climate change. In gathering these insights, this volume highlights lessons for the future of participatory development in communication for development and social change processes. This volume will be of great interest to students and scholars of agricultural and rural development, communication for development, digital communication, and sustainable development more broadly.

**Ataharul Chowdhury** is an Associate Professor in the School of Environmental Design and Rural Development, University of Guelph and the President of the Association for International Agricultural and Extension Education. He has led various collaborative projects with national, regional, and international partners to facilitate sustainable agriculture and rural development in remote and resource-poor communities. His research focuses on topics like agricultural advisory services and knowledge mobilization, climate change, communication and innovation studies, digital development, technology adoption, and misinformation.

**Gordon A. Gow** is Director of the Media and Technology Studies unit at the University of Alberta, where he is cross-appointed with the Department of Sociology. He has led various collaborative research projects in digital leadership and literacy with community-engaged projects, including a technology stewardship initiative with Canadian and international partners.

# Earthscan Food and Agriculture

For more information about this series, please visit: www.routledge.com/books/series/ECEFA/

# Digital Communication for Agricultural and Rural Development

Participatory Practices in a Post-COVID Age

**Edited by**
**Ataharul Chowdhury and Gordon A. Gow**

Routledge
Taylor & Francis Group
LONDON AND NEW YORK

earthscan
from Routledge

First published 2025
by Routledge
4 Park Square, Milton Park, Abingdon, Oxon OX14 4RN

and by Routledge
605 Third Avenue, New York, NY 10158

*Routledge is an imprint of the Taylor & Francis Group, an informa business*

*British Library Cataloguing-in-Publication Data*
A catalogue record for this book is available from the British Library

*Library of Congress Cataloging-in-Publication Data*
Names: Chowdhury, Ataharul, editor. | Gow, Gordon A., editor.
Title: Digital communication for agricultural and rural development :
participatory practices in a post-COVID age / edited by Ataharul
Chowdhury and Gordon Gow
Other titles: Earthscan food and agriculture
Description: New York, NY : Routledge, 2024 | Series: Earthscan food and
agriculture | Includes bibliographical references and index
Identifiers: LCCN 2024005482 (print) | LCCN 2024005483 (ebook) |
ISBN 9781032252087 (hardback) | ISBN 9781032252094 (paperback) |
ISBN 9781003282075 (ebook)
Subjects: LCSH: Communication in agriculture—21st century. | Communication
in rural development—21st century. | Digital communications—21st century.
Classification: LCC S494.5.C6 D56 2024 (print) | LCC S494.5.C6 (ebook) |
DDC 630.1/4—dc23/eng/20240410
LC record available at https://lccn.loc.gov/2024005482
LC ebook record available at https://lccn.loc.gov/2024005483

ISBN: 9781032252087 (hbk)
ISBN: 9781032252094 (pbk)
ISBN: 9781003282075 (ebk)

DOI: 10.4324/9781003282075

Typeset in Times New Roman
by codeMantra

# Contents

# Acknowledgments

Writing a book is a journey that involves the support and encouragement of numerous individuals and organizations. We are immensely grateful to all those who have contributed to the creation of this work.

First and foremost, we would like to express our heartfelt gratitude to the Social Science and Humanities Research Council (SSHRC) of Canada for their generous funding through the Connection Grant for the Global Networks Forum on Communication for Agriculture, Rural Development, and Environment. This initiative laid the foundation for the ideas presented in this book, which fostered a global exchange of knowledge on the evolving role of communication and digital technology in development and social change.

We would also like to express my appreciation to the Rural Communication Group of the International Association of Media and Communication Research (IAMCR) for their invaluable network support and small grants that played a crucial role in shaping the content of this book.

The challenges posed by the remote communication of the COVID-19 pandemic served as a catalyst for the ideas presented in the book. The resilience and adaptability demonstrated through the digital participation of various stakeholders of the SSHRC-funded initiative during these times have inspired much of the content. We are grateful to Dr. Helen Hambly, Dr. Katherine Reilly, Dr. Ricardo Ramírez, Dr. Sarah Cardey, Dr. Stella Tirol, Fred Campbell, and Dr. Habib M. Ali, for their support and sharing their valuable insights during webinars, which have inspired and shaped the content of the book.

A special thanks to the School of Environmental Design and Rural Development, Collaborative Change Communication for Development (CCComdev), Association for International Agricultural and Extension Education (AIAEE), and the Canadian Communication Association (CCA) for their organizational support in organizing the series of webinars and discussions. Their commitment to the advancement of knowledge in these fields has been instrumental.

We extend our deepest appreciation to the chapter authors, each of whom brought unique and valuable insights—whether practical or theoretical—to this book. Your contributions have enriched the content and provided diverse perspectives on the evolving role of communication in development.

We acknowledge J. Lynn Fraser for her copy editing service.

Last but not least, our sincere appreciation to the anonymous reviewers who provided constructive feedback, contributing to the refinement of the book's content.

This book would not have been possible without the collective efforts and support of these individuals and organizations. Thank you for being an integral part of this journey.

# Contributors

**Abdul-Rahim Abdulai** is currently a Postdoctoral Fellow of Political Economy of Food Systems Transformation with The Alliance Bioversity and CIAT on the CGIAR project on Sustainable Healthy Diets through Food Systems Transformation. His research centers on rural/agricultural change through the lenses of political economy, social practices, and socio-technical transitions.

**Murat Akcayir** is a Research Associate at the University of Alberta. His research interests focus on the intersection of digital literacy, telecommunications policy, and the use of technology for pedagogical purposes. He worked on a large-scale DigitalNWT project and is currently involved in a research project on telecommunication services in northern Canada.

**Isabel Pavez Andonaegui** is an Associate Professor in the School of Communication, University of Los Andes. She is a Principal Researcher in the Millennium Nucleus MEPOP and IMHAY. She has participated in numerous research projects on digital inclusion in vulnerable populations and currently leads a project on digital inequality in rural populations.

**Maria Bakardjieva** is a Professor and Chair in Communication and Media Studies, University of Calgary. Her research examines the social construction of communication technologies and the use of digital media in various cultural and practical contexts with a focus on user agency, critical reflexivity, and emancipation. She engages in knowledge mobilization and community outreach.

**Julia Bello-Bravo** is an Assistant Professor in the Department of Agricultural Sciences, Education and Communication, Purdue University, and co-founder and co-director of Scientific Animations Without Borders. Her research interests lie at the intersection of effective communication and education, specializing in informal education and communication strategies to reach low- or non-literate learners in developing and developed countries.

**Samantha Blais** is a PhD candidate at the University of Alberta. Her research focuses on intersections among settler colonialism, development, the environment, and health within Indigenous landscapes with Indigenous understandings of health and well-being, and the ways settler colonial development projects ignore or impact Indigenous health and well-being.

**Ataharul Chowdhury** is an Associate Professor in the School of Environmental Design and Rural Development, University of Guelph and the President of the Association for International Agricultural and Extension Education. He has led various collaborative projects with national, regional, and international partners to facilitate sustainable agriculture and rural development in remote and resource-poor communities. His research focuses on topics like agricultural advisory services and knowledge mobilization, climate change, communication and innovation studies, digital development, technology adoption, and misinformation.

**Teresa Correa** is a Full Professor in the Faculty of Communication and Literature, Universidad Diego Portales. She is Director of the research center CICLOS and the Alternate Director of the Millennium Nucleus NUDOS. Her research, published in over 50 articles and book chapters, focuses on digital inequality, media sociology, gender and minority representation, communication, and health.

**Uvasara Dissanayeke** is a Senior Lecturer in the Department of Agricultural Extension at the University of Peradeniya, Sri Lanka. Dr. Dissanayeke has extensive experience studying agricultural digitalization in Sri Lanka and teaching courses on technology and development in conjunction with the Postgraduate Institute of Agriculture (PGIA). Her publications include studies on the use of m-Learning and social media in agriculture education.

**Uduak Edet** is a doctoral student in the School of Environmental Design & Rural Development, University of Guelph, Canada, where she provides support in research, scientific writing, data reporting, and visualization. Her research focuses on interdisciplinary approaches that reduce the impact and improve resilience to climate change. She is interested in the influence of information disorder on climate change adaptation practices within the agri-food sector as well as the adoption of digital agricultural tools in rural communities.

**Catalina Farías** is a PhD student in Northwestern University's Media, Technology, and Society Program. Her research focuses on digital inclusion in vulnerable populations, access and use of new technologies, networked social movements, and gender dynamics. Her international projects were centered on feminist digital activism, and her Chilean projects focused on digital inclusion and technology use.

**Wayne Ganpat** is a Professor of Agricultural Extension. He is retired Dean (2021) of the Faculty of Food and Agriculture at the University of the West Indies (UWI), Trinidad. He is widely published with over 100 scholarly articles in international journals and has co-authored seven books on various topics, such as agricultural extension, ICTs, and climate change. He also served as president of the Association for International Agricultural and Extension Education (AI-AEE), for the 2019–2020 period.

**Leanne Goose** is a master's student in Community Engagement program in the School of Public Health, University of Alberta. She has degrees in Arts and Cultural Management and Professional Communications. She is the President of the Northern Arts and Cultural Centre, a member of the Indigenous Juno Committee, and an advocate in Indigenous arts and traditional knowledge.

**Gordon A. Gow** is Director of the Media and Technology Studies unit at the University of Alberta, where he is cross-appointed with the Department of Sociology. He has led various collaborative research projects in digital leadership and literacy with community-engaged projects, including a technology stewardship initiative with Canadian and international partners.

**Byron Hauck** is a Professor at Okanagan College, Department of Communication, and Term Lecturer in the School of Communication, Simon Fraser University. He investigates social productivity of listening combined with intercultural insights on mutual incompleteness of cultures to facilitate cultural translation. He focuses on historically peripheralized peoples to amplify their academic and public voices, and to dewesternize communication theory.

**Khondokar Humayun Kabir** is a Banting Postdoctoral Fellow in the School of Environmental Design and Rural Development at the University of Guelph. His research revolves around capacity development and extension. Recently, he has been involved in a project concerning climate change misinformation and tactics to tackle it.

**James Kamuye Kataru** is a computer professional, blogger, and content creator at www.kataruconcepts.com. He is a content manager for Scientific Animations Without Borders in Africa. He is also a social development advocate and heads a social media-based network of experts, farmers, health workers, and extension service providers.

**Judy Lawry**, until recently, was teaching in the Master of Communication Department in the School of Media and Communications at the Royal Melbourne Institute of Technology. Her specialty is strategic communications and organizational leadership. She is currently consulting in community engagement within local government and is still affiliated with RMIT.

**Anne Namatsi Lutomia** is a Postdoctoral Research Associate in the Department of Entomology, Purdue University, with Scientific Animations Without Borders. Her research focuses on scientific knowledge production and networking, science communication, learning technologies and adult learning, and centers on developing effective scientific communication educational content for communities to learn using mobile phones in local languages.

**Noel Iminza Lutomia** has a Bachelor's of Science in Business Management from Masinde Muliro University and a master's degree in project management from Nairobi University. She works with HIV/AIDS clients to provide access health services through digital health outreach, psychosocial support, prompt communication on their appointments, and adherence to medication.

**Linje Manyozo** is a communication for development scholar and practitioner based in the School of Media and Communication, College of Design and Social Context, RMIT University. His praxis has involved collaborating with communities to understand how media and communication breaks down structures and institutions of inequality.

**Fally Masambuka-Kanchewa** is an Assistant Professor of Agricultural Communication at Iowa State University. Her research focuses on exploring the role of communication as a science for understanding people's behaviors and amplifying voices of the marginalized. The goal of her research is to identify ways for incorporating needs of various stakeholders.

**Rob McMahon** is an Associate Professor in media and technology studies and political science at the University of Alberta and co-director of DigitalNWT project. It focuses on digital literacy and research on digital technologies in the Northwest Territories. His research focuses on development, adoption, and broadband and Internet technologies use by rural, Northern, and Indigenous communities.

**Michael B. McNally** is an Associate Professor in the Faculty of Education, University of Alberta. His research spans the areas of rural connectivity, radio-spectrum management, digital literacy, and open education. He is a contributor to the Opening Up Copyright Instructional Module Series, and the 3D Heart Project.

**John W. Medendorp** is an Associate Director of the Urban Center, Department of Entomology, Purdue University. He has managed and supported capacity development projects for USAID and USDA in 40 countries and in all of Feed the Future's target countries. His research focuses on Agricultural Education and Training learning system development.

**Kyle Napier** is a doctoral student of Education Policy Studies at the University of Alberta. He is a board member with Native Land Digital and is an adjunct faculty member with the University of Victoria and several other post-secondary institutions. He is a member of the Northwest Territory Métis Nation and is an academic, researcher, and language revitaliationist.

**Trang Pham** (PhD in Communication Studies, University of Calgary) freelances as a co-lead, user experience research, at IncluCity and as a researcher at non-profit organizations. The projects she has been working on bring voices of various groups of users to service providers to improve the users' digital experiences.

**Barry R. Pittendrigh** is the Director of Purdue University's Urban Center. His research centers on molecular mechanisms behind pesticide resistance, structural and functional genomics of lice, termite genomics, and creation of cowpea pest management systems in Africa. He is co-founder and co-director of Scientific Animations Without Borders, a research initiative for disseminating scientific information.

**Ricardo Ramírez** is an Adjunct Professor in Capacity Development and Extension in the School of Environmental Design and Rural Development, University of Guelph. His current work includes the design and implementation of utilization-focused and developmental evaluation approaches; capacity development in evaluation and communication, multi-stakeholder, and collaborative approaches for natural resource management; and participatory planning in complex adaptive contexts.

**Jeet Ramjattan** is an experienced Agricultural Extension agent with 25 years of experience in Trinidad & Tobago. He holds a PhD in Agricultural Extension from the University of the West Indies, blending theoretical insight with practical proficiency. Dedicated to advancing agriculture and sustainable development, Jeet excels in e-Extension, rural advisory services, and climate-smart agricultural techniques, contributing to operational efficiency and sustainable growth.

**Katherine Reilly** is an Associate Professor in the School of Communication, Simon Fraser University, Vancouver, Canada. Works at the intersection of social change and digital transformation where she studies how data and information systems mediate social relations. Her most recent project explored 'citizen data audits' as a means to build data literacy.

**Mary T. Rodriguez** is an Associate Professor at Ohio State University. She focuses on supporting communities in change processes at individual, household, and community levels. She strives to develop contextual, research-based solutions to build sustainable and resilient communities through exploration of behavior change and leadership development.

**Kevin Zhu** has a Bachelor's in Computing Science from the University of Alberta. He is currently pursuing a Master's degree in Advanced Computing at Tsinghua University. His research interests include the use of technology to address digital literacy.

# Acronyms and abbreviations

## A

| | |
|---|---|
| ADS | Anglican Development Services |
| AEAS | agricultural extension and advisory service |
| AES | agricultural extension services |
| AI | artificial intelligence |
| AIAEE | Association for International Agricultural and Extension Education |
| AIS | agricultural innovation systems |
| APC | Association for Progressive Communications |
| ASM | alternative social media |

## C

| | |
|---|---|
| C4D | communication for development |
| CBPR | community-based participatory action research |
| CCComdev | Collaborative Change Communication for Development |
| CCA | Canadian Communication Association |
| CDSC | College of Design and Social Context |
| CE | Community engagement |
| CEO | Chief Executive Officer |
| CfSC | Communication for Social Change |
| CIAT | International Center for Tropical Agriculture |
| CIRA | Canadian Internet Registration Authority |
| CoP | communities of practice |
| COVID | coronavirus disease |
| COVID-19 | coronavirus disease |
| CSM | corporate social media |

## D

| | |
|---|---|
| DA | digital agriculture |
| DAES | Department of Agricultural Extension Services |
| DE | digital engagement |
| DEAP | Developmental Educational Assistance Program |

| | |
|---|---|
| DET | digital extension tools |
| DIA | digitalization in agriculture |
| DLT | Digital Learning Team |
| DOI | Diffusion of Innovation |
| DP | digital platforms |
| DTPs | digital transformation pathways |

**E**

| | |
|---|---|
| EAS | extension and advisory services |
| ECLAC | Economic Commission for Latin America and the Caribbean |

**F**

| | |
|---|---|
| FAO | Food and Agriculture Organization |
| FAO-UN | United Nations Food and Agriculture Organization |
| FFS | Farmer Field Schools |
| FRT | Farm Radio Trust |

**G**

| | |
|---|---|
| GAP | Guide to Agricultural Production |
| GFRAS | Global Forum for Rural Advisory Services |
| GHGs | greenhouse gas emissions |
| GIS | geographic information system |
| GM | genetically modified |
| GMO | genetically modified organisms |
| GSMA | Global System for Mobile Communications Association |

**I**

| | |
|---|---|
| IAMCR | International Association of Media and Communication Research |
| IAP | individual action plan |
| IAP2 | International Association of Public Participation |
| ICs | informational capabilities |
| ICT4D | information and communication technology for development |
| ICTs | information and communication technologies |
| IFC | International Finance Corporation |
| IP | Innovation Project |
| IPT | Internet Performance Test |
| ISPs | Internet service providers |
| IT | information technology |
| IVR | Interactive Voice Response |
| IXPs | Internet Exchange Points |

**J**

JETI                      Joint Education Training Initiative

**L**

LocNet                    Local Access and Community Networks

**M**

MaFAAS                    Malawi Forum for Rural Advisory Services
MRT                       Media Richness Theory
MSN                       mobile social networking

**N**

NACDC                     National Content Development Committee
NGO                       non-government organization
NRM                       natural resource management
NWT                       Northwest Territories

**O**

OCAP                      ownership, control, access and possession
OCoPs                     Online Communities of Practice
ODK                       Open Data Kit
OECD                      Organization for Economic Cooperation and Development

**P**

PAR                       participatory action research
PPP                       power, profit, and propaganda

**R**

RAPID                     Responsive, Adaptive, Participatory Information
                          Dissemination
RMIT                      Royal Melbourne Institute of Technology

**S**

SARS-CoV-2                Severe acute respiratory syndrome coronavirus 2
SAWBO                     Scientific Animations Without Borders
SAWBO RAPID               Scientific Animations Without Borders
SDGs                      sustainable development goals
SM                        social media
SMC                       social media channels
SMGs                      social media groups

| | |
|---|---|
| SMPs | social media platforms |
| SMS | Short Message Service |
| SPSS | Statistical Package for Social Services |
| SSHRC | Social Science and Humanities Research Council |

## T

| | |
|---|---|
| TaCA | technology augmented capability approach |
| TAM | Technology Acceptance Model |
| TPB | Theory of Planned Behavior |
| TRA | Theory of Reasoned Action |
| TSP | technology stewardship program |
| TSTP | technology stewardship training program |
| T&T | Trinidad and Tobago |

## U

| | |
|---|---|
| UN | United Nations |
| UNESCAP | United Nations Economic and Social Commission for Asia and the Pacific |
| UNESCO | United Nations Educational, Scientific and Cultural Organization |
| UN-FAO | United Nations Food and Agriculture Organization |
| UNICEF | United Nations International Children's Emergency Fund |
| UofA | University of Alberta |
| USAID | United States Agency for International Development |
| USB | Universal Serial Bus |

## V

| | |
|---|---|
| VAGO | Victoria Auditor-General Office |

## W

| | |
|---|---|
| WiFi | wireless fidelity |
| WIL | work-integrated learning |

## Numbers

| | |
|---|---|
| 2G | second-generation cellular network |
| 3D | three dimensional |
| 3G | third-generation cellular network |

# 1 Introduction

*Ataharul Chowdhury and Gordon A. Gow*

Participatory research and development processes are central to the fields of communication for development (C4D) and communication for social change (CfSC). Scholars have explored various theoretical approaches, practice-based studies, and knowledge bases over the past decades including works by Gumucio-Dagron (2001), Gumucio-Dagron and Tufte (2006), Manyozo (2012), Thomas (2014), Wilkins et al. (2014) Melkote and Steeves (2015), Servaes (2018), and Waisbord (2018). These works emphasized the importance of participation as an integral component of inclusive development processes. In recent years, many successful practices, principles, and experiences have emerged, thanks in part to digital tools, such as video, radio, Internet, collaborative, and social media. These tools have provided a means for mediated participation, enabling marginalized groups to have a voice and make choices in development projects.

The COVID-19 pandemic has highlighted the significance of digital technology and mediated participation in managing ongoing C4D projects (Baffoe-Bonnie et al., 2021; Davis et al., 2021; Food and Agricultural Organization [FAO], 2020; Grove et al., 2020; Lendel & Meier, 2020). However, it has also exposed the challenges and gaps in knowledge associated with digital participatory practices. For instance, the swift shift to online communication has made it difficult to engage marginalized groups and those with limited access to digital technology, leading to further exclusion (Chadwick et al., 2022; Lamiño Jaramillo et al., 2022). As remote communication became the norm during the pandemic, researchers and practitioners had to rethink the concept of participation and explore new ways of engaging communities within C4D and CfSC initiatives.

This edited volume draws on contributions from participants in the Global Networks Forum on Communication for Agriculture, Rural Development, and Environment. The forum was a three-year-long project, funded by the Social Science and Humanities Research Council (SSHRC) of Canada, between 2020 and 2022, to foster a global exchange of knowledge on the evolving role of communication and digital technology in development and social change. The editors have invited formal contributions from a series of webinars with established and emerging scholars and practitioners of communication for development and social change from regional and global networks, such as Collaborative Change Communication for Development (CCComdev), the Rural Communication Group of

DOI: 10.4324/9781003282075-1

the International Association of Media and Communication Research (IAMCR), Association for International Agricultural and Extension Education (AIAEE), and Canadian Communication Association (CCA).

Initially, the Global Networks Forum was to focus on various activities related to knowledge mobilization, including academic practice networks and partnerships. However, with the onset of the COVID-19 pandemic in 2020, there were forced lockdowns and remote-only interactions that were to have a profound effect on participatory research and development. A new reality had taken hold, in which many of the long-held assumptions and established practices in the fields of C4D and CfSC had to be re-examined. As such, it was felt there would be value in compiling a collection of discussions that reported on pandemic-related insights set against some of the persistent challenges for participatory research in agriculture and rural development.

Communication for development and social change originated in the early twentieth century with a critical and reflective view that communication and media play a vital role in addressing fundamental human needs like agriculture, natural resources, and rural and Indigenous community development. Today, analysts continue to remind us that one must not lose sight of the rural and agricultural roots of communication for development and social change, as evidenced in practice-informed scholarship (e.g., Bessette, 2018; Quarry & Ramírez, 2009, 2012; Servaes & Lie, 2014; Van de Fliert et al., 2017). Thus, it is crucial to continue the conversation on the founding networks of academia and practice that intersect in communication for development and social change, particularly in the context of global agriculture and food security, the well-being of rural communities, and evolving environmental challenges, such as climate change.

As the field has evolved, the values and principles associated with participatory practice and human rights continue to remain central to it and may even be considered by some as non-negotiable foundations for the field (Melkote & Steeves, 2015). Nevertheless, researchers and their community-based counterparts are exploring new approaches to address sustainability challenges. Although some Sustainable Development Goals (SDGs), such as gender equality (SDG 5) and partnerships (SDG 17), have touched on communication for development, there is a need to be creative in how one is responsive to these foundational values. This could involve integrating and co-producing knowledge in new ways (as suggested by Chowdhury, 2018). To achieve this, one must reimagine "development" as an evolutionary adaptive process, rethink communication as a complex adaptive system, and embrace cross-disciplinary perspectives that build connections to other related fields. Ultimately, one must challenge development communication agencies to transform into convivial institutions responsive to the communities they serve (as recommended by Acunzo et al., 2014; Chowdhury, 2018; FAO, 2006; Jacobson, 2016; Lennie & Tacchi, 2013; Patel, 2018).

Going forward, a significant concern is how to engage in participatory processes that benefit marginalized communities, while factoring in both the opportunities and risks of mediated communication in an era of widespread and sometimes controversial digital transformation (Bradshaw & Howard, 2019). In this regard, the matter of communication for social change and development continues to be extensively debated, with discussions that confront us with questions about coerced participation and participation resistance (Manyozo, 2012; Muturi & Mwangi, 2009; Ramírez et al., 2015; Waisboard, 2014). For some, the focus of communication for social change and development is not the degree of involvement but rather the action and ownership of decision-making processes (Acunzo et al., 2014; Lennie & Tacchi, 2013). The rapid growth of social and Internet-based collaborative media has opened up opportunities for fostering collective innovation and amplifying voices at the grassroots level. Recent studies have shown, in fact, that social and digital media can be beneficial for strengthening agricultural and rural communities (e.g., Chowdhury & Hambly, 2013; Kaushik et al., 2018; Molyneaux et al., 2014; Munthali et al., 2021; Riley & Robertson, 2021). However, there are also ever-present risks of these mediated processes further marginalizing communities. This has been noted by researchers such as Albu and Etter (2018), Couldry et al. (2018), Hintz et al. (2018), and Narula et al. (2018). For instance, media corporations and non-government organizations have driven the delivery of community programs through satellite-based technologies that tend to reinforce centralized communications processes. The sustainability of community radio is uncertain in light of shrinking resources and competition from the private sector (Ali & Conrad, 2015).

Another area of concern is the spread of misinformation through these platforms, which can have negative consequences such as creating political polarization (Azzimonti & Fernandes, 2018) and hindering disease prevention and community health (Sommariva et al., 2018; Tan & Bigman, 2020). This was particularly evident during the COVID-19 pandemic, when remote engagement and online participation became the norm, as stakeholders searched for information on various topics (Albu & Etter, 2018). For example, over the past few decades, agri-food systems have faced numerous controversies regarding climate change, animal welfare, environmental stewardship, technology in food production, and digitization of food chains (Eenennaam, 2022; Ji et al., 2019; Klerkx, 2021; Maye et al., 2021; Treen et al., 2020). These issues have led to polarizing opinions, which can spread through online communities. To combat this threat, agricultural, rural, and environmental groups will require an enhanced capacity to participate in processes whereby they can critically examine the information and data shared in the digital context (Chowdhury & Firoze, 2022; Cunliffe-Jones et al., 2021; Mabaya et al., 2021; Tan & Bigman, 2020).

The need to respond to threats of misinformation is one aspect of examining how participatory processes are essential in building the capacity of rural and marginalized communities to leverage new and emerging communication technologies to improve their livelihoods (Chowdhury et al., 2023; Esteban-Navarro et al., 2020; Gow et al., 2020; Klerkx, 2021; Seto, 2022). There is also a growing awareness of the risk that digital platforms pose for state and corporate surveillance, censorship, and social control (Castells, 2015; Couldry et al., 2018) in contrast to their potential for supporting participation in critical debates and discussions affecting agriculture and rural development. Several studies (Beaunoyer et al., 2020; Lamiño Jaramillo et al. 2022; Mathrani et al., 2021) have highlighted these concerns. Though digital tools have rapidly emerged for communication, participation, and creating virtual communities of practices (CoPs), they have not necessarily created democratic socio-cultural fabrics. In fact, digital platforms have shown a tendency toward centralized control and influence (Baruah & Borborah, 2021; De Blasio et al., 2020). Rather than creating more opportunities for participation, new digital platforms may, in fact, result in greater market concentration that marginalizes the voices of smallholder farmers and their communities. Therefore, it is important to be cautious while welcoming digital agriculture and environmental services (Birner et al., 2021; Kruk et al., 2021).

The contributions of this book are set against the backdrop of the aforementioned theoretical and critical discussions. The book's intent is to offer a collection of theoretical perspectives and practical experiences relevant to participatory research in a post-pandemic world. Contributions to this edited collection are organized into three sections, beginning with reflections on the participatory paradigm in C4D and CfSC. In this section, the contributors examine the concept of "participation" while also considering the impact of digital transformation on equity and inclusivity. In the second section, two complementary critical perspectives on the unintended effects of depending on commercial social media as a platform for participatory engagement with agricultural and rural development stakeholders are presented. The third section offers a set of contributions that include findings from empirical studies and related experiences on digitally mediated participation, including several that respond directly to disruptions caused by the COVID-19 pandemic. In the concluding chapter, the lessons learned for participatory methods and practices are reflected upon as we look ahead to what we anticipate will be an increasingly digitally mediated world for those working in C4D and CfSC.

## Part 1: Reflecting on the participatory paradigm in C4D

In Part 1, the book has included contributions that focus on the historical and theoretical debate of participation rooted in the C4D paradigm. The part also focuses on conceptualizing participation in a digital setting.

In his analysis in Chapter 2, Ricardo Ramírez delves into the theoretical approaches of C4D and explores the concept of participation. The chapter emphasizes the need to understand four types of access, including mental, material, skill, and use, and the meaning of effective use of digital media in understanding participation in the Digital Age. It identifies four discourses of participation in C4D, such as stakeholder discourse emphasizing top-down interaction, networked discourse focusing on more collaborative interaction, mobilization discourse highlighting bottom-up efforts, and oppositional discourse that stresses conflict among actors. Chapter 2 highlights the potential of social and internet based collaborative media to amplify marginalized voices and encourage collective innovation, while also acknowledging the systemic risks it poses for further exclusion. Using two cases, one before the advent of social media and broadband and one involving digital technologies, the chapter identifies key components of defining participation in the Digital Age that Ramírez contends are "non-negotiable." Ramírez stresses the emergent and unexpected outcomes that new media can produce, especially in terms of negative consequences, and calls for a greater understanding of the complex adaptive system between new media and different stakeholders.

In Chapter 3, Katherine Reilly approaches the topic of participation as a means of exerting power through a multidisciplinary lens. Although numerous studies have tackled the subject of power within the context of social change, there is a lack of exploration of power dynamics within digital spaces in C4D literature. This chapter offers a compelling and critical analysis of communication's role in participatory development in the new era of platform capitalism. By presenting participation as a powerful tool, Reilly provides readers with a useful framework for understanding the motivations behind participating in digital spaces. The concept of "digital participation" as a rational, informed, and independent process is a valuable contribution to both C4D and CfSc scholarship. The chapter distinguishes between the dominant development narrative and digital participation, helping readers comprehend the critical perspectives that have remained understudied in digital spaces. Using the concepts of datafication, logistics brokerage, and heteromation, the chapter argues that the adoption and use of digital technologies and platforms can limit or even eliminate meaningful contributions to decision-making and social change. Reilly describes how platformization is reshaping social relations and economic structures, transforming the subjectivities of social agents irreversibly. The chapter advocates for new participation approaches that empower alternative visions for using digital platforms to advance development objectives that align with equity goals.

## Part 2: Critical perspectives on digital participation

The chapters in Part 2 examine contemporary and emerging challenges, including but not limited to the ethical and long-term implications of social media for participation in online communities. It considers increasingly prevalent concerns about the propagation of misinformation with user-generated content while also drawing attention to questions of Internet governance that influence possibilities for digital self-determination.

Chapter 4 introduces the idea of information disorder in relation to agricultural and rural communities. Khondokar Kabir, Ataharul Chowdhury, and Uduak Edet argue that this phenomenon poses significant challenges for farmers and the public alike, as they are constantly exposed to misleading information in the Digital Age. Using historical and current cases and examples, the authors discuss why scholars and practitioners of C4D and CfSc need to engage with this issue. Since a vast array of their service targets are agricultural and rural communities, it is crucial to address this problem. The authors also introduce the concept of Power, Profit and Propaganda (PPP), which can help shed light on the complex web of misinformation within the agri-food domain. The PPP framework is founded on the interplay between power dynamics, economic incentives, and deliberate propaganda. It provides insight into the underlying forces driving the spread of false information. In addition, the authors examine various factors that amplify misinformation, including social media platforms, scientific literacy gaps, confirmation bias, political interests, the media landscape, and globalization. According to the authors, understanding these contextual nuances will help stakeholders develop strategies to combat information disorder and foster a more discerning and informed society in the Digital Age.

In Chapter 5, Gordon A. Gow argues that agricultural extension and rural development should expand their use of ICT tools and platforms beyond corporate social media as part of a digital diversification strategy. The author uses the concept of path dependency to demonstrate how organizations have become increasingly "locked-in" to corporate social media platforms as a primary means of communication. He points out some of the implications that this form of path dependency has for participation in online spaces, suggesting it may be an emerging form of digital colonialism. In order to disrupt the path dependency dilemma, he points to the possibilities for non-commercial social media alternatives as a digital diversification strategy. However, Gow argues that implementing such a strategy will be challenging for extension services, and they will need to consider research alliances, ICT infrastructure, service resources, and digital skills training.

## Part 3: Practices, experiences, cases, and tools

In Part 3, we turn to examine empirical studies on digital participation and the digitalization processes in rural and agricultural communities. The chapters in this section cover a wide range of topics, including how C4D activities have continued during the COVID-19 pandemic, lessons learned from these experiences, and potential changes to participatory practices as we move beyond the pandemic. The cases and experiences presented come from both developed and developing regions covering Australia, Canada, Chile, China, Ghana, Kenya, Malawi, Trinidad and Tobago, and Vietnam. They shine a light on how rural and agricultural communities utilize Internet technologies and participate in the Digital Age, taking into account social, cultural, political, and economic factors that shape the process. Taken together, these chapters invite readers to reflect on how participation actually happens in a digital context, highlighting the diverse experiences and perspectives of those who are often left out of the conversation.

In Chapter 6, Rob McMahon, Michael B. McNally, Samantha Blais, Murat Akcayir, Kyle Napier, Leanne Goose, and Kevin Zhu share their insights on how they had to adapt their approach to participatory research when confronted with travel restrictions imposed by pandemic. In doing so, they describe their community-led initiatives for digital inclusion and literacy within rural Indigenous communities in the Northwest Territories, Canada. This chapter emphasizes the importance of trust and reciprocity among team members and participants when implementing digital tools for mediated participation. Reflecting on the interview process with local researchers and digital innovators, their chapter suggests that while digital participation can be effective, it remains crucial to have opportunities for in-person contact.

Chapter 7 features a research study conducted by Jeet Ramjattan, Ataharul Chowdhury, and Khondokar Kabir on how agricultural extension communities in

Trinidad and Tobago interact on Facebook and WhatsApp platforms. The study focused on the information shared, misconceptions, and concerns that arise among members. Findings revealed that most respondents were worried about the amount of false information they encountered in social media groups. This was due to individuals sharing materials that aligned with their own opinion or ideology, or following misleading posts from family members. These concerns highlight the challenges of promoting inclusive participation as people tend to join groups of like-minded individuals to avoid misinformation.

Gordon A. Gow, Ataharul Chowdhury, Uvasara Dissanayeke, Jeet Ramjattan, and Wayne Ganpat in Chapter 8 discuss the important role played by agricultural extension and advisory services in fostering inclusive digital transformation pathways in Sri Lanka as well as Trinidad and Tobago. The authors suggest that by "reframing" "digital agriculture" as "digitalization in agriculture," we can imagine a variety of transition pathways that are more inclusive of smallholders and marginalized groups. The chapter proposes a logic model for micro-level innovation that combines situated learning theory and a normative framework for human development, based on the Capabilities Approach. This model is used to guide the design and evaluation of a participatory research education program that has been implemented and evaluated with extension practitioners in the two countries. Finally, the chapter reports on the initial results of the program while highlighting its immediate impacts, and identifies important insights for future research.

In Chapter 9, Isabel Pavez and Catalina Farías investigate the impact of school closures caused by the pandemic on Internet access, usage, and perception among 12 dyads of rural mothers and their seventh-grade children in rural Chile. These interviews were conducted one year after school closure due to COVID-19 and shed light on the role of digital technology in a new learning situation. The findings imply that children tend to feel more powerful due to device accessibility, and that their mothers do not play a significant role in addressing any problems that may arise. The authors argue that the two groups have different perspectives on the primary purpose of smartphones, but they both agree on the educational function of computers.

In Chapter 10, Abdul-Rahim Abdulai takes a mixed-method approach by conducting surveys of 386 farmers and organizing three focus groups in Northern Ghana. The chapter delves into the obstacles that inhibit farmers from utilizing digital advisories, as well as the political and economic implications of these impediments. According to the farmers surveyed, the primary barrier is the lack of digital literacy. Since most smallholders have limited formal education, they lack the basic skills required to perform online tasks such as browsing the Internet, following Interactive Voice Response (IVR), using a computer, social media, and sending text messages. The chapter contends that the absence of (digital) literacy is a result of political and economic realities, which will continue to limit smallholders' adoption of digital ICTs in future. The argument presented suggests that the current initiatives may result in increased structural inequality, and thus recommend implementing strategies that align with the capabilities and circumstances of smallholders in Ghana. The authors advocate for integrating digital literacy training and low-tech methods alongside the digitalization process.

In Chapter 11, Fallys Masambuka-Kanchewa and Mary T. Rodriguez share insights gathered from interviews with agricultural extension and rural development experts in Malawi. They applied Rogers' diffusion of innovation (DOI) and media richness theories to show that digital tools are currently being used to merely transfer information to farmers, without engaging them in the creation of innovative practices or taking ownership of their own development. The authors emphasized the need for co-creation and co-designing messaging to make digital tools more effective in addressing the needs of the most vulnerable people. To support this, they provided a framework that incorporates Rogers' DOI decision-making stages and reciprocal communication outlined in the media richness theory.

In Chapter 12, Maria Bakardjieva, Isabel Pavez, Teresa Correa, and Trang Pham reflect on separate studies conducted in rural areas of Canada, Chile, and Vietnam. They introduce the concept of "Internet use genres" and trace these to the participants' relevance systems and socio-biographical situations. These situations are shaped by the geographical, socio-economic, and cultural conditions of their communities. Despite distinct social and political profiles in each case, there are also commonalities observed among rural residents in each country. This suggests that Internet use genres respond to recurring local situations and are highly malleable and subject to user agency. Understanding the evolution of these genres provides researchers and policymakers with a more nuanced understanding of Internet adoption processes in rural communities. It also opens up possibilities for exploring the inclusion of end-users' choices in technological development.

Since the 1950s, agricultural extension has prioritized developing effective communicators. This involved North American and South East Asian land grant universities experimenting with more interactive and participatory communication methods with farming communities. Chapter 13 of this book addresses how higher education teaches effective community engagement officers in the post-COVID-19 digital era. Judy Lawry and Linje Manyozo discuss their experiences in building a Community Engagement Course and Stream as part of the Master of Communication program at Royal Melbourne Institute of Technology (RMIT). The chapter also reviews community engagement pedagogy across universities in Australia and New Zealand, finding a cohesive body of knowledge that guides community engagement teaching, paving the way for standardization in community engagement pedagogy. The authors define community engagement as a participatory and interventionist form of communication. Overall, the chapter focuses on institutionalizing community engagement teaching at the graduate level and its impact on agriculture and rural extension pedagogy and practice.

Byron Hauck posits, in Chapter 14, that participation when viewed through the lens of universalized actors, tends to perpetuate a masculine perspective on knowledge, action, and accomplishment. To delve deeper into this issue, the chapter takes a transcultural political economy of communication approach to examine masculinity as displayed through participation in village development in China, with a particular focus on marginalized perspectives. Through this investigation, the chapter seeks to shed light on the insights and potentials that such perspectives can bring to the table. The chapter also explores the prospects of finding a sense of belonging in today's society based on socialist ideals, as seen from the vantage point of rural men who have been marginalized by China's current urban and market-centric development trajectory. It compares the disparities between communist state mechanisms and digital platforms in observing, interpreting, and acting on the practices of the populace, with the aim of supporting the argument that participation theories should not generalize opportunities for those with dominant power to practice. Rather, participation theories should aid in specifying what enables individuals, from the "ideal" individual to government officials and technology entrepreneurs, to act. This helps to dismantle assumptions of free autonomy and facilitate a more equitable and just society.

A research study on the effectiveness of the Scientific Animations Without Borders (SAWBO) program in Kenya is presented in Chapter 15 by Anne Namatsi Lutomia, Julia Bello-Bravo, Noel Iminza Lutomia, John W. Medendorp, and Barry R. Pittendrigh. SAWBO creates educational video animations, translates them into local languages, and distributes them free of charge through digital devices, including mobile phones. The study focused on the use of SAWBO videos by a WhatsApp network that engages with rural farmers across the country. Using adult learning theory, the chapter explores the interactions between the educational content in the animated videos and the WhatsApp network. The study uses a case study method to capture the experiences of the WhatsApp network members and the outcomes of their interactions with the animated videos. Finally, the chapter concludes with a discussion of the lessons learned from the program and its implications for future practices in the use of digital platforms for educating agriculture and rural communities.

## References

Acunzo, M., Pafumi, M., Torres, C. S., & Tirol, M. S. (2014). *Communication for rural development source book*. Food and Agriculture Organization.

Albu, O. B., & Etter, M. A. (2018). How social media mashups enable and constrain online activism of civil society organizations. In J. Servaes (Ed.), *Handbook of communication for development and social change* (pp. 891–909). Springer.

Ali, C., & Conrad, D. (2015). A community of communities? Emerging dynamics in the community media paradigm. *Global Media and Communication, 11*(1), 3–23. https://doi.org/10.1177/1742766515573970

Azzimonti, M., & Fernandes, M. (2018). Social media networks, fake news, and polarization. NBER Working Paper Series no. 24462. National Bureau of Economic Research. http://www.nber.org/papers/w24462

Baffoe-Bonnie, A., Martin, D. T., & Mrema, F. (2021). Agricultural extension and advisory services strategies during COVID-19 lockdown. *Agricultural & Environmental Letters, 6*(4). https://doi.org/10.1002/ael2.20056

Baruah, C., & Borborah, P. (2021). A re-imagined community: Pandemic, media, and state. *India Review, 20*(2), 176–93. https://doi.org/10.1080/14736489.2021.1895562

Beaunoyer, E., Dupere, S., & Guitton, M. J. (2020). COVID-19 and digital inequalities: Reciprocal impacts and mitigation strategies. *Computers Human Behavior, 111*, 106424. https://doi.org/10.1016/j.chb.2020.106424

Bessette, G. (2018). Participatory development communication and natural resources management. In J. Servaes (Ed.), *Handbook of communication for development and social change* (pp. 1141–1154). Springer.

Birner, R., Daum, T., & Pray, C. (2021). Who drives the digital revolution in agriculture? A review of supply-side trends, players and challenges. *Applied Economic Perspectives and Policy, 43*(4), 1260–85. https://doi.org/10.1002/aepp.13145

Bradshaw, S., & Howard, P. N. (2019). The global disinformation order: 2019 Global inventory of organised social media manipulation. The Computational Propaganda Project at the Oxford Internet Institute, University of Oxford. https://demtech.oii.ox.ac.uk/wp-content/uploads/sites/93/2019/09/CyberTroop-Report19.pdf

Castells, M. (2015). *Networks of outrage and hope: Social movements in the Internet Age.* Wiley.

Chadwick, D., Ågren, K. A., Caton, S., Chiner, E., Danker, J., Gómez-Puerta, M., Heitplatz, V., Johansson, S., Normand, C. L., Murphy, E., Plichta, P., Strnadová, I., & Wallén, E. F. (2022). Digital inclusion and participation of people with intellectual disabilities during COVID-19: A rapid review and international bricolage. *Journal of Policy and Practice in Intellectual Disabilities.* https://doi.org/10.1111/jppi.12410

Chowdhury, A. (2018, November 28–29). *Rethinking capacity development for communication for development.* [Paper presentation.] Roundtable discussion, development communication theorizing: New perspectives, innovative practices, and future directions, University of The Philippines, Los Baños, Philippines.

Chowdhury, A., & Firoze, A. (2022, April 4–7). Combatting online agriculture misinformation (OAM): A perspective from political economy of misinformation. In *Proceedings of 2022 Conference of the Association for International Agricultural and Extension Education*, Thessaloniki, Greece.

Chowdhury, A., & Hambly Odame, H. (2013). Social media for enhancing innovation in agri-food and rural development: Current dynamics in Ontario, Canada. *Journal of Rural and Community Development, 8*(2), 97–119.

Chowdhury, A., Kabir, K. H., Abdulai, A.-R., & Alam, M. F. (2023). Systematic review of misinformation in social and online media for the development of an analytical framework for agri-food sector. *Sustainability, 15*(6), 4763. https://doi.org/10.3390/su15064753

Couldry, N., Rodriguez, C., Bolin, G., Cohen, J., Volkmer, I., Goggin, G., Kraidy, M., Iwabuchi, K., Qiu, J. L., Wasserman, H., Zhao, Y., Rincón, O., Magallanes-Blanco, C., Thomas, P. N., Koltsova, O., Rakhmani, I., & Lee, K.-S. (2018). Media, communication and the struggle for social progress. *Global Media and Communication, 14*(2), 173–91. https://doi.org/10.1177/1742766518776679

Cunliffe-Jones, P., Diagne, A., Finlay, A., Gaye, S., Gichunge, W., Onumah, C., Pretorius, C., & Schiffrin, A. (2021). *Misinformation policy in Sub-Saharan Africa: From laws and regulations to media literacy.* University of Westminster Press.

Davis, K. S., Archibald, A., Thomas, Grove, B., & Babu, S. (2021). Organizational innovation in times of crises: The case of extension and advisory services. *Journal of International Agricultural and Extension Education, 28*(1), 6–14. https://doi.org/10.5191/jiaee.2021.28101

De Blasio, E., Kneuer, M., Schünemann, W., & Sorice, M. (2020). The ongoing transformation of the digital public sphere: Basic considerations on a moving target. *Media and Communication, 8*(4), 1–5. https://doi.org/10.17645/mac.v8i4.3639

Eenennaam, A. V. (2022). The history and impact of misinformation in the agricultural sciences. CAS initiative on conspiracy, misinformation, and the infodemic. https://mediaspace.illinois.edu/media/t/1_k0b1s1mh

Esteban-Navarro, M.-Á., García-Madurga, M.-Á., Morte-Nadal, T., & Nogales-Bocio, A.-I. (2020). The rural Digital Divide in the face of the COVID-19 pandemic in Europe— Recommendations from a scoping review. *Informatics*, 7(4), 54. https://doi.org/10.3390/informatics7040054

Food and Agricultural Organization (FAO) (2006). Communication for sustainable development. Background paper for World Congress on Communication for Development, 2006. Communication for Development Group, FAO.

Food and Agricultural Organization (FAO, April 17). Morris, A. (Ed.) (2020). *Extension and advisory services: At the frontline of the response to COVID-19 to ensure food security.* https://www.fao.org/3/ca8710en/CA8710EN.pdf

Gow, G., Chowdhury, A., Ramjattan, J., & Ganpat, W. (2020). Fostering effective use of ICT in agricultural extension: Participant responses to an inaugural technology stewardship training program in Trinidad. *The Journal of Agricultural Education and Extension*, 26(4), 335–50. https://doi.org/10.1080/1389224x.2020.1718720

Grove, B., Archibald, T., & Davis, K. (2020). Extension and advisory services: Supporting communities before, during and after crises. https://globalagriculturalproductivity.org/wp-content/uploads/2019/01/International-Extension_2020_GAP.pdf

Gumucio-Dagron, A. (2001). *Making waves: Participatory communication for social change*. The Rockfeller Foundation.

Gumucio-Dagron, A., & Tufte, T. (Eds.) (2006). *Communication for social change anthology: Historical and contemporary readings*. Communication for Social Change Consortium.

Hintz, A., Dencik, L., & Wahl-Jorgensen, K. (2018). *Digital citizenship in a datafied society*. Polity.

Jacobson, T. L. (2016). Amartya Sen's capabilities approach and communication for Development and Social Change. *Journal of Communication*, 66(2016), 789–810. https://doi.org/10.1111/jcom.12252

Ji, J., Chao, N., & Ding, J. (2019). Rumor mongering of genetically modified (GM) food on Chinese social network. *Telematics and Informatics*, 37, 1–12. https://doi.org/10.1016/j.tele.2019.01.005

Kaushik, P., Chowdhury, A., Hambly Odame, H., & van Passen, A. (2018). Social media for enhancing stakeholders' innovation networks in Ontario, Canada. *Journal of Agricultural & Food Information*, 19(4), 331–53. https://doi.org/10.1080/10496505.2018.1430579

Klerkx, L. (2021). Digital and virtual spaces as sites of extension and advisory services research: Social media, gaming, and digitally integrated and augmented advice. *The Journal of Agricultural Education and Extension*, 27(3), 277–86. https://doi.org/10.1080/1389224x.2021.1934998

Kruk, S. R. L., Kloppenburg, S., Toonen, H. M., & Bush, S. R. (2021). Digitalizing environmental governance for smallholder participation in food systems. *Earth System Governance*, 10. Article no. 100125. https://doi.org/10.1016/j.esg.2021.100125

Lamiño Jaramillo, P., Boren-Alpizar, A., Morales Vanegas, S., & Millares-Forno, C. (2022). Training, trust, and technology: A mixed-methods study of Latin American extension workers' experiences during COVID-19. *Journal of International Agricultural and Extension Education*, 29(1), 40–56. https://doi.org/10.4148/2831-5960.1017

Lendel, N., & Meier, C. (2020). Responding in a time of crisis: Assessing extension efforts during COVID-19. *Advancements in Agricultural Development*, 1(2), 12–23. https://doi.org/10.37433/aad.v1i2.35

Lennie, J., & Tacchi, J. (2013). *Evaluating communication for development*. Routledge.

Mabaya, E., Nsofor, I. M., & Evanega, S. (2021). Battling misinformation wars in Africa: Applying lessons from GMOs to COVID-19. https://theconversation.com/battling-misinf ormation-wars-in-africa-applying-lessons-from-gmos-to-covid-19-156183

Manyozo, L. (2012). *Media, communication and development: Three approaches.* SAGE.

Mathrani, A., Sarvesh, T., & Umer, R. (2021). Digital divide framework: Online learning in developing countries during the COVID-19 lockdown. *Globalisation, Societies and Education, 20*(5), 625–40. https://doi.org/10.1080/14767724.2021.1981253

Maye, D., Fellenor, J., Potter, C., Urquhart, J., & Barnett, J. (2021). What's the beef?: Debating meat, matters of concern and the emergence of online issue publics. *Journal of Rural Studies, 84*, 134–46. https://doi.org/10.1016/j.jrurstud.2021.03.008

Melkote, S. R., & Steeves, H. L. (2015). *Communication for development: Theory and practice for empowerment and social justice.* Sage.

Molyneaux, H., O'Donnell, S., Kakekaspan, C., Walmark, B., Budka, P., & Gibson, K. (2014). Social media in remote First Nation communities. *Canadian Journal of Communication, 39*(2014), 275–88.

Munthali, N., Lie, R., Van Lammeren, R., Van Paassen, A., Asare, R., & Leeuwis, C. (2021). Intermediation capabilities of information and communication technologies (ICTs) in Ghana's agricultural extension system. *The African Journal of Information and Communication, 28*, 1–37. https://doi.org/10.23962/10539/32212

Muturi, N., & Mwangi, S. (2009). The theory and practice gap in participatory communication. *Journal of International Communication, 15*(1), 74–91. https://doi.org/10.1080/132 16597.2009.9674745

Narula, S., Rai, S., & Sharma, A. (2018). *Environmental awareness and the role of social media.* IGI Global.

Patel, F. (2018). Glocal development for sustainable social change. In J. Servaes (Ed.), *Handbook of communication for development and social change* (pp 501–517). Springer.

Quarry, W., & Ramírez, R. (2009). *Communication for another development: Listening before telling.* Zed Books.

Quarry, W., & Ramírez, R. (2012). The limits of communication. *Nordicom Review, 33*(Special Issue), 121–34. https://doi.org/10.2478/nor-2013-0030

Ramírez, R., Quarry, W., & Guerin, F. (2015). Can participatory communication be taught? Finding your inner phronēsis. *Knowledge Management for Development Journal, 11*(2), 101–11.

Riley, M., & Robertson, B. (2021). #farming365 – Exploring farmers' social media use and the (re)presentation of farming lives. *Journal of Rural Studies, 87*, 99–111. https://doi.org/10.1016/j.jrurstud.2021.08.028

Servaes, J. (2018). *Handbook of communication for development and social change.* Springer.

Servaes, J., & Lie, R. (2014). New challenges for communication for sustainable development and social change: A review essay. *Journal of Multicultural Discourses, 10*(1), 124–48. https://doi.org/10.1080/17447143.2014.982655

Seto, A. (2022). Hallway pedagogy and resources loss: Countering fake news in rural Canadian schools. In K. R. Foster & J. Jarman (Eds.), *The right to be rural* (pp. 51–68). University of Alberta Press.

Sommariva, S., Vamos, C., Mantzarlis, A., Đào, L. U-L., & Martinez Tyson, D. (2018). Spreading the (Fake) news: Exploring health messages on social media and the implications for health professionals using a case study. *American Journal of Health Education, 49*(4), 246–55. https://doi.org/10.1080/19325037.2018.1473178

Tan, A. S. L., & Bigman, C. A. (2020). Misinformation about commercial tobacco products on social media-implications and research opportunities for reducing tobacco-related health disparities. *American Journal of Public Health*, *110*(S3), S281–83. https://doi.org/10.2105/AJPH.2020.305910

Thomas, P. N. (2014). Development communication and social change in historical context. In K. G. Wilkins, T. Tufte, & R. Obregon (Eds.), *The handbook of development communication and social change* (pp. 7–19). Wiley & Sons.

Treen, K. M. d'I., Williams, H. T. P., & O'Neill, S. J. (2020). Online misinformation about climate change. *WIREs Climate Change*, *11*(5). https://doi.org/10.1002/wcc.665

Van de Fliert, E., Cooper, T., Sam, S., Cardey, S., Lie, R., Witteveen, L., Torres, C., Tirol, S., Dagli, W., Hambly, H., & Perez, M. (2017). *Inclusive rural communication services: Building evidence, informing policy*. Report. Food and Agricultural Organization of the United Nations. https://www.fao.org/3/i6535e/i6535e.pdf

Waisboard, S. (2014). The strategic politics of participatory communication. In K. G. Wilkins, T. Tufte, & R. Obregon (Eds.), *The handbook of development communication and social change* (pp. 145–67). Wiley & Sons. https://doi.org/10.1002/9781118505328

Waisbord, S. (2018). Family tree of theories, methodologies, and strategies in development communication. In J. Servaes (Ed.), *Handbook of communication for development and social change* (pp 93–132*)*. Springer,

Wilkins, K. G., Tufte, T., & Obregon, R. (Eds.) (2014). *The handbook of development communication and social change*. Wiley & Sons.

# Part 1

# Reflecting on the participatory paradigm in C4D

# 2 Non-negotiable components of participation in the digital age of communication for development and social change

*Ricardo Ramírez*

## Introduction

Marshall McLuhan's famous book titled *The Medium Is the Massage* was the result of a typographical error, one that the author immediately adopted because it fit his book's message. The question arises: Does being immersed or massaged make us real participants? The promise of social media is fading. Two newspaper headlines that appeared days apart in November 2021 illustrates this, "The more hate the more viral: How Twitter is plagued with hate with the rising of the ultra-right" (translated from Spanish) (Fanjul, 2021, n.p.), and "'I'm happy to lose £10m by quitting Facebook,' says Lush boss" (Wood, 2021, n.p.).

In the first, there is reference to the algorithms in social media that prioritize postings that capture attention to maximize user attention through 'likes' and directs users exclusively to like-minded opinions. The louder or more extreme the posting—the more 'likes' it captures while casting a shadow on differences of opinion and values. Now the ultraright in many Western societies are using the medium to increase polarization as opposed to the democratic promise of these platforms in their early days. This result is increased participation among the like-minded while antagonizing those with different views. This is exacerbated by the challenge of misinformation and disinformation, for instance, in climate change communications, and in pandemic communications about vaccinations (Conway, 2020; Treen et al., 2020; Wardle & Derakhshan, 2017).

The second article was about the pervasive pressure among youth to portray an image that is shaped by social pressures that lead to abuse and bullying. As the CEO (chief executive officer) of Lush noted, the algorithm's "generate addictive, mindless scrolling" (Wood, 2021, n.p.). From an ethical standpoint, this predicament reached the limit for this cosmetic corporation. Their concern was teenagers' mental health.

These are articles examples of the challenges to the meaning and significance of 'participation' in the digital age, with specific attention to rural and remote areas. Salemink et al. (2017, p. 360) observed that "The paradox is that rural communities are most in need of improved digital connectivity to compensate for their remoteness, but they are least connected and included." Klerkx (2021) expands on the digital and virtual spaces emerging as part of the agricultural service provision, from social

DOI: 10.4324/9781003282075-3

media to gaming and digitally integrated and augmented advice, where the value proposition remains known. There are concerns regarding the promises of technology in agriculture that relegate farmers to being a 'cog' in industrialized agriculture removing their power to choose how they wish to participate (Wang, 2020).

This chapter focuses on digital media where the user choses to engage as an active participant, with the expectation of foreseeable and positive outcomes in agricultural and rural development. As Conway (2020) emphasizes, communication involves cultural translation. This means that stakeholders must engage with each other to arrive at shared interpretations. This chapter revisits some of the foundations of participation with references to Communication for Development and Social Change concepts. The main contribution of this chapter is a set of non-negotiable components of participation and a critical lens to assess the potential of new technologies.

## Conceptual framework and relevant literature

### The meaning of access

Jan Van Dijk (2001) offers four different types of access that people can experience when attempting to use new media: mental, material, skills, and use.

### Mental access

Refers to anxiety or the motivation to use technology. In other words, either fearing it or wanting it. Some of us experience reservations, while others cannot wait to access a new technology. Behavioral scientists and marketers focus on awareness raising as the starting point for a shift in attitudes about a new medium.

Take the example of Ontario farmers in the late 1990s. They faced the end of publicly funded farm advisors or extension specialists and their replacement in the form of printed fact-sheets. Many struggled with the loss of the social interaction and the end of tailored and timely advice the advisors offered. These were all critical to farming decisions.

### Material access

Refers to having physical access to the hardware, be it privately owned or shared in a public library, tele-center, or private cybercafé. Globally, rural and remote households continue to lag behind urban ones in material access and affordability, due to their lower tele-density. The challenge of rural households left behind due to low bandwidth during the COVID-19 lockdown is a case in point demonstrating unequal material access, one that is associated with higher costs and poorer service.

### Skills access

Is concerned with the different types of training that are necessary to master a new technology. Instrumental skills are about operating a digital device whereas informational skills are about searching, selecting, processing information from a device and from a network or the Internet. Strategic skills are about integrating, valuing and applying information to ones' job or task. Only when these are successfully integrated can we talk about people using technology for different applications, or use access.

During the tele-center era, much of the training that was offered to rural and remote households centered on computer skills. These days, mobile devices including smartphones and tablets are owned individually. Training often happens across generations in a household. Van Dijk's model is 'recursive' in that it shows that a sequence of access and skills is needed for each new innovation. As each new technology arrives, some people often go through some of the same steps to become familiar with it and put it to effective use while others refrain as they do not find that the new technology adds value.

*Effective use*

In 2003, Gurstein published an article entitled "Effective use: A community informatics strategy beyond the digital divide." At the time, the digital divide was a trending issue, with concerns about the growing inequality in access to technology. The divide had many definitions and connotations, most of which remain a concern two decades later. This chapter's discussion focuses on 'effective use.' This is where Gurstein's work connects with the last stage of Van Dijk's model. Gurstein proposed that effective use might be defined as "The capacity and opportunity to successfully integrate ICTs into the accomplishment of self or collaboratively identified goals" (Gurstein, 2003, n.p.).

Gurstein argued that achieving effective use requires a number of conditions: carriage facilities, input / output devices, tools and supports, content services, service access / provision, social facilitation, and governance. Beyond these categories of access, he argued for more attention as to how such access could respond to broader concerns, strategies, and applications that would benefit individuals and communities. Gurstein advocated for participatory design where users, i.e., the demand side, and providers, i.e., the supply side, would agree on the following appropriate infrastructure, available and usable output devices, local availability of any peripherals (for example, a digital thermometer for an e-health application), appropriate and relevant content and language, service offerings that are suited to the users' needs, the provision of capacity development, and the appropriate structure of finance and governance. Gurstein (2003, n.p.) observed:

> What this means in practice is that for, for example, a community e-health application would be designed with the active participation of the local community including the local health care professionals and para-professionals and one would expect that the framework for their design activity would be something similar to what has been presented as the "effective use" framework, i.e. addressing from a local ICT application design perspective each of the identified categories in turn, and making provision for these in the ultimate application design.

Not surprisingly, Gurstein advocated for "effective use" initiatives in information and communication technology (ICT) to make use of participatory action research methodologies. Gurstein and others were inviting the participatory communication field (Bessette, 2004; Gumucio-Dagron, 2001; White, 1999; White et al., 1994) to join the information and communication technology for development (ICT4D) field. At the same time, reported field experiences confirmed the value of such approaches for 'effective use' in community development in Canada and globally (Hudson et al., 2017; Ramírez, 2001; Richardson, 1999).

## The dimensions of 'participation'

### Participation discourses in the field of ICT4D

Singh and Flyverbom (2016) proposed four types of discourses on participation in ICT4D efforts. The four types are based on two main domains (1) consensual (state-dominated) versus conflictual (societal pressures), and (2) structured (hierarchical) versus horizontal (grassroots):

1  **Stakeholder discourses.** Emphasize mainly top-down interactions and consensus among participants. They tend to be state-dominated, including World Bank projects. "At the practical level," as noted by Singh and Flyverbom (2016, p. 694), "this discourse, perhaps reflecting the bureaucratic jargon of international agencies, represents participation as a seemingly consensual activity."
2  **Networked discourses.** Focus on more horizontal interactions and collaborations based on shared interests, such as crowdsourcing, and other networks of service delivery in ICT4D projects.
3  **Mobilization discourses.** Highlight bottom-up efforts that involve contestation over the meanings and effects of participation such as the World Summit for Information Society and social movements' use of ICTs.
4  **Oppositional discourses.** Stress conflict among actors and grassroots and other bottom-up and agent-driven forms of participation examples of which include community radio and community content creation.

These four discourses are a useful anchor to our discussion in real-world situations where the forces of the market and government fail to provide adequate services for rural and remote communities (McMahon et al., 2021; Salemink & Strijker, 2018; Strover et al., 2021).

### Origins of the participation debate

Debates about 'participation' in international development began in the 1960s–1970s. By the 1980s some attention shifted with community participation becoming a hallmark of projects aiming at assisting the poor. According to Cornwall (2002), the rationale shifted from an initial focus on empowerment and capacity development to one emphasizing project efficiency and effectiveness. Since then, participation has been associated with decentralization efforts with more attention being given to beneficiaries as 'clients' who are receiving services. In the late 1980s and early 1990s, the literature on participation was associated with action research and social change movements (Selener, 1997). Practitioners of participatory research talked about making the voices of the poor heard (Chambers, 1997). Participation became a 'must have' approach. Recently, participation is associated with calls for the decolonization of aid and the localization of decision-making (McKegg, 2019).

One feature of the 'participation' phenomenon has been the creation of 'spaces and places' for interaction. For McGee (2002) these spaces are either offered by the powerful, such as by implementing organizations or generated from the bottom-up. Of interest to this chapter is the role of media in creating such spaces, especially to engage communities in rural and remote areas that are otherwise left at the margin.

In addition to the two types of spaces, there is a temporal dimension as some spaces are one-off consultations while others constitute regularized processes (Cornwall, 2002). Policy spaces are those instances in which intervention or events bring about new opportunities, reconfiguring relationships between actors, or bringing in new actors, and opening the possibility of new directions (Grindle & Thomas, 1991).

There is also a growing recognition that there are a number of challenges and paradoxes with participatory approaches and that we need to be more rigorous and transparent with the use of the terms and concepts (Kanji & Greenwood, 2001). Lotz (1998), for example, examines the use of the word 'community.' He observes, "Like every other word, 'community' has a history of effective use and simplistic misuse." In turn, Apthorpe (1997, pp. 53–54) notes the following:

> Participation, like community, 'Community' (and 'participation') are other sure-fire winning words ... living blameless lives of their own in language, policy and analysis of whatever hue .... It is the one type of term, along with, for example, 'cooperation' and 'participation' which has never been used in a negative sense ... One crucial characteristic of these sorts of keywords is that they do not require an opposite word to give or enhance their meaning. They acquire much of their winning warmth from their popular meanings in everyday usage. A further characteristic is that ... they are not ever put to serious empirical test—or if they are, and they fail, they continue to circulate in good currency nevertheless. The projects they herald may be evaluated, and whether they are winner or not is another matter.

Critics of participation argue that it has become a tyranny. Ironically, participatory approaches are seen as imposing decision-making and control as well as group dynamics and methods onto communities and groups. Critics like Cooke and Kothari (1998) argue that participatory development's tyrannical potential is systemic. They contend the problem goes beyond how the practitioner operates or the specificities of the techniques or tools employed. They argue 'participation' remains as a system of representation within a project. Rather than the incorporation of people's knowledge, they propose that participatory planning is mostly about the acquisition and manipulation of planning knowledge (Ramírez, 2008).

Critiques of participation are present across many elements such as watershed management (Rhoades, 1997, 1998), information systems (Heeks, 1999), participatory technology development (Biggs, 1995), participatory communication (Gumucio-Dagron, 2001), forest management (Hildyard et al., 1998a, 1998b; Van Dam, 2000; Vira et al., 1998), and gender and power (Mosse, 1993).

Hickey and Mohan (2004), in response to the *tyranny* critique, suggest that for participatory approaches to be transformative they would need to be explicit about their ideological and theoretical motivations. They added that a more radicalized citizenship must follow along with the fulfillment of associated rights.

*Qualifying the meaning of engagement and participation*

'Community engagement' has gained acceptance at the institutional level. Many institutions and practitioners are embracing the term and seeking ways of applying it to program requirements. Some call for caution as there is a difference between what is advocated for and what actually happens on the ground as there may be some endorsement of informal institutions while the actual focus remains on the formal (Cooke & Kothari, 1998). There is merit in the notion that as bureaucracies seek to adopt the terms in stakeholder discourse (Singh & Flyverbom, 2016), internal spaces for change emerge within those organizations that may have long-term consequences. As these approaches become mainstream, organizational response mechanisms eventually catch up. Take the example of how Nike's organizational learning went through different stages from a defensive stand to an active advocacy one as the sweatshop attacks 'matured' from an activist agenda to a consolidated, mainstream issue (Zadek, 2004).

*Definitions*

Kanji and Greenwood (2001, p. 7) observe that "definitions of participation range from assisting people to exercise their democratic rights to a means of obtaining views from different stakeholders." They offer this definition:

> Participation: enabling people to realize their rights to participate in, and access information relating to, the decision-making processes which affect their lives. Democratic institutions and access to information about governments' policies and performance are necessary to enable people to participate in the decisions that affect their lives. They also need to be able to form an organization, such as unions, women's groups or citizens' monitoring groups, to represent their collective interests.
>
> (DFID, 2000, p. 24, as quoted in Kanji & Greenwood, 2001, p. 9)

According to Kanji and Greenwood (2001), stakeholder engagement constitutes one of several participatory methods and can take place at different stages of a project:

* defining the agenda
* development of a proposal
* preparatory phase
* implementation
* analysis of results
* dissemination and action

These stages are worth keeping in mind in the context of Van Dijk's recursive model of access, and Gurstein's notion of 'effective use.' At each one of these projects' stages, there can be a different level of engagement. The most commonly referred to is the ladder of participation that was presented in the late 1960s. It suggested that participation ranged from manipulation all the way to citizen control (Arnstein, 1968). Since that time other ladders have been published, and Table 2.1 compares three of them (Ramírez, 2008).

Table 2.1 shows a common pattern from self-mobilizing at the top to totalitarian control at the bottom. A more recent variation gave more attention to the shifting roles and relationships between outsiders and local people who are able to gradually take ownership over a space or process, and a shift upward along the ladder toward more self-control (Chambers, 2005). The aforementioned direct our attention on the importance of local people and their organizations taking ownership over a process.

The idea of people and communities 'owning' a problem places importance on identifying stakeholders who are affected by an issue and can influence how it is addressed. The original reference to this concept comes from the business management literature where there has been recognition of the importance of consultation among stakeholders affected by an issue (Checkland & Scholes, 1990). It also emphasizes that those who 'own' a problem should be involved in resolving it (Ramírez, 2008).

*Table 2.1* Three ladders of participation

| *Arnstein (1968)* | *Pretty (1994)* | *Kanji and Greenwood (2001)* |
| --- | --- | --- |
| Citizen control | Self-mobilization | Collective action |
| Delegated power | Interactive participation | Co-learning |
| Partnership | Functional participation | Cooperation |
| Placation | Participation for material incentives | |
| Consultation | Participation by consultation | Consultation |
| Informing | Participation in information giving | |
| Therapy | Passive participation | Compliance |
| Manipulation | | |

A common pattern in rural and agricultural development has been for third parties to become intermediaries between those who own the problem and the government services or service providers that could help improve it (see the top-down discourses mentioned by Singh and Flyverbom [2016]). These mediating organizations have an important role to play, especially if they are committed to enabling the grassroots to take on an active role in each stage of project development. This challenge is as common to community development in developing countries as it is in industrialized settings (Lotz, 1998).

A case worth mentioning is Farm Radio International's Participatory Radio Campaign. Here farmers are involved in selecting the improvement or innovation to be featured, discussing the strength and weaknesses, making a decision about whether or not to take up the innovation, and providing advice during the adoption process (Hudson et al., 2017). In this instance, the process is facilitated by a non-government organization (NGO) and the medium is pre-determined.

### *Digital media—Emerging properties in complex systems*

Responding to community needs, the demand side, and matching it with accessible and affordable services, the supply side, has become more complex since the early attention to 'access.' For rural and remote communities, the success of mediating organizations that represented their interest has been documented, both in rural and remote Canada by Hambly and Rajabiun (2021) and Ramírez (2001), and internationally by Gumucio-Dagron (2001), Futch and McIntosh (2009), and Richardson et al. (2000). Mediating organizations lobbied governments to provide incentives for private carriers to reach into high cost serving areas. Today, the gap between rural and urban communities remains a challenge, something that the COVID-19 pandemic has made more urgent as remote education and work have become a necessity.

As technological platforms have become more sophisticated, their potential impact has become less predictable (Bar et al., 2000; Myers, 2004). This is caused in part by the fact that they affect multiple dimensions of every day life of individuals, organizations, and businesses (Totolo et al., 2015; Van Mele et al., 2010). As the technology unfolds, it creates new services that were unimagined before (Ramírez, 2003). The notion of unexpected outcomes is common in the literature on complex systems. There has been much exploration in the field of natural resource management that is relevant to ICTs. This notion makes multi-stakeholder engagement more challenging as different interests and expectations need to be reconciled, along with the notion that not all outcomes will unfold as expected.

There is much to learn from the field of natural resource management especially with regards to common property areas, conservation areas, and watersheds where there is a need to accommodate multiple interests. It follows that each stakeholder will defend his or her own interests and will interpret opportunities differently. The relationship among the different parties often shifts from periods of collaboration to periods of conflict (Ramírez, 1999; Singh & Flyverbom, 2016). Understanding the complexity of this situation is helped by referencing legal frameworks, natural sciences and social sciences research, as well as participatory planning and learning processes. This represents a significant challenge, hence the need for an epistemology that embraces pluralism (Anderson et al., 1998; Wollenberg et al., 2001).

In natural resource management (NRM), a new epistemology has emerged to add clarity to these features. It evolved into a range of action-research methodologies that include collaborative management (Borrini-Feyerabend, 1996), collaborative learning (Daniels & Walker, 2001), adaptive collaborative management (Buck et al., 2001; Röling & Jiggins, 1998), rapid appraisal of agricultural knowledge systems (Engel & Salomon, 1997), and linked local learning (Lightfoot et al., 2001a, 2001b).

The similarities between ICTs and NRM suggest that some of the lessons from natural resource management are applicable to ICT research. Planning is seen as an interactive process. The goals among the stakeholders are multiple and at times contradictory with outcomes that are only partly predictable. What has become evident in recent critiques of social media is that some of the negative outcomes were not foreseen. Gurstein's words about 'effective use' ring true, "The capacity and opportunity to successfully integrate ICTs into the accomplishment of self or collaboratively identified goals" (2003, n.p.). This means that stakeholders need to have continued engagement to co-manage and adjust new media as new unexpected outcomes become evident. Sadly, with much digital media this is not the case as users are seen as customers not as co-participants. Worse, customers are seen as 'currency' in this age of surveillance capitalism (Deibert, 2020).

In summary:

- There are four discourses in which to locate each context stakeholder, networked, mobilization, and oppositional discourses.
- Different types of access lead to effective use.
- There are many meanings of participation and their possible expression during the stages of a project or initiative.
- Multiple stakeholders are engaged with different goals leading to new media producing some outcomes that are emergent.
- The interactions between new media and different stakeholders are a complex adaptive system.

## Case studies

The following are two contrasting cases: one is from before social media and broadband access, while the other took place two decades after and includes the introduction of digital technology.

### Case 1

In the mid-1990s, a project by the Food and Agriculture Organization (FAO) of the United Nations in the Philippines introduced a participatory approach to agricultural communication and extension (Coldevin, 1995). Farmers were consulted in both the choice of agricultural technologies and the media to disseminate them. The farm communities organized to co-manage the media distribution system to the extent that it was still functioning a decade after the project ended (personal communication). A school on air approach using multiple media was implemented. A full case study was prepared showing pre- and post-campaign levels of awareness and adoption of the technologies. This case fits with the stakeholder discourse (as it was government and United Nations-led) with an intent to shift to a mobilization discourse.

**Different types of access leading to effective use.** The project engaged farmers in agricultural technology selection that led to a shared interest in accessing the extension and communication services. Farmers were able to select media combinations that they felt were appropriate, affordable, and accessible. For instance, they chose to use audio towers distributed throughout the community and they decided on a broadcasting schedule that enabled them to listen before heading out to the fields. Their engagement ensured that the different types of access to the media technology were confirmed.

**There are many meanings of participation and their possible expression during the stages of a project or initiative.** The level of participation during most of the project's stages was closer to the 'interactive' and 'functional' levels as shown in Table 2.1 as the project team was able to consult and respond to community priorities. For the national organization coordinating the project, this was a departure from previous top-down approaches that had been imposed on communities. As the farm communities decided to contribute funding for the generator and broadcast equipment, they took ownership over the sustainability of the media.

**Multiple stakeholders are engaged with different goals, leading to new media producing some outcomes that are emergent.** There were three main types of stakeholders involved in the farm communities; the public extension system (connected to the agricultural research system); and FAO advisors. While the farmers were active in the definition of agricultural topics and media selection, the public extension organization took the lead in packaging the communication materials. The project did not actively engage with other stakeholders such as those involved in marketing of inputs or agricultural produce, nor other rural residents.

**Interactions between new media and different stakeholders are a complex adaptive system.** The pre- and post-campaign data confirmed very positive outcomes in the adoption of the agricultural innovations. It was evident that the audio towers and broadcast system was operational at least a decade after the project ended (personal communication), and overall it was a success. However, in that encounter a decade later it became evident that there had been an unexpected outcome. The project had not analyzed land tenure during its assessment phase, and as it turned out, most of the land was rented and the landowners increased the rental price as soon as they noticed increases in productivity (Quarry & Ramírez, 2009).

## Case 2

A team of two human rights NGOs piloted a project to support health access for tea garden workers in Assam, India. The workers belonged to the indigenous Adivasi community that suffered from high rates of maternal and infant mortality with minimal access to legal and health services. A short message service (SMS) mobile technology was introduced to allow the women to report health rights violations, which would be linked to a mapping platform. As reported by Zaveri et al. (2016, p. 17), the project faced several challenges:

> While the geo-mapping of health rights violations was taking place, the tea garden owners and government were unaware of it. The women volunteers were afraid to report health rights violations as some of them were employed by the government (as nurse, etc.). Discussions with the [first NGO] staff indicated that some women volunteers were not texting the infringements, they faced problems in selecting the right code, and did not seem to be motivated. Many women had never reported while others were very active. It was also clear that without good community level data, [other NGO] would not be able to confront the government. The face-to-face discussions enabled a better understanding of the project under review, the challenges being faced and the context in which the App was being used.

With regards to the four discourses, this case belonged between the mobilization discourse (NGO-led process) with the intent to challenge the status quo as well as in the oppositional discourse.

**Different types of access lead to effective use.** It was evident that different types of access to the new media had not been explored. This was the result of inadequate pre-project consultations to understand media access preferences and barriers by the women that the project sought to support. There were unexpressed assumptions about how tea workers would access the technology. This, in turn, led to an initial lack of effective use. The barriers were identified and addressed when the project revisited its theory of change and changed its approach.

**There are many meanings and expressions of participation in a project or initiative.** The project's needs assessment had not consulted with the stakeholders sufficiently. It had not engaged with the women it sought to protect nor with the government authorities charged with supplying health services. A communication audit uncovered this weakness and this led to closer engagement.

**Multiple stakeholders mean there are different goals, leading to new media producing some outcomes that are emergent.** The project would have benefited from a stakeholder analysis to confirm who it needed to consult with, and what linkages and relationships needed to be strengthened. While at first there was an apparent contrast between tea workers' needs and the health authorities' services, the project ended up helping align them. This led the project to become a success.

**Interactions between new media and different stakeholders are a complex adaptive system.** Following an external review, the NGOs were able to put the project back on track. This included better communication with health authorities, addressing the access barriers among the tea workers, and press releases announcing the plan to establish a "Citizen Grievance Forum" to address maternal health violations. A video was produced and disseminated with testimonies about how the app was used as a tool to report such cases (Zaveri et al., 2016).

## Conclusion

The two cases show a contrast in the extent to which project stakeholders were consulted. However, one can argue that both cases share a problem in terms of weak or incomplete needs assessment. While one could argue that a more thorough needs assessment with more dedicated participation by affected stakeholders would have helped the last concepts, emergent properties and complex adaptive systems, it is suggested that this may not be a real solution.

The issue of consultation brings us back to the type of participation sought (Table 2.1). Singh and Flyverbom (2016) conclude that a distinction is necessary between consensus and conflict, what they refer to as "'cooperation without consensus'—organized and regularized interactions among actors with very different and often incompatible goals," and, they argue "that participation increasingly allows for difference and multiplicity, as well as experiments with procedures and forms of engagement beyond technocratic and expert-driven processes" (p. 701).

In the digital age, there has been a fascination with the promise of technology, for society as a whole and for those working in rural and remote community development. Projects such as the Assam example have often been led by organizations and individuals that are familiar with the technology. They have pushed ahead making assumptions about access and participation, and their expected outcomes have often not come to fruition. Despite these promises, digital extension tools (DET) continue to be introduced, and they have numerous problems (Coggins et al., 2022: Abtract):

> (unawareness of DET, inaccessible device, inaccessible electricity, inaccessible mobile network, insensitive to digital illiteracy, insensitive to illiteracy, unfamiliar language, slow to access, hard to interpret, unengaging, insensitive to user's knowledge, insensitive to priorities, insensitive to socio-economic constraints, irrelevant to farm, distrust).

The negative consequences of social media mentioned in this chapter illustrate emergent properties. Who would have expected that the focus on gaining acceptance through 'likes' would lead to a rise in hate? What is missing is a means for user engagement as early satisfaction has led to ineffective, damaging use. Social media corporations' focus on profit has demonstrated that self-regulation does not work. When they are challenged, they shift to other types of technology, seeking to lure users into accessing their services, regardless of users' needs. This is further exacerbated by information pollution, as noted by Wardle and Derakhshan (2017, p. 4):

> complex web of motivations for creating, disseminating and consuming these "polluted" messages; a myriad of content types and techniques for amplifying content; innumerable platforms hosting and reproducing this content; and breakneck speeds of communication between trusted peers.

Wardle and Derakhshan (2017, p. 7) argue that we need to understand the ritualistic function of communication. This means moving beyond thinking about communication as messaging from one person to another. Instead, we need to "recognize that communication plays a fundamental role in representing shared beliefs. It is not just information, but drama."

There are emerging initiatives that seem to address the concepts covered in this chapter. These examples belong to the mobilization discourse proposed by Singh and Flyverbom (2016). The Association for Progressive Communications (APC) hosts a Local Access and Community Networks (LocNet) (https://www.apc.org/en/node/34231/) initiative, it seeks to address un-met connectivity needs in least developed countries by engaging with stakeholders from the start. Similarly, the Mradi Africa initiative by Mozilla seeks to "build with and not for" a healthier Internet (https://blog.mozilla.org/netpolicy/2021/10/26/mozilla-and-omidyar-network-launch-new-reimagine-open-initiative-powering-local-innovation-in-the-global-south/). The value of community networks in terms of improvements to local community benefits, financial savings, and improved connectivity for communities left behind by market forces has been confirmed in multiple countries (see netCommons project in Europe [https://netcommons.eu]). These grassroots examples seek to make local needs and local content a priority, especially in those areas where market forces or rural policies have failed to create an even playing field (Salemink & Strijker, 2018; Strover et al., 2021).

The concepts listed in this chapter can be seen as 'non-negotiable' principles that call for a more collaborative design as they:

- Acknowledge which one of the four discourses of participation is involved (stakeholder, networked, mobilization, or oppositional).
- Attend to the different types of access that lead to effective use.
- Be clear and honest about the type of participation sought, and its expression during the stages of a project or initiative.
- Engage with multiple stakeholders and understand their goals.
- Acknowledge that new media will produce emergent, unexpected outcomes.
- Appreciate the interactions between new media and different stakeholders as a complex adaptive system that will need ongoing monitoring and course correction.

Several of these non-negotiables are about situating oneself along discourses of participation, types of access, and purpose of an intervention or initiative. This calls for a significant understanding of contextual and strategic thinking, while embracing the odds of unexpected outcomes to emerge.

In agricultural and rural development, many of the new technologies are being developed with an urban footprint, an industrial mindset, and without a sense of the social, political, and dependency consequences that can occur (Wang, 2020). This contrasts with Gurstein's remarks about effective use calling for collaboratively identified goals.

For new media to achieve its intended purposes, stakeholders need to have meaningful engagement (participation) throughout the stages of development, types of access, and a say in course correction when emergent outcomes become negative or only favor the interests of a few.

## References

Anderson, J., Clement, J., & Crowder, L. (1998). Accommodating conflicting interests in forestry—concepts emerging from pluralism. *Unasylva, 49*(194), 3–10.

Apthorpe, R. (1997). Writing development policy and policy analysis plain or clear: On language, genre and power. In C. Shore & S. Wright (Eds.), *Anthropology of policy: Critical perspectives on governance and power* (pp. 43–58). Routledge.

Arnstein, S. (1968). A ladder of citizen participation. *AIP Journal, 35*(4), 216–24. https://doi.org/10.1080/01944366908977225

Bar, F., Cohen, S., Cowhey, P., DeLong, B., Kleeman, M., & Zysman, J. (2000). Access and innovation policy for the third-generation internet. *Telecommunications Policy, 24*, 489–518. http://repositories.cdlib.org/brie/BRIEWP137

Bessette, G. (2004). *Involving the community: A guide to participatory development communication*. Southbound/IDRC.

Biggs, S. (1995). Contending coalitions in participatory technology development: Challenges for the new orthodoxy. [Unpublished paper.] Norwich, UK, December, 1995.

Borrini-Feyerabend, G. (1996). *Collaborative management of protected areas: Tailoring the approach to the context*. IUCN. https://portals.iucn.org/library/efiles/documents/1996-032.pdf

Buck, L., Geisler, C., Schelhas, J., & Wollenberg, E. (2001). *Biological diversity: Balancing interests through adaptive collaborative management.* CRC Press.

Chambers, R. (1997). *Whose reality counts? Putting the first last.* IT Publications.

Chambers, R. (2005). *Ideas for development.* Earthscan.

Checkland, P., & Scholes, J. (1990). *Soft systems methodology in action.* John Wiley and Sons.

Coggins, S., McCampbell, M., Sharma, A., Sharma, R., Haefele, S. M., Karki, E., Hetherington, J., Smith, J., & Brown, B. (2022). How have smallholder farmers used digital extension tools? Developer and user voices from Sub-Saharan Africa, South Asia and Southeast Asia. *Gobal Food Security, 32,* 100577. https://doi.org/10.1016/j.gfs.2021.100577

Coldevin, G. (1995). *Farmer-first approaches to communication: A case study from the Philippines.* Communication for Development, Research, Extension and Training Division. FAO.

Conway, K. (2020). *The art of communication in a polarized world.* AU Press.

Cooke, B., & Kothari, U. (1998). *Participation: The new tyranny?* Zed Books.

Cornwall, A. (2002). Making spaces, changing places: Situating participation in development. *IDS Working Paper 170.*

Daniels, S., & Walker, G. (2001). *Working through environmental conflict: The collaborative learning approach.* Praeger.

Deibert, R. J. (2020). Reset: Reclaiming the Internet for civil society. CBC Massey Lectures. https://www.masseycollege.ca/cbc-massey-lectures/

Gumucio-Dagron, A. (2001). *Making waves: Stories of participatory communication for social change.* The Rockefeller Foundation. https://www.ircwash.org/sites/default/files/Gumucio-2001-Making.pdf

Engel, P., & Salomon, M. (1997). *Facilitating innovation for development.* Royal Tropical Institute.

Fanjul, S. C. (2021). 'Cuanto más odio, más viral; por qué Twitter se ha llenado de saña en pleno auge de la ultraderecha'. Icon. https://elpais.com/icon/actualidad/2021-11-26/cuanto-mas-odio-mas-viral-por-que-twitter-se-ha-llenado-de-sana-en-pleno-auge-de-la-ultraderecha.html

Futch, M. D., & McIntosh, C. T. (2009). Tracking the introduction of the village phone product in Rwanda. *Information Technologies and International Development, 5*(3), 54–81.

Grindle, M. S., & Thomas, J. W. (1991). *Public choices and policy change: The political economy of reform in developing countries.* John Hopkins University Press.

Gumucio-Dagron, A. (2001). *Making waves: Stories of participatory communication for social change.* The Rockefeller Foundation. https://www.ircwash.org/sites/default/files/Gumucio-2001-Making.pdf

Gurstein, M. (2003). Effective use: A community informatics strategy beyond the Digital Divide. *First Monday, 8*(12). https://firstmonday.org/article/view/1107/1027

Hambly, H., & Rajabiun, R. (2021). Rural broadband: Gaps, maps and challenges. *Telematics and Informatics, 60,* 101565. https://doi.org/10.1016/j.tele.2021.101565

Heeks, R. (1999). The tyranny of participation in information systems: Learning from development projects. Development Informatics Working Paper, no. 4. https://papers.ssrn.com/sol3/papers.cfm?abstract_id=3477771

Hickey, S., & Mohan, G. (2004). Towards participation as transformation: Critical themes and challenges. In S. Hickey & G. Mohan (Eds.), *Participation: From tyranny to transformation? Exploring new approaches to participation in development* (pp. 3–24). Zed Books.

Hildyard, N., Hegde, P., Wolverkamp, P., & Reddy, S. (1998a). Pluralism, participation and power. *Forests, Trees and People Newsletter, 35*(March), 31–35. https://eurekamag.com/research/003/234/003234077.php

Hildyard, N., Hegde, P., Wolverkamp, P., & Reddy, S. (1998b). Same platform, different train: The politics of participation. *Unasylva, 194*(49), 26–34. https://hdl.handle.net/10535/8415

Hudson, H. E., Leclair, M., Pelletier, B., & Sullivan, B. (2017). Using radio and interactive ICTs to improve food security among smallholder farmers in Sub-Saharan Africa. *Telecommunications Policy, 41,* 670–84. https://doi.org/10.1016/j.telpol.2017.05.010

Kanji, N., & Greenwood, L. (2001). *Participatory approaches to research and development in IIED: Learning from experience.* IIED.

Klerkx, L. (2021). Digital and virtual spaces as sites of extension and advisory services research: Social media, gaming, and digitally integrated and augmented advice. *Journal of Agricultural Education and Extension, 27*(3), 277–86. https://doi.org/10.1080/1389224X.2021.1934998

Lightfoot, C., Fernandez, M., Noble, R., Ramírez, R., Groot, A., Fernandez-Baca, E., Shao, F., Muro, G., Okelabo, S., Mugenyi, A., Bekalo, I., Rianga, A., & Obare, L. (2001a). A learning approach to community agroecosystem management. In C. Flora (Ed.), *Interactions between Agroecosystems and Rural Communities* (pp. 115–30). CRC Press.

Lightfoot, C., Ramírez, R., Groot, A., Noble, R., Alders, C., Shao, F., Kisauzi, D., & Bekalo, I. (2001b). Learning our way ahead: Navigating institutional change and agricultural decentralisation. *Gatekeeper Series,* no. 98. IIED.

Lotz, J. (1998). *The lichen factor: The quest for community development in Canada.* CBU Press.

McGee, R. (2002). Conclusion: Participatory poverty research: Opening spaces for change. In K. Brock & R. McGee (Eds.), *Knowing poverty: Critical reflections on participatory research and policy* (pp. 189–05). Earthscan.

McKegg, K. (2019). White privilege and the decolonization work needed in evaluation to support indigenous sovereignty and self-determination. *Canadian Journal of Program Evaluation, 34*(2), 357–67. https://doi.org/10.3138/cjpe.67978

McMahon, R., Akcayir, M., McNally, M. B., & Okheena, S. (2021). Making sense of digital inequalities in remote contexts: Conceptions of and responses to connectivity challenges in the Northwest Territories, Canada. *International Journal of Communication, 15*(1), 5229–51. https://ijoc.org/index.php/ijoc/article/view/18213

Mosse, D. (1993). *Authority, gender and knowledge: Theoretical reflections on the practice of participatory rural appraisal.* Network Paper, 44. ODI.

Myers, M. (2004). *Evaluation methodologies for information and communication for development (ICD) programmes.* DFID.

Quarry, W., & Ramírez, R. (2009). *Communication for another development: Listening before telling.* Zed Books.

Ramírez, R. (1999). Stakeholder analysis and conflict management. In D. Buckles (Ed.), *Conflict and collaboration in Natural Resource Management* (pp. 101–26). IDRC/World Bank.

Ramírez, R. (2001). A model for rural and remote information and communication technologies: A Canadian exploration. *Telecommunications Policy, 25*(5), 315–30. https://doi.org/10.1016/S0308-5961(01)00007-6

Ramírez, R. (2003). Bridging disciplines: The natural resource management kaleidoscope for understanding ICTs. *Journal of Development Communication, 14*(1), 51–64. https://citeseerx.ist.psu.edu/viewdoc/download?doi=10.1.1.525.8517&rep=rep1&type=pdf

Ramírez, R. (2008). A 'meditation' on meaningful participation and engagement. *The Journal of Community Informatics, 4*(3). https://openjournals.uwaterloo.ca/index.php/JoCI/article/view/2948/3810

Rhoades, R. (1997, July 25–26). *The participatory multipurpose watershed project: Nature's salvation or Schumacher's nightmare?* [Conference presentation.] Global Challenges in Ecosystem Management in a Watershed Context, Toronto, ON, Canada.

Rhoades, R. (1998). *Participatory watershed research and management: Where the Shadow Falls.* Gatekeeper Series, 81. IIED.

Richardson, D. (1999). Facilitating participation: Accessing Internet services for development: The art of facilitating participation: Releasing the power of grassroots communication. In S. White (Ed.), *The art of facilitating participation: Releasing the power of grassroots communication* (pp. 259–80). Sage.

Richardson, D., Ramírez, R., & Haq, M. (2000). *1 - Grameen Telecom's Village Phone Programme: A multi-media case study.* https://www.researchgate.net/publication/246338707_Grameen_Telecom"s_Village_Phone_Programme_in_Rural_Bangladesh_A_MultiMedia_Case_Study_TeleCom

Röling, N., & Jiggins, J. (1998). The ecological knowledge system. In N. Röling & M. Wagemakers (Eds.), *Facilitating sustainable agriculture: Participatory learning and adaptive management in times of environmental uncertainty* (pp. 283–11). Cambridge University Press.

Salemink, K., & Strijker, D. (2018). The participation society and its inability to correct the failure of market players to deliver adequate services levels in rural areas. *Telecommunications Policy, 42,* 757–65.

Salemink, K., Strijker, D., & Bosworth, G. (2017). Rural development in the digital age: A systematic literature review on unequal ICT availability, adoption, and use in rural areas. *Journal of Rural Studies, 54,* 360–71. https://doi.org/10.1016/j.jrurstud.2015.09.001

Selener, D. (1997). *Participatory action research and social change.* Global Action Publications.

Singh, J. P., & Flyverbom, J. (2016). Representing participation in ICT4D projects. *Telecommunications Policy, 40*(7), 692–703. https://doi.org/10.1016/j.telpol.2016.02.003

Strover, S., Riedl, M. J., & Dickey, S. (2021). Scoping new policy frameworks for local and community broadband networks. *Telecommunications Policy, 45,* 102171.

Totolo, A., Renken, J., & Sey, A. (2015). The impact of public access venue information and communication technologies in Botswana public libraries. *Evidence Based Library and Information Practice, 10*(3), 64–84. https://doi.org/10.18438/B8NP5F

Treen, K. M. d'I., Williams, H. T. P., & O'Neill, S. J. (2020). Online misinformation about climate change. *WIREs Climate Change, 11*(5). https://doi.org/10.1002/wcc.665

Van Dam, C. (2000). Two decades of participatory development but how participatory? *Forests, Trees and People Newsletter, 42,* 11–17.

Van Dijk, J. A. G. M. (2001). *The ideology behind 'Closing Digital Divides': Applying static analysis to dynamic gaps.* Paper presented at *IAMCR/ICA Symposium on the Digital Divide.* Nov. 16, Austin, TX, USA.

Van Mele, P., Wanvoeke, J., Akakpo, C., Dacko, R. M., Ceesay, M., Beavogui, L., Souman, M., & Anyang, R. (2010). Videos bridging Asia and Africa: Overcoming cultural and institutional barriers in technology-mediated rural learning. *Journal of Agricultural Education and Extension, 16*(1), 75–87. https://doi.org/10.1080/13892240903533160

Vira, B., Dubois, O., Daniels, S., & Walker, G. (1998). Institutional pluralism in forestry: Considerations of analytical and operational tools. *Unasylva, 49*(194), 3, 35–42.

Wang, X. (2020). *Blockchain chicken farm and other stories of tech in China's countryside.* FSG/LOGIC.

Wardle, C., & Derakhshan, H. (2017). *Information disorder: Toward an interdisciplinary framework for research and policy making.* Council of Europe.

White, S. (Ed.) (1999). *The art of facilitating participation: Releasing the power of grassroots communication*. Sage.

White, S., Nair, K. S., & Ashcroft, J. (Eds.) (1994). *Participatory communication: Working for change and development*. Sage.

Wollenberg, E., Edmunds, D., & Anderson, J. (2001). Pluralism and the less powerful: Accommodating multiple interests in local forest management. *International Journal of Agricultural Resources, Governance and Ecology, 1*(3/4), 199–222. https://doi.org/10.1504/IJARGE.2001.000012

Wood, Z. (2021, November 26). 'I'm happy to lose £10m by quitting Facebook,' says Lush boss. Ethical Business section. *The Guardian*. https://www.theguardian.com/business/2021/nov/26/im-happy-to-lose-10m-by-quitting-facebook-says-lush-boss

Zadek, S. (2004). The path to corporate responsibility. *Harvard Business Review*, December, 125–32.

Zaveri, S., Ramelan, V., Ramírez, R., & Brodhead, D. (2016). Mentoring three ISIF-funded project in evaluation and research communication. DECI case study. https://evaluationandcommunicationinpractice.net/wp-content/uploads/2017/08/Zaveri-Ramelan-Ramirez-Brodhead_CaseStudyISIF_201608.pdf?9023db&9023db

# 3 Datafication, aggregation, and heteromation

## Participation in a digitally mediated world

*Katherine Reilly*

## Introduction

In development studies, participation is typically understood as either engagement in a process or event, or power sharing among stakeholders (Carpentier, 2020). Participation in the planning and implementation of development initiatives is seen as a means to enhance power sharing around decision-making with the hope that it will also improve the allocation of benefits (Melkote, 1991). There is also a commitment to democratization (Mohan, 2001) including transparency, accountability, or equity, since these are thought to create a favorable context for inclusive decision-making and power sharing. Development studies positions participation as a normative goal, and much of the associated literature explores whether and how people can contribute meaningfully to decision-making and social change.

In this research, communications are seen as a prerequisite for participation, or a way to facilitate associated goals such as access, inclusion, dialogue, or negotiation. Critical works explore how communication channels or practices create barriers to achieving these goals (Gumucio-Dagron, 2009). For example, a classic debate in development communications pits access to the means of communication (e.g., radio towers) against control over the content of communications (i.e., control over the broadcast industry, or algorithms) as the primary barriers to participation, freedom, development, or other normative measures. These same discussions are reflected in recent takes on participation in the Digital Age, for example, when access to digital technologies is contrasted against use and appropriation (Morales, 2009).

These approaches to participation and communication tend to be strategic or instrumental. That is, they are based on the search for practical responses to inequality and marginalization given cultural experience and the allocation of the material benefits of development. However, if we dig a little deeper, we encounter important prior questions about the role of participation and communication in development. For example, Bodley's classic work *Victims of Progress* (2014) detailed the high costs of incorporation into modernization projects for Indigenous and rural communities. Similarly, the dual forces of COVID-19 and the platform economy have led to expanded connectivity in rural communities, which have long faced digital divides. These shifting communicative horizons raise questions about how new forms of participation shape rural incorporation into development processes.

DOI: 10.4324/9781003282075-4

Meanwhile, recent work by Dutta (2015) argues that development studies literature concerning participation has actually served to pave the way for cultural acceptance of neo-liberalism in the Global South. As he explains, in response to critiques from dependency theory in the 1970s, the United Nations Educational, Scientific and Cultural Organization (UNESCO) introduced the 'Another Development' framework:

> A substantive body of work on participatory communication started taking root in the development communication literature as well as in the mainstream structures of the WB, FAO, and USAID. Within these hegemonic structures that were key players in catalyzing the neoliberal transformation of global spaces, participatory communication played a pivotal role as a strategic tool for enhancing reach, effectiveness, and efficiency of top-down development innovations designed by experts.
>
> (Dutta, 2015, n.p.)

Dutta's observations about participation build on the observation that neo-liberalism depends on creating "an efficient and harmonious consumer capitalist society" (Laurie & Bonnett, 2002, p. 31). The expansion of communicative facilities and platformization into rural communities can be viewed as a new deepening of these processes.

In these examples, communication for participation is regarded as a vector of power, which can serve to advance hegemonic goals including the reproduction of inequality or marginalization. Departing from this perspective, this chapter focuses on how digital communication is reshaping the context for development and social change. It asks, how does the digital platform economy reshape the strategic role of participation in development? The term *digital platform economy* refers to technologically mediated spaces of networked interaction between producers and consumers that structure both cultural and market realities in the contemporary moment. I follow Srnicek (2017) in highlighting the role of data and algorithms in regulating these spaces, and therefore embrace van Dijck's (2014) assertation that new institutions and cultural processes have arisen to facilitate the processes of datafication that support digital mediation.

In what follows, the concept of *societies of control* (Deleuze, 1992) offers a critical frame for thinking about participation in the Digital Age. Having established this, I explore three mechanisms through which coercive forces mediate the digital platform society and enact societies of control: datafication, or the transformation of social reality into data points (Mayer-Schönberger & Cukier, 2013); logistics brokerage, or the commodification of the process of organizing social interactions and transactions (Thompson, 2015); and heteromation, or the extraction of economic value from free labor in digital networks (Ekbia & Nardi, 2017). The chapter concludes with suggestions about the types of interventions required to build capacity for alternative forms of participation in the digital realm.

### Regulation of digital participation

If participation is to be a means of decentralizing power, then we need to know how power is expressed so that participation can be organized appropriately.

We have experienced a fundamental change in how power is articulated with the rise of digital technologies. We see this, for example, in the idea of societies of control (Deleuze, 1992). This concept forms part of a body of work that explores how social institutions come to be normalized through regimes of thought or rationality. Winiecki (2007) explains that "codes of actuarial accounting for and evaluating the actions of individuals give rise to institutionalised forms that serve to control access and privileges throughout modern societies" (p. 1). Deleuze argues that in previous time periods, power was articulated through analogue mechanisms of discipline, but with the rise of the information age power is being re-organized around digital forms of control. Thus, "as we move from analog [*sic*] to digital, module to modulation, and from the barracks of a prison to stacks of code, Deleuze prompts us to pay attention to the changing forms of power from discipline to control" (Choi, n.d., n.p.). Bogard (2007, n.p.) argues:

> This shift can be framed historically and economically as a problem of capitalist governance involving the limits of enclosure as a tool of capitalist accumulation . . . . After WWII, information technologies make it possible to release populations more into the open. Rather than pack them into closed spaces, Capital begins a new strategy to disperse them. Network controls, like remote surveillance and electronic passwords, allow it to keep its grip on bodies . . . . The new controls promise to counter the resistance of populations to confinement by instituting a kind of mobile and free form of enclosure.

Deleuze argues that in either systems of discipline or systems of control, there is a sense of freedom around life choices. These freedoms are constrained by the parameters of the system we occupy. Choi (n.d.) observes, "subjects (or for that matter objects) are liberated as long as they adhere to a variety of prescribed comportments" (n.p.). A core example of this in the platform economy is the idea that you can access an online service if and only if you agree to the terms of service (Thatcher et al., 2016). Winiecki (2007, p. 1) explains, "When faced with such constraints and potentials actors apply existing knowledge and understanding, and what that knowledge and understanding allows them to see, in activating only a limited range of what might actually be possible."

This idea undergirds a common observation of Big Data and platform studies that individualization and personalization create an illusion of freedom even as systems gather data to train the algorithms that play on our preferences. These processes are driven by our addiction to always on communications in which "Social media networks produce value through monetizing our attention span and by emphatically blurring the boundary between work and leisure," and "Our participation in these systems, apps and services, leads to the financialization of social interaction" (Choi, n.d.).

Similarly, Martín-Barbero (1993) recognized a new organization of power with the rise of digital technologies. He argued that different time periods are marked by cultural mutations that reflect how the sensorium, or culture of knowing, manifests. The sensorium reflects arrangements of power and resistances to them, and, importantly, communications technologies change how these processes are mediated and expressed. In earlier works, Martín-Barbero (1993) mapped the expression of power between logics of mass media production, and communitarian cultures of consumption within fixed spaces of hegemonic struggle. He argued that audiences often received messages differently from how they were intended, and from these creative acts produced new expressions of culture and knowledge. These acts of resistance constituted an expression of cultural sovereignty in the face of international or national hegemonic projects. Here power is expressed in *disciplinary* institutions that attempt to force the unification of diverse populations.

In later works, Martín-Barbero identified a new sensorium that consisted of two synchronic factors, temporality and spatiality, and two diachronic factors, technicity and sensoriality. *Technicities* describe how new media and digital technologies meld technique, language, expression, and apparatus into one socio-cultural-technical system. Martín-Barbero was concerned about how new media or information and communications technologies were coming to inform the possibility for meaning-making, knowledge production, and especially how we know and express ourselves. He argued that "If, during centuries technology was considered a mere instrument, today it is already well on its way to becoming reason [itself]" (Martín-Barbero, 2011, p. 110). *Sensoriality*, meanwhile, refers to the act of signifying the world from our senses—the process of interpretation that contemplates both our biological perception and socio-cultural standing (Valquiria et al., 2019). This concept captures the interlacing of meaning-making, techniques, and apparatuses in the contemporary moment. However, in this case, it is focused on recovering what is human in processes of meaning-making. The contemporary sensorium, following the logic of Martín-Barbero, emerges from the articulations of the media apparatus, human processes of meaning-making, and specific temporal and spatial realities.

The platformization and datafication (technicity) plus social experience (sensoriality) that characterize the contemporary sensorium have raised concerns among communications scholars because of the potential for technicities to colonize sensorialities. Rincón (2017, 2018) fears that contemporary media establishes the conditions for a zombie democracy, where citizens are guided by the logics of continuous, unthinking consumption. Couldry (2019) documented a conversation that had occurred in 2017 in which he and Martín-Barbero discussed how digital platforms were reshaping "the very spaces where culture and social life are lived out" (p. 187). Both Rincón and Couldry identified a new pattern of *control* through the normalization of consumeristic comportments, which in turn shut down the possibility for human sympathy and connection. It is not that the apparatuses are emotionless, but rather that their particular formulation contributed to a sensorium devoid of community and care. Martín-Barbero did not share this view. As Couldry (2019) recounts:

> I felt from Jesús Martín Barbero a certain caution towards the pessimism with which these developments undoubtedly leave me: there must, he seemed to be saying, surely be some paths towards resistance to this, the deepest commercialization imaginable of human life.
>
> (p. 187)

In earlier works during the mass media era, Martín-Barbero relied on audiences' inherent affective, cognitive, creative, and interpretive power as a boundary to, or perversion of, hegemonic media flows. Today, given how information systems actively colonize spaces for perception, emotion, meaning-making, and relationality, we cannot rely on affect as an automatic source of emancipatory potential. Ahmed (2010) offers valuable insights in this regard. She argues that social truths arise out of networks of affective relations, and we tend to go along with these processes of social construction because we rely on these networks for a sense of affinity. For example, Ahmed argued that to feel a sense of belonging, we accept the notion that family is a social good and actively participate in its reproduction. To reject *the family* is to also reject the solidarity of the family construct.

Extending this notion to the media sphere, to reject the digital is also to reject the sociality of platformized processes, *so our very participation in digital spaces is an act of prosuming consumerism's disconnected and uncaring affective relations*. As expressed by Martín-Barbero and Couldry, the challenge is that the very articulations that hold this map together are precisely the forces that disarticulate our communities and destabilize collective projects. Galloway's (2014, pp. 109–10) observations further complicate the scenario by suggesting how always-on communications undermine the potential of participation to counter the power structures that controlled access to decision-making with implications for resource distribution:

> The ultimate significance of control society is not so much the continuous encroachment of the border checkpoint on the passport control, not so much data mining or facial recognition algorithms, but that it has eviscerated history, not by banning dissent but by accelerating the opportunities and channels for critical thought to infinity and therefore making it impossible to think historically in the first place. *Thus the central challenge within control society will be not simply to resist the various new nefarious control apparatuses, but to rescue history from its own consummation.*
>
> [Emphasis added]

Thus, where participation may have served to decentralize hegemonic power during a time of totalizing hegemony, in the contemporary moment, to participate can actually fuel the apparatuses of control in ways that undermine the objectives of communities. With this in mind, development communications should not be undertaken unthinkingly or a-historically. We need new ways of participating that help ensure development is grounded. We must ask: "How can we rethink and re-orient participation for this new time period, as a tool to help advance these goals?"

## Questioning assumptions about participation

This question can be addressed by thinking through common assumptions about participation. These are the views that participation is an agnostic technique or best practice, that it is a discrete process, that it is laddered, and that stakeholders are clearly defined, rational, and autonomous. In this section, each of these assumptions is challenged and alternative thinking offers a possible avenue toward new practices.

The first and most obvious challenge to the literature on participation is the idea that it is a set of best practices for community intervention that exist in a relation of objectivity and impartiality to social realities. From a Deleuzian point of view, best practices are comportments that structure social processes. This means that participation needs to be understood as a social process that (re)produces both relations of power and also our experience of freedoms. We can also say that the comportments prescribed for participatory activities are a form of mediation (Livingstone, 2009) or "the cultural processes by which power is negotiated between dominant institutions and popular or resistance movements" (Livingston citing Martín-Barbero, 2003, 2009, p. 12). These observations are further complicated when participatory processes are shifted into digitally mediated realms given the recursive nature of the latter. That is to say, "media today mediates the conditions of mediation" (Hansen, 2010, p. 181). It can also be said that "Information and data do not just capture but now actually constitute everyday life" (Beer & Burrows, 2013, p. 56). In other words, the affordances of digital media structure a set of comportments, which themselves structure participatory comportments, which in turn condition social relations.

These observations suggest that rather than understand participation as a set of best practices for power sharing, it needs to be understood as a form of social mediation that conditions flows of power and attendant access to resources. This suggests the need to pay close attention to whether and how the comportments created by either analogue or digital forms of participation benefit the community or not. Social media is widely thought to generate social isolation, addictive behaviors, antipathy, and unthinking engagements. This is not to say that social media or participation, as in the case of any technology, cannot be used productively. But care must be taken to appropriate these resources in ways that respond to contemporary expressions of power.

Secondly, the literature on participation generally assumes that stakeholders are rational and autonomous. For Mitchell et al. (1997), stakeholders are defined as any actors who have at least two out of the three factors that link them to a clearly defined issue: power to assert influence over decision-making, recognized legitimacy within the affected community, and an immediate and urgent need to be included. To be included in participatory decision-making, therefore, stakeholders must be able to clearly articulate their interest in an issue, and they must come to the table ready to represent their needs and desires.

These assumptions are challenged by the idea of societies of control and social mediation. On social media, legitimacy is based on the power to convene networks, which calls into question the very concept of stakeholders. Meanwhile, immediacy or urgency can be manipulated through algorithmic forms of control. In such a context, feelings often trump reason, as they become the basis for mobilizing followers, and emotions may come to supersede substantive outputs. At the same time, apathy takes on a new significance, eroding the quality of participation and contributing to decisions that reflect popular mood rather than clearly articulated interests, need, or desires. In this context, participatory processes may be perverted as they become a means to take control rather than a means to decentralize it. These processes may lead to outcomes that are far from just, both in how decisions are made and in how resources are allocated.

These observations raise questions about how best to pursue power sharing or resource distribution when communications channels undermine their achievement. This brings us to the third challenge, which is a hallowed icon of the literature on participation: Sherri Arnstein's (1969) well-known ladder of participation. This is a model of decentralization or delegation of control, such that decision-making shifts from central axes of power toward the grassroots. When the Internet first appeared, it arguably offered the perfect mechanism to facilitate this goal. However, what promised to be the definitive architecture of the public sphere has descended into a morass of asocial behavior (Deb et al., 2017), thus illustrating the point that social media mediates participation, often in undesirable ways.

Social media reveals that it is possible to have decentralization and still produce unjust or unequal relations within communities. Delegation of control will not necessarily overcome and may in fact reinforce comportments that uphold inequality or marginalization. Coming back to the mediation of participatory processes, this suggests the importance of performing versions of *the good* that attend to diverse positionalities and intersectionalities. This means that participation is not only about securing a measure of control in the context of highly centralized and centralizing processes, but it is also about articulating meaningful processes based in recognition and respect, against the backdrop of atomizing and centrifugal forces.

The fourth assumption about participation is that it is a discrete process, with a clear and legitimate convener such as a municipal office, private sector actor, or community leader; a beginning, middle, and end that takes place within the lifecycle of a project and a common goal such as planning or decision-making. This thinking makes sense when generating stakeholder support for initiatives that align with capitalist processes of enclosure or imperialist processes of colonization. Here, participation allows actors to vie for 'a slice of the pie' given a finite resource base.

But as we move into a networked world, the logic of participation is greatly altered. Digitally mediated social networks work to synchronize technicities and sensorialities around the shifting spatial and temporal experience of diverse individuals. *Participants*, therefore, come together in shifting moments and spaces of confluence. These processes of articulation are layered and multiple, consisting of diverse conveners and objectivities. In the hyper-competitive space of social media, conveners vie for the limited attention of individuals, often using tactics that debase the quality of dialogue or engagement, and participants become adept at code switching (Tatum, 2017) as a means to cope with the differing expectations of multiple spaces of confluence (Benjamin, 2019, p. 119). Here the *quantity* of engagement becomes a measure of *the good*.

In this context, activists search for new ways to create a basis for dialogue, often exploring the politics of care, or slow engagements, as means to enact alternative versions of *the good* grounded in the *quality* of connection. Participants might still flow through spaces freely, but they will encounter alternative enactments of technicities and sensorialities that provoke a different set of comportments more in line with critical reflection, active listening, or collaboration. There is also a focus on creating spaces that obviate the need for code switching, by embracing difference as a starting point for dialogue and engagement (Benjamin, 2019). The goal of participation shifts from influencing a plan or a decision to altering the dominant comportments of sociability so that projects can emerge organically from below, in ways that constitute alternatives to dominate mediations of power.

### Datafication, logistics brokerage, heteromation, and participation

This section looks at how the digital platform economy can prefigure and foreclose meaningful participation in decision-making and social change. The digital platform economy consists of technologically mediated spaces of networked interaction between producers and consumers. For Srnicek (2017), these platform spaces are governed by algorithms that establish the rules for engagement within spaces of interaction in ways that surveil and control those interactions. *Platformization* describes the process of making Web-based spaces and their meta-data platform ready (Helmond, 2015), which in turn relies on new institutions, social processes, and discourses that support datafication (van Dijck, 2014).

For Srnicek (2017), this new approach to capitalist-driven accumulation arose as a response to declining profits from manufacturing. He notes, "capitalism has turned to data as one way to maintain economic growth and vitality in the face of a sluggish production sector" (2017, p. 5). Data has opened up new terrains of accumulation by allowing capitalism to trade in, and therefore colonize our attention, ideas, and preferences (Couldry & Mejias, 2019). For Benjamin (2019), the social processes that are normalized by the platform economy are particularly insidious, because they masquerade as impartial, personalized, merit-based, and forward looking (p. 109). This implies they are, in fact, the architecture of free and equal participation. But it is important to reiterate that the platform economy was designed as a mechanism of surplus accumulation. The architecture of platformization was designed to commoditize our interactions. Our participation in these spaces feeds this logic. When we leverage these spaces to engage in participatory activities, our interactions can be influenced by platform logics. Platformization unfolded through waves of digital transformation in which "digital technologies create disruptions triggering strategic responses from organizations that seek to alter their value creation paths while managing the structural changes and organizational barriers that affect the positive and negative outcomes of this process" (Vial, 2019, p. 118). The resulting entities included familiar social media companies like Google and Facebook as well as disruptive companies like Uber or Airbnb, and new approaches to organizing production and distribution, as with the rise of Amazon. But these tendencies also extended behind platform spaces. Van Dijck and Poell (2013) argue that the algorithmic programmability, popularity seeking, social networking, and datafication that shape social media spaces enter into our work processes, incentive structures, and social tendencies in the offline world. For example, as we became used to the highly transactional nature of digital spaces, we find ourselves expecting a transactional experience in our face-to-face encounters.

Since the primary objective of platformized spaces is capitalist accumulation, their contribution to social development is necessarily subordinated and curtailed. Participation in these spaces, as well as participation through these spaces, reproduces social and economic relations in ways that necessarily shut down possible avenues for societal development. But it is extremely difficult to escape these spaces and their logic. Digital transformation causes established institutions to adjust their activities around the new logics introduced by platformization so as to protect investments of time, energy, or resources and continue the stream of benefits that emerges from historical legacies and cultural relevance. For example, farmers will feel pressure to adopt data intensive platformized production processes as the other players within the farm ecosystem shift toward new production, shipping, and processing standards. They will feel pressure to 'participate' in the new logics that permeate agricultural industries. This will reshape what it means to participate in governance of these new economic processes.

A brief examination of mechanisms of platformization can illustrate how it reshapes and / or curtains participation. Van Dijck argues that datafication is upheld by a set of ideological frameworks that she calls "dataism," which she defines as a "belief in the objectivity of quantification and in the potential of tracking all kinds of human behavior and sociality through online data" (2014, p. 201). These are based on the normalization of a set of social institutions, such as the terms of service agreements we click on when we sign up for a new online space (Thatcher et al., 2016). These agreements are backed by new legal frameworks in the form of personal data protection laws, which normalize corporate collection of personal data. Datafication is not just the transformation of cultural information into machine-readable form (Mayer-Schönberger & Cukier, 2013), but the establishment of a set of institutions that regulate comportments and controls access or privileges (O'Neil, 2016). It enables an architecture that allows us to participate in platform spaces. In doing so, it transforms human interactions into virtual transactions.

When we engage in datafication—when we, for example, share personal data at our pharmacy—we enact these comportments and contribute to the assertion of these controls. We do this not only by legitimating the categories for data gathering but also by legitimating the social processes that surround data gathering, such as the conversion of a human interaction into a digital transaction. These categories and practices come to shape how we know ourselves and our reality. For example, when human interactions become digital transactions, our sense of community and attachment is transformed. Submission to these processes produces valuable insights for loyalty programs, but it creates atomization and isolation for users. These systems have both inherent flows and deliberate traits that sort people into categories, which shape people's access to essential services, such as health care.

In the pursuit of feeling well and efficient access to health care, our well-being is negatively affected because we lose social ties. As Carroll et al. (2019, p. 501) point out in their work on community and information systems,

> we need to think about what it would mean to reduce the potential for conversations and, potentially, to do away with certain roles in the community. It could be the case that efficiency and access are not worth these costs.

A second example comes from logistics brokerage or aggregation theory (Thompson, 2015). This is the idea that platforms enable new ways to distribute goods and services based on helping producers and consumers find each other. For example, Uber offers the service of helping a driver find a customer, and so positions itself as a logistics company. Each time the Uber algorithm facilitates a transaction, Uber earns a small return. In logistics brokerage, both workers and consumers are the nodes linked together by platformized interaction. As nodes in these networks, our rich human relations are reduced to the signaling of needs and receipt of services. Fulfilment in this context is defined as service delivery, rather than, say, producing a benefit for society or forming sustained relationships. Logistics brokerage generates efficiencies and conveniences, but also the expectation of immediacy, and while producers and consumers coincide with each other in time and space for the brief time during which a service is rendered, they do not actually coincide with each other at the level of affective relationships. This undermines the potential for the production of relations of care. Momentary connections are prioritized over solidarity, social supports, or civil society.

A final example is provided by Ekbia and Nardi's (2017) concept of heteromation. It has been suggested that platformization will lead to automation by supporting artificial intelligence (AI) and new services, such as driverless cars. Heteromation is offered as an alternative thesis to automation as a transformative force for production and labor. Where automation argues that computers are taking over the work of producing human necessities, heteromation argues that, in fact, participation in social media is transforming humans into part of the machine. For example, each time we fill a CAPTCHA, we are doing the free labor of training AI how to read type, or how to interpret what it sees on a roadway. Meanwhile, Irani (2015) outlines AI's dependencies on the work of low paid data janitors who moderate content or calibrate information systems.

When Facebook users see violent content, they can flag it to the moderators so that it can be reviewed for removal. In this instance, users are doing the work of moderating the content of a media platform, while also training AI how to recognize unsuitable content. This act of citizenship makes the network space safe for everyone, while producing knowledge and monetary value for the platform itself. Meanwhile, the user has been exposed to a violent act, which is displaced in time and space since it is being experienced virtually. What circulates as a result is a sense of outrage, with the platform, or with its content moderation software, or with the content of the message, but perhaps not with the act of violence itself. The expectation is that this moderation will take place during online participation.

In the fluid spaces of articulation that characterize socially mediated participation, atomized users will bump up against unpleasantness and are expected to address this as part of their participatory process.

These three aspects of the digital platform economy each normalize new social institutions and new regimes of thought or rationalities that assert order and controls in society. Rather than a disciplinary form of order that taught people their *place* in society, these new forms of control prioritize freedom of movement and the exchange of goods. But this freedom is not actually free given that it is mediated by spaces of capitalist accumulation. Because the primary objective of platformized spaces is capitalist accumulation, any contribution they make to facilitating participation for social development is necessarily curtailed. The mechanisms that enable digital transformation toward a platformized society not only enable institutions but also spill over into our work processes, incentive structures and social tendencies in the offline world, reshaping our communities, and the distribution of resources.

## Conclusion: implications for development

As long as we live in a pro-growth economy, then we are always participating in development in one way or another. The nature of that participation changes over time with changes to the means of production and associated mechanisms of power, social institutions, and regimes of thought. These shifts have implications for who wins and who loses, who is a beneficiary of development, and who is marginalized from these gains. Therefore, it makes sense to ask how our participation affects the distribution of goods, and to seek out forms of participation that will alter the unequal terms of growth.

When the allocation of benefits is determined through centralized processes, then gaining access to those processes—being able to participate in them—is an obvious strategy. This can be a difficult thing to achieve when people have been *disciplined* to accept their lot, and in such a context deconstructing those processes of disciplinary control is a necessary aspect of pursuing greater equity.

However, when the allocation of benefits is shaped by networked forms of communication and commerce, when there is no clear center of decision-making, then different strategies are in order. It becomes necessary to rethink what it means to participate, and to facilitate new and different skills that redress inequities and marginalization. In the context of platformization and digital capitalism, the challenge is to articulate stakeholders around interventions that reassert control over the resources that enable contemporary digital forms of control.

If the goal is to respond to inequality and marginalization by reallocating goods and services, then traditional forms of participation will no longer work. They relied on best practices that could bring together clear stakeholders from fixed places who would participate in specific processes of power sharing. But when power organizes itself through distributed channels of control, new responses are required. They must overcome the always-on comportments, the immediacy and outrage that allow contemporary systems of control to keep us in check. We must also slow down, take the time to listen to one another, and perhaps relearn the skills of community engagement. Participation needs to be rethought, not for decentralizing control but rather as a means to construct alternatives that are more human and more caring.

One approach is to use the features of the system in your own favor. For example, data intermediaries give groups or communities power over their data so that they can demand concessions from those who would like to use it. Or, for example, algorithm auditing allows people to make the terms of resource distribution more transparent. These are examples of how power sharing might be engineered through novel forms of participation in a system that was originally designed or leveraged to accrue benefits in uneven or biased ways. Another approach is to reject certain features of the system, such as its disarticulation, immediacy, outrage, and displacement by pursuing small, local, slow processes of social engagement.

New forms of participation are required, and they present many challenges. In the former, participation faces the challenge of articulating diffuse users around a common project that serves concrete needs. Thus, the challenge is to take control of the sustenance of the network so that it can be brought to heel. Researchers are addressing this through the question of data governance (Ruhaak, 2017). They need to figure out models that offer benefits that are greater than the freedom that users experience through open networking. In the latter, participation faces the challenge of overcoming the dominant culture of the platform age, with its always on, never available ethos, by creating spaces that foster quality community connections.

## Acknowledgments

This work was supported by a grant from the Social Sciences and Humanities Research Council of Canada (SSHRC). Many thanks to Esteban Morales, Ataharul Chowdhury, and Ricardo Ramirez for their intellectual support of this paper.

## References

Ahmed, S. (2010). Happy objects. In M. Gregg & G. Seigworth (Eds.), *The affect theory reader*. Duke University Press. https://read.dukeupress.edu/books/book/1469/chapter/170328/Happy-Objects

Arnstein, S. R. (1969). A ladder of citizen participation. *Journal of the American Institute of Planners, 35*(4), 216–24. https://doi.org/10.1080/01944366908977225

Beer, D., & Burrows, R. (2013). Popular culture, digital archives and the new social life of data. *Theory, Culture & Society, 30*(4), 47–71. https://doi.org/10.1177/0263276413476542

Benjamin, R. (2019). *Race after technology: Abolitionist tools for the new Jim Code*. Polity.

Bodley, J. H. (2014). *Victims of progress* (6th ed.). Rowman & Littlefield Publishers.

Bogard, W. (2007). The coils of a serpent: Haptic space and control societies. *CTheory*, Sept. https://journals.uvic.ca/index.php/ctheory/article/view/14513

Carpentier, N. (2020). Media and participation. In J. Servaes (Ed.), *Handbook of communication for development and social change* (pp. 195–216). Springer. https://doi.org/10.1007/978-981-15-2014-3_47

Carroll, J. M., Beck, J., Boyer, E., Dhanokar, S., & Gupa, S. (2019). Empowering community water stakeholders. *Interacting with Computers*, *31*(4), 492–506. https://doi.org/10.1016/j.patter.2022.100449

Choi, T. (n.d.). Notes on the control society. *Poetic Computation.* http://taeyoonchoi.com/poetic-computation/control-society/

Couldry, N. (2019). Afterword. In H. C. Stephansen & E. Treré (Eds.), *Citizen media and practice: Currents, connections, challenges* (pp. 181–89). Routledge.

Couldry, N., & Mejias, U. A. (2019). Data colonialism: Rethinking Big Data's relation to the contemporary subject. *Television & New Media*, *20*(4), 336–49. https://doi.org/10.1177/1527476418796632

Deb, A., Donahue, S., & Glaisyer, T. (2017). *Is social media a threat to democracy?* The Omidyar Group. https://fronteirasxxi.pt/wp-content/uploads/2017/11/Social-Media-and-Democracy-October-5-2017.pdf

Deleuze, G. (1992). Postscript on the societies of control. *October*, *59*, 3–7. https://www.jstor.org/stable/778828

Dutta, M. J. (2015). Decolonizing communication for social change: A culture-centered approach. *Communication Theory*, *25*(2), 123–43. https://doi.org/10.1111/comt.12067

Ekbia, H. R., & Nardi, B. A. (2017). *Heteromation, and other stories of computing and capitalism.* MIT Press.

Galloway, A. R. (2014). *Laruelle: Against the digital.* University of Minnesota Press.

Gumucio-Dagron, A. (2009). Playing with fire: Power, participation, and communication for development. *Development in Practice*, *19*(4–5), 453–65. https://doi.org/10.1080/09614520902866470

Hansen, M. B. N. (2010). New media. In W. J. T. Mitchell & M. B. N. Hansen (Eds.), *Critical terms for media studies* (pp. 172–85). University of Chicago Press.

Helmond, A. (2015). The platformization of the Web: Making Web data platform ready. *Social Media + Society*, *1*(2), 2056305115603080. https://doi.org/10.1177/2056305115603080

Irani, L. (2015). Justice for 'Data janitors.' *Public Books.* https://www.publicbooks.org/justice-for-data-janitors/

Laurie, N., & Bonnett, A. (2002). Adjusting to equity: The contradictions of neoliberalism and the search for racial equality in Peru. *Antipode*, *34*(1), 28–53. https://doi.org/10.1111/1467-8330.00225

Livingstone, S. (2009). On the mediation of everything: ICA Presidential Address 2008. *Journal of Communication*, *59*(1), 1–18. https://doi.org/10.1111/j.1460-2466.2008.01401.x

Martín-Barbero, J. (1993). *Communication, culture and hegemony: From the media to mediations.* Sage.

Martín-Barbero, J. (2003). Cultural change: The perception of the media and the mediation of its images. *Television & New Media*, *4*(1), 85–106. https://doi.org/10.1177/1527476402239435

Martín-Barbero, J. (2011). La pertenencia en el horizonte de las nuevas tecnologías y de la sociedad de la comunicación. In M. Hopenhayn & A. Sojo (Eds.), Abrahamson, P., Arditi, B., Costa, S., Courtis, C., Gargarella, R., Molina, G., Güell, P., Jaquette, J., Marramao, G., Barbero, J. M., Arauco, V. P., Richard, N., Stavenhagen, R. (Authors), *Sentido de pertenencia en sociedades gragmentadas: América Latina desde una perspectiva global* (pp. 105–27). Siglo Veintiuno Editores/Asdi/CEPAL. https://repositorio.cepal.org/bitstream/handle/11362/2027/1/S306983S4782011_es.pdf

Mayer-Schönberger, V., & Cukier, K. (2013). *Big data: A revolution that will transform how we live, work, and think* (Reprint edition). Mariner Books.

Melkote, S. R. (1991). *Communication for development in the Third World: Theory and practice*. Sage.

Mitchell, R. K., Agle, B. R., & Wood, D. J. (1997). Toward a theory of stakeholder identification and salience: Defining the principle of who and what really counts. *Academy of Management Review*, *22*(4), 853–86. https://doi.org/10.2307/259247

Mohan, G. (2001). Participatory development. In V. Desai & R. Potter (Eds.), *The Arnold companion to development studies* (pp. 49–54). Hodder. http://www.hoddereducation.co.uk/Title/9780340760512/The_Companion_to_Development_Studies.htm

Morales, S. (2009). La apropiación de TIC: una perspectiva. In S. Morales & M. I. Loyola (Eds.), *Los jóvenes y las TIC. Apropiación y uso en educación* (pp. 99–120). Escuela de Ciencias de la Información. https://apropiaciondetecnologias.com/wp-content/uploads/2017/05/Los_jóvenes_y_las_TIC.pdf

O'Neil, C. (2016). *Weapons of math destruction: How big data increases inequality and threatens democracy*. Crown.

Rincón, O. (2017). De celebrities, pop y premodernos: hacia una democracia zombie. *Contratexto*, *27*, 135–47. https://hdl.handle.net/20.500.12724/5105

Rincón, O. (2018). Mutaciones Bastardas de La Comunicación. *MATRIZes*, *12*(1), 65–78.

Ruhaak, A. (2017, October 27). Getting started with data governance. *Mozilla Foundation*. https://foundation.mozilla.org/en/blog/getting-started-with-data-governance/

Srnicek, N. (2017). *Platform capitalism*. Polity Press.

Tatum, B. D. (2017). *Why are all the Black kids sitting together in the cafeteria?: And other conversations about race*. Basic Books.

Thatcher, J., O'Sullivan, D., & Mahmoudi, D. (2016). Data colonialism through accumulation by dispossession: New metaphors for daily data. *Environment and Planning D*, December 30, 2015. https://papers.ssrn.com/sol3/papers.cfm?abstract_id=2709498

Thompson, B. (2015, July 21). Aggregation theory. *Stratechery*. https://stratechery.com/2015/aggregation-theory/

Valquiria, M. J., Ribeiro, R. R., & Da Silva, G. H. (2019). Sensorialidad, la mediación que siempre estuvo presente. In O. Rincón (Ed.), *Un Nuevo Mapa Para Investigar La Mutación Cultural: Diálogo Con La Propuesta de Jesús Martín-Barbero* (pp. 117–36). CIESPAL.

van Dijck, J. (2014). Datafication, dataism and dataveillance: Big data between scientific paradigm and ideology. *Surveillance & Society*, *12*(2), 197–208. https://doi.org/10.24908/ss.v12i2.4776

van Dijck, J., & Poell, T. (2013). Understanding social media logic. *Media and Communication*, *1*(1), 2–14. https://doi.org/10.17645/mac.v1i1.70

Vial, G. (2019). Understanding digital transformation: A review and a research agenda. *The Journal of Strategic Information Systems*, *28*(2), 118–144. https://doi.org/10.1016/j.jsis.2019.01.003

Winiecki, D. (2007). Accidental participation in control, in the small of society. *Transformations*, 14. https://scholarworks.boisestate.edu/ipt_facpubs/50

# Part 2

# Critical perspectives on digital participation

# 4 Data-driven digital participation in agri-food context

## Why should C4D and CfSC scholars and practitioners pay attention to information disorder?

*Khondokar Humayun Kabir, Ataharul Chowdhury and Uduak Edet*

## Introduction

The emergence of social and online media platforms has presented both opportunities and challenges for sustainable agriculture in the modern era (Bianco, 2022; Yadav et al., 2023). These platforms have enabled data-driven digital engagement in the agri-food space, providing information sharing and participation opportunities. However, rural communities often face limited connectivity and inclusion despite efforts to improve digital access (Salemink et al., 2017). Additionally, while these platforms were initially seen as promoting democratic values, they have led to issues such as polarization and marginalization due to conflicting and political topics. This chapter argues that addressing information disorder on social and online media platforms is critical to achieving sustainable agricultural development. It's a unique era in human history as we now inhabit an information-based society. We have access to an overwhelming amount of data, thanks to the democratization of information through the Internet. Shockingly, about 90% of the world's information has been generated within just two years, and it doubles every two years (Jayagopal & Basser, 2022). As a result, we have little time for careful editing, testing, and reflection on the information we produce. Unfortunately, this polluted information ecosystem shapes our world and distorts our perception of reality. Helfand (2016) argues that our access to excess information is becoming a burden because it undermines the integrity of the information we consume. Our current information ecosystem is severely polluted, resulting in an information disorder that has a negative impact on science and society (Wardle & Derakhshan, 2017). Experts in communication for development (C4D) and communication for social change (CfSC) have raised concerns about providing accurate and relevant information on agriculture. This is a crucial area for analyzing and understanding new challenges related to sustainable development and social change (Servaes & Lie, 2014). C4D and CfSC scholars and practitioners have been working for a long time on ways to use information and communication services, including digital media, to empower rural and agri-food communities (see Acunzo et al., 2014; Food and Agriculture Organization FAO], 2017; Gumucio-Dagron & Tufte, 2006; Servaes, 2018).

DOI: 10.4324/9781003282075-6

It is vital to address the complexities of information disorder on social and online media platforms to achieve sustainable agricultural development (Chowdhury et al., 2023; Demestichas et al., 2020).

The rise of the information age has brought to light the credibility crisis in the information we consume. Experts in the field refer to this as information disorder, which encompasses mis-, dis-, and mal-information that can cause harm due to its falseness (Wardle & Derakhshan, 2017). Misinformation is incorrect information without malicious intent, whereas disinformation is deliberately misleading and harmful. Malinformation, conversely, is the dissemination of accurate information taken out of context to cause harm. Research on misinformation has grown recently, especially after the 2016 election in the United States and during the pandemic, covering topics such as political communication, public health, and climate change (De Paor & Heravi, 2020). However, there remains a lack of research on agri-food misinformation. It is crucial to address information disorder to foster trust, ensure the integrity of decision-making processes, and promote ethical and sustainable farming practices (Chowdhury et al., 2023).

Misinformation has affected the agri-food sector, as evidenced by historical examples. Debates, controversies, and doubts regarding new practices and technologies have long plagued the agricultural industry. The Green Revolution, for instance, has faced criticism for causing environmental degradation, income inequality, unequal asset distribution, and unnecessary mechanization (IFPRI, 2002; John & Babu, 2021). This history has demonstrated how communities of scholars and practitioners have become divided on various issues despite a shared vision for development. Similarly, the development of genetically modified organisms (GMOs) for crops, livestock, and fisheries have been disrupted due to information disorder (Mintz, 2017). The Villejuif Leaflet is another historical example of misinformation in the European agri-food industry, which spread fear through flyers about toxic and cancer-causing substances in various foods and drinks (Kapferer, 1989). Recently, Sri Lanka's food and economic crisis has been attributed to misinformation, magical thinking, ideological delusion, propaganda, and short-sighted policies, polarizing people around organic farming practices (Bhowmick, 2022).

Although agri-food misinformation has been an issue in the past, the current challenges are unparalleled due to the rise of data-driven digital capitalism. The expansion of global digital infrastructure has enabled the collection of user data for the benefit of digital capitalism. This has resulted in the vast majority of the world population being able to access the Internet. This has led to a hyper-connected global society, where misinformation can rapidly spread causing significant socioeconomic consequences. Information scholars have recognized misinformation as a severe hazard to both science and society (Lee, 2020).

This chapter discusses the global problem of information disorder and its devastating impact on the agri-food industry. This chapter highlights the crucial need to tackle this issue in online and social media, especially in the context of sustainable agricultural development and data-driven digital participation. As a result, it is suggested that scholars and practitioners in C4D and CfSC must focus on information disorder to protect the livelihoods of agri-food communities.

Additionally, it is proposed as a fresh perspective for studying information disorders within the agri-food sector.

## Why should C4D and CfSC scholars and practitioners pay attention to information disorder while facilitating agricultural development?

Information plays a vital role in decision-making, knowledge sharing, and innovation adoption in sustainable agricultural development. Accurate and reliable information is crucial for identifying, implementing, and evaluating sustainable development strategies at all levels (Kelly, 1998). Farmers, policymakers, researchers, and other stakeholders require precise information to make informed decisions about agricultural practices, technologies, and market opportunities. However, the emergence of social and online media has revolutionized the speed and reach of information dissemination, leading to the rapid spread of both accurate and false information globally (Omoregie & Ryall, 2023). Considering this, it is important to re-evaluate how information is generated, processed, and accepted in agriculture and food systems.

The use of information systems in the agri-food sector has undergone a significant transformation in the last three decades. Previously, agricultural extension services supported farmers through front-line extension workers and back-office activities (Belay & Abebaw, 2004). They employed various methods and tools to help farmers at different stages of their development by disseminating necessary information and knowledge. During the post-Green-Revolution period, farmers relied on traditional information and capacity development sources, such as face-to-face training, posters, radio, and television to meet their information needs (Dejene, 1989; Quizon et al., 2001; Zalom, 1993). However, the digital revolution has changed how agri-food actors search for relevant knowledge and information (Bentley et al., 2016). The recent COVID-19 pandemic has made it challenging for extension and advisory services to continue their work in traditional ways, leading to a shift toward digital extension, online events, and agricultural training videos to enhance local innovation (FAO, 2020: Sekabira et al., 2023; Van Mele et al., 2010).

In the digital world, misinformation and disinformation can mislead agri-food actors, leading to sub-optimal decisions and harmful practices. Social and online media platforms use algorithms to shape how content reaches the target audience based on their personal data and participation. However, the untransparent use of these algorithms, which are corporate secrets, raises concerns about information manipulation and the promotion of specific interests (Fuchs, 2018). Using user data without explicit consent and the control exerted by these platforms over algorithms is a particular issue.

Social and online media platforms gather information from users' interactions, such as messaging, posting, and liking, to personalize content delivery to individual users. This is made possible by algorithms and Big Data, which enable precise interest-based targeting (Berry, 2019; Fuchs, 2018). However, this process exposes the workings of social media capitalism (also known as digital capitalism), where different actors manipulate these algorithms to serve their own agendas.

As a result, users are often presented with a distorted version of reality, leading to misinformation on various topics. In the realm of agricultural development, the impact of algorithms on information disorder is of great importance (Tripoli & Schmidhuber, 2018). Social media platforms and search engines rely on these algorithms to determine the visibility and accessibility of agricultural information. This leaves users vulnerable to misinformation regarding farming practices, food safety, and environmental impacts, which can distort their understanding of these topics. Therefore, it is imperative to acknowledge the influence of these algorithms and the motivations behind their implementation. Social and online media platforms must increase their transparency and accountability to combat information disorder in agriculture. The negative effects of information manipulation can be mitigated by advocating for algorithmic transparency and making their processes and criteria more explicit (Reviglio & Agosti, 2020).

Additionally, promoting digital media literacy among users is vital to empowering individuals to critically evaluate the information they encounter. To achieve goals of sustainable agricultural development, stakeholders must come to a consensus on critical issues like environmental concerns, greenhouse gas emissions (GHGs) from the livestock sector, and the use of fertilizers in crop farming. Unfortunately, these stakeholders often have polarized perspectives on scientific knowledge, often fueled by misinformation. Forming a common consensus based on accurate and reliable information makes it challenging. Agri-food actors may spread misinformation for reasons other than their ability to identify and understand false information. Cognitive scientists suggest that rational thinking and psychological training can help individuals overcome misinformation hazards (Benkler et al., 2018; Roozenbeek & Van der Linden, 2019). However, we believe that addressing misinformation requires a deeper understanding of its root causes and the larger information ecosystem. Focusing solely on cognitive training ignores the contradictions of capitalist production, including the role of algorithms in shaping information. By recognizing this, we can move beyond the moral panic surrounding false information and address the underlying systemic factors that perpetuate misinformation in agricultural practices (Morozov, 2017).

In the world of digital capitalism, the spread of misinformation is a major issue due to the availability of user data, especially their preferences. This allows corporations and politicians to manipulate people's emotions for their gain, hindering efforts to combat information inaccuracies (Boler & Davis, 2020). The concept of surveillance capitalism (Zuboff, 2015) further complicates the issue by treating human experiences as raw materials. Additionally, the decline in public trust in science and institutions, driven by inequality and political polarization, has made agri-food stakeholders vulnerable to misinformation, regardless of their political beliefs.

In what follows, the impact of information disorder on agricultural development cannot be ignored. The use of social and online media platforms and algorithms has disrupted traditional information channels, amplifying misinformation in the agri-food sector. To address this issue, transparency, accountability, and algorithmic transparency from platforms are needed, along with promoting digital media literacy. Recognizing the broader context of digital capitalism and its impact on information disorder is essential, acknowledging the inherent contradictions and challenges posed by the capitalist mode of production.

### Agricultural misinformation: case studies

As researchers and practitioners in C4D and CfSC, it is important to remember the devastating consequences of agricultural misinformation throughout history. Despite being extensively documented by historians, it often gets overshadowed by sensationalist clickbait in today's media landscape. It is important to acknowledge that agricultural misinformation has had the most severe human toll in the twentieth and twenty-first centuries, making it one of the deadliest forms of misinformation. Case studies (Table 4.1) show that both the Global North and South are equally vulnerable to misinformation.

These examples serve as a reminder to scholars in the fields of C4D and CfSC to consider the impact of information disorder on agricultural development. These instances demonstrate the severe consequences of disinformation on agricultural practices, food supplies, and community cohesion. C4D and CfSC experts must address the threats of agricultural misinformation, especially in the current information era driven by social and online media. We will further turn into this discussion in the next section. These platforms have extensive reach and rapid dissemination capabilities, making them a prime breeding ground for the spread of misinformation among rural farmers (Chowdhury et al., 2023; Ogbette et al., 2019). Rural farmers are not immune to the disruptive effects of misinformation, as it infiltrates their screens, influencing their decision-making, and shaping the future of sustainable agricultural development. To combat the challenges of misinformation in the constantly evolving media landscape, researchers, and practitioners in C4D and CfSC must adopt holistic approaches encompassing media literacy, community engagement, and innovative communication strategies. In a later section of this chapter, this issue will be re-examined drawing on the political economy of information.

### Areas of the agricultural / food system that are susceptible to information disorder

In agriculture, there are several controversial issues, such as GMOs, global warming, and the use of chemical fertilizers (Norwood et al., 2015). These issues make various aspects of the farming system susceptible to information disorder (Seed World Europe, 2020; Tsiboe & Turner, 2022), particularly when considering the risks and uncertainties surrounding changing weather conditions, fluctuating market prices, government policies, crop yields, and more (Tsiboe & Turner, 2022). Information disorder, including misinformation, disinformation, and malinformation, thrives when science fails to provide timely responses or when questions are left unanswered (Seed World Europe, 2020). Uncertainties can lead to biases in information-seeking and processing, leading society to choose what to believe or who to blame (Dugas et al., 2005; Rains & Tukachinsky, 2015; Seed World Europe, 2020). Recognizing this sheds light on how information disorder spreads in the agri-food context (see Figure 4.1).

*Table 4.1* Cases of agricultural misinformation and their implications

| Case | Description | Impacts | Lessons |
|---|---|---|---|
| Holodomor (Hunger Plague) in Ukraine (1932–1933) (Kolchinsky et al., 2017; Reznik and Fet, 2019; Soyfer, 1989) | The Holodomor, a man-made famine that ravaged Soviet Ukraine from 1932 to 1933, resulted in the catastrophic loss of millions of lives. While the primary causes of this tragedy lie in Stalin's brutal policies of forced collectivization and grain requisitioning, the role of Lysenkoism, a pseudo-scientific theory championed by Soviet botanist Trofim Lysenko, cannot be understated. Lysenkoism, rooted in neo-Lamarckian ideology, asserted that environmental influences on crop plants could be inherited through all cells of the organism. This flawed concept, devoid of any scientific basis, stood in stark contrast to established Mendelian genetics.<br><br>Lysenkoism's impact extended beyond theoretical debates; it became the official doctrine of Soviet agriculture. The implementation of Lysenkoist practices, such as vernalization (exposing seeds to cold temperatures), close planting, and the rejection of fertilizers and pesticides, had a disastrous effect on agricultural productivity. These practices, devoid of scientific merit, led to a decline in crop yields, soil degradation, and increased susceptibility to pests and diseases. | The dissemination of agricultural misinformation during the Stalin era by Lysenko resulted in catastrophic consequences, causing more deaths than any other scientist in history. His refusal to acknowledge genetics and natural selection led to incorrect farming practices, such as planting seeds in close proximity without using pesticides or fertilizers. These practices aggravated the agricultural problems and worsened the famine in Ukraine. | Although the scientific community discredited Lysenkoism in 1965, the damage it caused to agricultural science was permanent. This case highlights the link between political ideology and promoting unproven scientific theories. Lysenko's loyalty to Socialist ideology coincided with the ruling regime's interests, resembling political propaganda tactics. The Holodomor case stresses the significance of fighting agricultural misinformation and endorsing evidence-based practices to avoid comparable tragedies in the future. |

| The Great Leap Forward in China (1958–1962) (Peng, 1987) | In 1958, China embarked on the Great Leap Forward, a bold effort to modernize its rural and agricultural sectors by promoting collectivism and rapid industrialization. Unfortunately, the agricultural reforms were poorly implemented, leading to disastrous consequences. The misguided policies and practices of the time resulted in a significant decline in agricultural productivity, leading to one of the most catastrophic famines in human history. | China's Great Leap Forward caused immense food shortages and widespread famine, which resulted in the estimated deaths of 15–55 million individuals. The agricultural practices employed during this time, including communal farming and prioritizing backyard steel production, were misguided and disturbed the traditional agricultural systems, leading to significant disruptions in food production. The lack of expertise and knowledge in implementing these radical changes has had disastrous consequences for the Chinese population. | The Great Leap Forward serves as a poignant reminder of the hazards of uninformed agricultural policies and the necessity of adopting evidence-based methods in agricultural and rural development. This case emphasizes the importance of comprehending the intricacies of agricultural systems and the potential repercussions of hastily executed and inadequately planned reforms. It underscores the significance of informed decision-making, research-based practices, and integrating indigenous knowledge to promote sustainable agricultural development. Lessons learned from the Great Leap Forward are crucial in ensuring China's food security and rural prosperity, as well as that of other nations. |
| Misinformed Agricultural Policy in Sri Lanka (Bhowmick, 2022) | The negative attitudes toward GMOs in Sri Lanka have had a notable impact on how the public views agriculture and the policies surrounding it. Recently, the president of Sri Lanka announced a goal to become the world's first fully organic farming country, influenced by misguided agricultural policies. Unfortunately, this decision had dire consequences, resulting in a significant food shortage and posing a major threat to food security in the country. | In Sri Lanka, flawed agricultural policies have led to a shortage of food supply, causing food insecurity, affecting both farmers and the general population. The exclusive focus on organic farming disregarded the potential advantages of GMOs and modern agricultural methods in increasing productivity and combating issues like pests and diseases. Consequently, the country struggled to meet the food requirements of its people. | In Sri Lanka, the negative effects of an ill-informed agricultural policy have made it clear that agricultural decision-making must be based on reliable evidence. To make informed decisions, it is crucial to thoroughly understand the advantages and disadvantages of various farming practices, such as the use of chemical fertilizers and pesticides. This highlights the importance of a well-rounded approach that considers both scientific evidence and local circumstances while also considering the long-term effects on food security and sustainability. |

*(Continued)*

*Table 4.1* (Continued)

| Case | Description | Impacts | Lessons |
| --- | --- | --- | --- |
| Destruction of Golden Rice Field Trial in the Philippines (Alberts et al., 2013) | A group of activists attacked and destroyed a field trial of genetically modified (GM) golden rice in the Bicol region of the Philippines in 2013. The purpose of developing this rice variety was to combat the serious health risks associated with vitamin A deficiency, which is prevalent in the region and particularly affects children under five, often resulting in blindness and increased mortality rates. Golden rice was engineered to produce beta-carotene, a precursor to vitamin A. | The disruption of the golden rice field experiment had impeded the progress toward combating malnutrition and avoidable illnesses, leading to a substantial loss of human lives. | The vandalism of the golden rice field trial in the Philippines clearly illustrates how misinformation can negatively impact agricultural progress and public health. It highlights the importance of disseminating accurate information and engaging in evidence-based discussions to counteract the spread of false narratives. |
| Agricultural Misinformation Induced Vandalism in the United States (Fought, 2013) | In 2013, a group of vandals attacked Syngenta's 1,000 genetically modified sugar beet plants in Oregon, causing significant damage. Such acts of economic sabotage were attributed to anti-genetically modified crop organizations like Greenpeace and the Earth Liberation Front. | Syngenta had suffered significant financial losses, research disruption, and hampered the potential benefits of genetically modified crops, such as increased yields and reduced reliance on herbicides. | This example sheds light on the contentious debate surrounding GMOs, highlighting the impact of false information. It emphasizes the importance of effective communication, evidence-based decision-making, and constructive dialogue to bridge the gap between scientific research and public comprehension. |

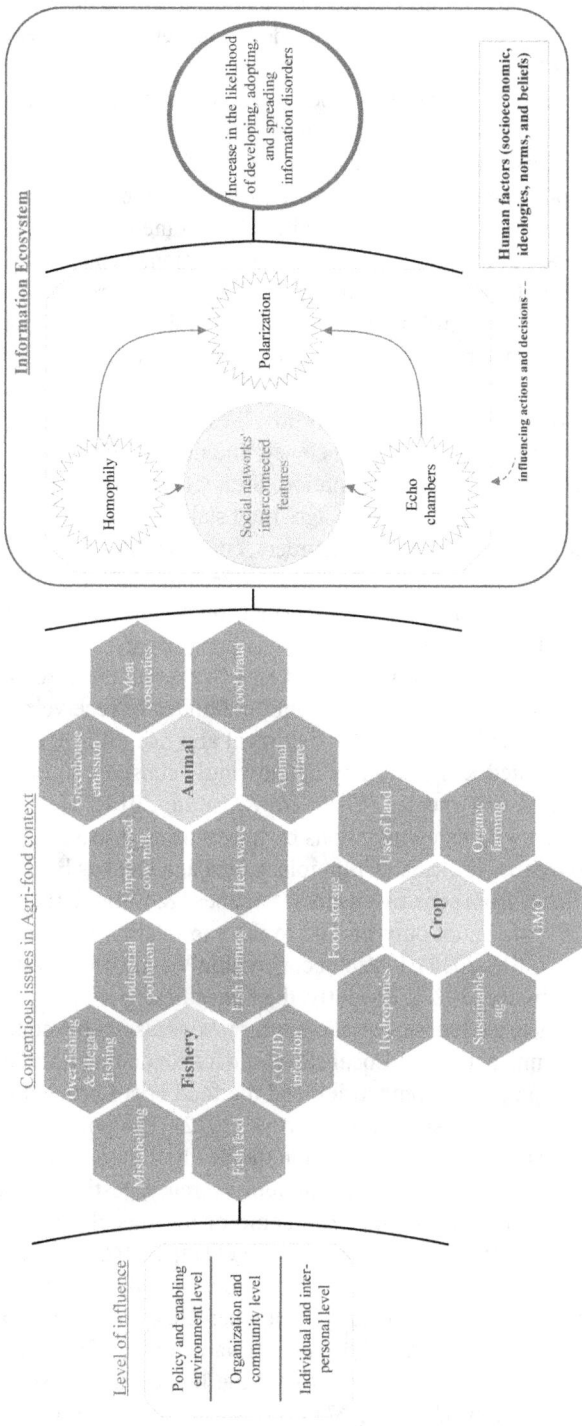

*Figure 4.1* Socio-technical factors influencing the development, adoption, and dissemination of mis-dis-mal-information.

Figure 4.1 shows how homophily, echo chambers, and polarization work together to create and spread information disorder (Treen et al., 2020). Homophily is when people prefer to spend time with those who share their beliefs. In this context, people tend to look for information supporting their existing opinions rather than seeking out different viewpoints. Echo chambers are spaces where people are exposed mainly to information that confirms their beliefs. These spaces do not offer diverse information sources and can create a cycle that amplifies misinformation. Polarization is when people with opposing beliefs become more divided. This can create echo chambers that further spread misinformation. Echo chambers, which are formed by homophily and polarization, provide a fertile ground for the spread of misinformation. This misinformation can spread rapidly within these chambers and solidify polarized perspectives. These dynamics are particularly relevant in the agri-food industry, where controversial topics like GMOs, organic agriculture, and animal welfare abound. Figure 4.1 portrays how these relationships affect individuals, organizations, and policies at various levels and underscores the importance of addressing the root causes of misinformation. Strategies that tackle these underlying causes are urgently needed. Various agri-food stakeholders are involved in the ongoing discussion about information disorders, contributing to a number of areas, as shown in Figure 4.1. In the fishery industry, enforcing a daily COVID-19 test for fishermen and fish in southern China has sparked controversy among scientists and on social media platforms (Cheung, 2022). This apprehension may stem from the first COVID-19 outbreak, which was linked to a live animal in a seafood market in Wuhan, China (Allen, 2022). Fish farming continues to receive criticism due to the overuse of marine ingredients for fish meal and fish oil in feed production. Consumers in the United Kingdom may unknowingly consume more wild fish by eating certain farmed fish species (Shepherd & Little, 2014). However, farmed fish can also threaten wild fish populations by transmitting diseases and pathogens such as piscine reovirus. In British Columbia, Canada, a ban has been imposed on phasing out fish farms from sensitive waters (Shepherd & Little, 2014), resulting in a backlash. Overfishing is not only limited to the exploitation of wild fish for fish feeds (Shepherd & Little, 2014), it also occurs within West African waters, most of which are illegal (Nwoye, 2022). Recently, the Senegalese government secretly issued fishing licenses to a Chinese vessel operator who is also believed to be associated with illegal fishing activities. Local stakeholders condemn this action because they struggle to keep up with competition and the food crises brought on by overfishing from European and Asian fleets (Nwoye, 2022). The pollution of the Oder River in Poland and Germany has resulted in the death of thousands of fish. Toxic substances, including mesitylene, lead, cadmium, mercury, pesticides, elevated salt levels, high temperature, and drought, are some of the speculated causes of this ecological disaster, but there is no confirmation yet (Ptak, 2022; Reuters, 2022).

Mislabeling of seafood is also an area prone to information disorder as the deliberate intent to deceive consumers is evident when cheap fish species are labeled as highly priced or farmed fish as wild species (Naaum et al., 2016). Almost 50% of fish examined in German restaurants found the less desired Japanese scallops mislabeled as king scallops (Leahy, 2021). The economic value of mislabeling is a major motivation. Unfortunately, this is enhanced by the porous seafood supply chain (Leahy, 2021).

The magnitude of greenhouse gas emissions from animal waste is enormous but overlooked in climate change discussions (Mojeed, 2022). Reported food fraud cases to indict China and India for processing dead animals and human bodies into corn beef for export (Akinpelu, 2019; Mohammad et al., 2015). Similarly, Bill Gates is held responsible for the death of over 2,000 cattle in Kansas City (U.S.A.), even though the government attributes the catastrophe to climate change (Kochi, 2022). Farmers' frustration over how to tail dock their sheep (New Zealand, Ministry for Primary Industries, 2021), and the Canadian government's resolve to prosecute those who distribute raw milk (Selick, 2018), are some of the areas prone to information disorder in animal farming.

In addition, meat color and appearance are significant factors in consumer-purchasing decisions, so for 50 years, food color manufacturers have supported the food industry giants with synthetic and organic food colorants for the prevention of microbial invasion and oxidation reactions but most especially for the cosmetics and appearance of food products including red meat (Uchegbu et al., 2020). In the United States, the Food and Drug Administration regulation ensures strict regulations before using these products. Generally, manufacturers claim to be well-informed and experienced about the regulatory requirements of administering approved dosages and specifications for food markets. Notwithstanding, consumers remain concerned about its health implications, leading to several opinions and research on the hazards of food colorants (Uchegbu et al., 2020). Ultimately, natural food colorants are recommended over synthetic food colorants (Uchegbu et al., 2020), although some consumers will avoid purchasing such meat products.

Using chemical fertilizers, pesticides, and genetically modified food may improve crop production (Norwood et al., 2015). Nevertheless, supporters of organic farming point to the harmful emission of nitric oxide, a greenhouse gas produced by manufacturing and applying fertilizer (Chai et al., 2019). Although the health effects from fertilizer usage are found to be relatively low (Nganchamung et al., 2017), health concerns from consuming GMO crops are still largely at the forefront of most debates and media discussions as there are a lot of uncertainties around the topic including the possible effects of genetically modified (GM) foods on health and environmental safety. The concerns remain speculative but should be taken in good faith (Zhang et al., 2016).

Hydroponic farming eliminates the use of soil by suspending plant roots in nutrient-rich water (Cordes, 2022; Maygar, 2022). This invention guarantees food quality and safety (Maygar, 2022). Still, a salmonella outbreak in Great Lakes region in the United States, which infected over 30 people, proves that hydroponics presents substantial contamination risks (Cordes, 2022). Vertical farming is also a form of hydroponics, and its groundbreaking innovation for higher crop yield is well established (Maygar, 2022). However, the claim that it plays a huge role in sustainable agriculture is countered by the fact that vertical farms use much more energy, making them less environmentally sustainable than traditional farming (Naus, 2018). The public and government have condemned the wrong application of organophosphates and gammalin to store and preserve beans in Nigeria because of its damage to the human kidneys and overall danger to health (Peters, 2018).

Cattle herders and sedentary farmers in West Africa have long been conflicted over using agricultural land and natural resources (Cabot, 2017). In Nigeria, this can be traced to the pre-colonial era, and the main source of the conflict is tied to its role in the survival of each group (Bello & Abdullahi, 2021; Cabot, 2017). Environmental and ecological concerns, such as resource scarcity, desertification, insecurity in northern Nigeria, population explosion, and urbanization, are some of the contributing factors (Cabot, 2017; Nwankwo, 2022), but apart from these, polarized views and opinions evolve from local disagreements and regional and geopolitical interests. Some of these views center on criticism of government policies, tribalism against the South, the corrupt media, and allegations that top government officials own the cattle. All these have contributed to information disorder.

### Power, profit, and propaganda: a renewed perspective for C4D and CfSC scholars to understand and counteract information disorders in the agri-food context

The scholars of C4D and CfSC have utilized different theories, including Rogers' Innovation Diffusion Theory, Technology Acceptance Model (TAM), Theory of Planned Behavior (TPB), and Theory of Reasoned Action (TRA), to comprehend decision-making and the process of adopting agricultural innovation. However, these theories do not encompass the intricacies of information disorders within the agri-food system. As a result, a new perspective is required, drawing from the political economy of information to shed light on these emerging issues. We suggest using the lens of power, profit, and propaganda (PPP) to understand better the information disorders occurring in the agri-food industry (Figure 4.2). Although there is no single established theory for PPP, its individual concepts have been widely explored in various fields. For instance, political science and psychology literature have studied misinformation and utilized PPP to offer insights into the distribution and consumption of misinformation and the mechanisms that contribute to its spread through networks (see Guess & Lyons, 2020; Kuo & Marwick, 2021). Additionally, events such as the 1973 oil crisis and the current dispute between fossil fuels and renewable energy exemplify the interconnected dynamics of PPP in shaping public perception and decision-making in the energy sector (Bini et al., 2016; de Kadt, 1983). By examining the effects of power dynamics, economic interests, and deliberate misinformation campaigns on information disorders, researchers can gain insight into how these factors influence decision-making and innovation adoption, ultimately shaping the information landscape in the agri-food system. Power distribution within the industry is a major factor in shaping information flows and narratives. Powerful actors, such as agrochemical companies, influential policymakers, and dominant market players, often use their position to mold information to their advantage and maintain their hold on power. Additionally, profit-driven agendas and economic motivations can contribute to the spread of misleading or biased information, particularly in industries like agrochemicals and large-scale agriculture. These motivations may overshadow environmental or health concerns, disseminating misinformation that favors certain products or practices. Finally, deliberate propaganda also plays a role in distorting information in the agri-food sector. Various actors, including biased media outlets, industry players, and online campaigns, intentionally spread misleading or false information in an effort to manipulate public perceptions, create confusion, and perpetuate information disorder within the industry.

This chapter asserts that various contextual factors influence the PPP framework (Figure 4.2) (see Van der Linden, 2023), including (1) social and online platforms, (2) lack of scientific literacy, (3) confirmation bias and cognitive biases, (4) political interests and ideological polarization, (5) media landscape and journalism, and (6) globalization and supply chain complexity (Brennan, 2017; Haas, 2012; Jolls & Johnsen, 2017; Kincaid & Cole, 2019; Waisbord, 2018). As mentioned in the first factor, the rise of digital technologies and social media platforms enables the spread of misinformation, allowing propaganda to reach a wider audience and facilitating its viral nature. This aligns with the profit-driven agendas of powerful actors in the agri-food sector, as outlined in the PPP framework. These actors may use digital platforms to disseminate misleading information and shape narratives that serve their interests.

Scientific literacy, or lack thereof, is another contextual factor that interacts with the propaganda aspect of the PPP framework. Farmers, consumers, and the general public, with a less extensive background in scientific principles, may be more likely to accept and transmit information without thorough evaluation. Farmers, consumers, and the general public with a limited understanding of complex scientific concepts are more susceptible to accepting and spreading misinformation. This reinforces power dynamics where influential actors exploit the lack of scientific literacy to manipulate information in their favor. The third factor highlights how confirmation bias and other cognitive biases can fuel the spread of

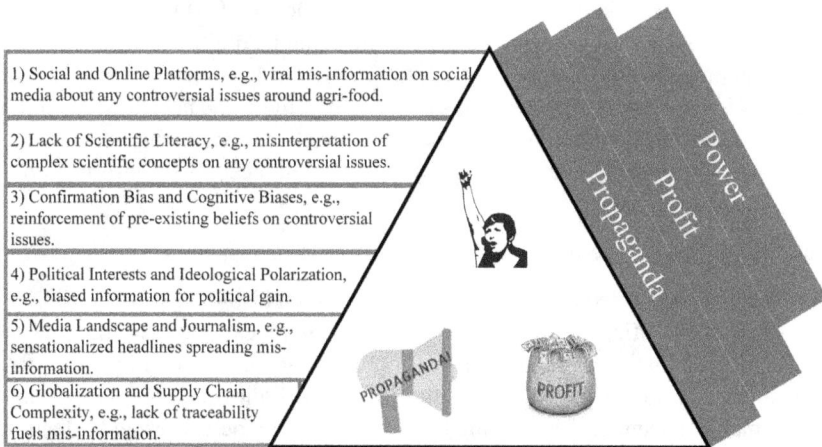

*Figure 4.2* Power, profit, and propaganda (PPP) with the various contextual factors that influence the dynamics of information disorder in the agri-food context.

misinformation, which aligns with the propaganda element of the PPP framework. People often seek out information that supports their existing beliefs, forming echo chambers that reinforce false information. This manipulation of information ultimately serves the agendas of those in power who use propaganda tactics to control public opinion and maintain their positions of influence.

The next factor highlights how political interests and ideological polarization intertwine with the power dynamics outlined in the PPP framework. Certain stakeholders may spread biased or inaccurate information to promote their agendas, contributing to the distorted information landscape in the agri-food system.

Touching on the media landscape and journalism, which are linked to the propaganda and contextual factors of the PPP framework, is the fifth factor. With decreasing trust in traditional media, partisan news outlets, and sensationalized headlines, misinformation can easily flourish, making disseminating precise and dependable information challenging. This, unfortunately, exacerbates the information disorder already present in the agri-food industry.

Finally, the sixth factor highlights the impact of globalization and the intricate supply chain on the PPP framework's profit-driven agendas. The interdependence and complexity of the global market can lead to false information about food safety, quality, or origin, influencing consumer choices and market dynamics. This can align with the profit-seeking behavior of dominant agri-food players. In the area of GMOs and the seed business, the negative effects of information disorders on decision-making and sustainable agricultural practices are evident (Gillam, 2020). Corporations with vested interests use disinformation campaigns to manipulate public opinion by spreading false claims about the safety or benefits of GM crops. These campaigns exploit digital technologies and social media to amplify their reach (Ryan et al., 2020). The lack of scientific literacy among farmers, consumers, and the general public further contributes to the acceptance and spread of misinformation. This misleading information may distort farmers' understanding of GMOs and hinder their ability to make informed choices (Daev et al., 2016). Power dynamics, profit-driven agendas, deliberate propaganda, digital technologies, scientific literacy, and ideological polarization collectively perpetuate information disorders in the agri-food system. This example highlights the need to study information disorders in agriculture with a new perspective, considering the complex contextual factors that influence the dissemination and reception of information.

## Conclusion and future initiatives

This chapter highlighted the importance of information disorder for C4D and CfSC scholars and practitioners to acknowledge and comprehend the existence and influence of information disorder systems in promoting participation in sustainable agricultural development in the face of the digital age. Policymakers, industry players, farmers, consumers, and the media must recognize the causes and effects of mis-, dis-, and mal-information. In analyzing information disorders in the agri-food industry, we uncover the intricate dynamics that impact the spread and consequences of misinformation, disinformation, and malinformation within the sector.

Topics such as GMOs, chemical fertilizers, fishery practices, seafood mislabeling, animal farming methods, food colorants, and genetically modified food are particularly prone to information disorders due to their controversial and uncertain nature. The chapter explored the relationship between homophily, echo chambers, polarization, and information disorder, which sheds light on how misinformation spreads and polarized views become deeply ingrained. We introduce a new perspective on PPP to study information disorders in the agri-food industry. Power dynamics, economic incentives, and deliberate propaganda within the industry all influence the flow of information, the narratives that emerge, and the decision-making processes. Additional contextual factors such as social and online platforms, limited scientific literacy, confirmation bias, political interests, media landscape, and globalization also contribute to information disorders. To combat misinformation, it is important to prioritize scientific literacy, critical thinking, and diverse perspectives. This can be achieved through transparency, fact-checking, and relying on reliable sources of information. All stakeholders must work together when addressing information disorders in the agri-food industry. Public-private partnerships, educational initiatives, media literacy programs, and ethical practices within the sector can contribute to a more accurate and reliable information landscape. By doing so, the agri-food system can make informed decisions, navigate complex issues, and adopt sustainable practices that benefit the industry and society.

## Acknowledgments

This work was supported by the insight grant from the Social Sciences and Humanities Research Council of Canada (SSHRC). Many thanks to Md. Firoze Alam for discussions about some initial ideas of this paper.

## References

Acunzo, M., Pafumi, M., Torres, C. S., & Tirol, M. S. (2014). *Communication for rural development source book*. Food and Agriculture Organization. https://www.fao.org/3/i3492e/i3492e.pdf

Akinpelu, Y. (2019). False! China is not importing dead bodies as corned beef into Africa. *Dubawa*. https://dubawa.org/false-china-is-not-importing-dead-bodies-as-corned-beef-into-africa/amp/

Alberts, B., Beachy, R., Baulcombe, D., Blobel, G., Datta, S., Fedoroff, N., Kennedy, D., Khush, G. S., Peacock, J., Rees, M., & Sharp, P. (2013). Standing up for GMOs. *Science, 341*(6152), 1320. https://doi.org/10.1126/science.1245017

Allen, K. (2022, August 22). Covid in China: Fish tested amid Xiamen outbreak. *BBC.com*. https://www.bbc.com/news/world-asia-china-62593217

Belay, K., & Abebaw, D. (2004). Challenges facing agricultural extension agents: A case study from South-Western Ethiopia. *African Development Review, 16*(1), 139–68. https://doi.org/10.1111/j.1467-8268.2004.00087.x

Bello, B., & Abdullahi, M. M. (2021). Farmers–herdsmen conflict, cattle rustling, and banditry: The dialectics of insecurity in Anka and Maradun Local Government Area of Zamfara State, Nigeria. *SAGE Open, 11*(4), 1.

Benkler, Y., Faris, R., & Roberts, H. (2018). *Network propaganda: Manipulation, disinformation, and radicalization in American politics.* Oxford University Press.

Bentley, J. W., Mele, P. V., Harun-ar-Rashid, M., & Krupnik, T. J. (2016). Distributing and showing farmer learning videos in Bangladesh. *Journal of Agricultural Education and Extension, 22*(2), 179–97. https://doi.org/10.1080/1389224X.2015.1026365

Berry, D. M. (2019). Against infrasomatization: Towards a critical theory of algorithms. In D. Bigo, E. Isin, & E. Ruppert (Eds.), *Data politics: Worlds, subjects, rights* (pp. 43–63). Routledge. https://doi.org/10.4324/9781315167305

Bhowmick, S. (2022, July 2033). *Understanding the economic issues in Sri Lanka's current debacle.* Observer Research Foundation. http://20.244.136.131/research/understanding-the-economic-issues-in-sri-lanka-s-current-debacle

Bianco, A. (2022). Agriculture and new technologies: A basic challenge for the twenty-first century. In C. Facioni, G. Di Francesco, & P. Corvo (Eds.), *Italian studies on food and quality of life* (pp. 113–24). Springer International Publishing.

Bini, E., Garavini, G., & Romero, F. (Eds.) (2016). *Oil shock: The 1973 crisis and its economic legacy.* Bloomsbury Publishing.

Boler, M., & Davis, E. (Eds.) (2020). *Affective politics of digital media: Propaganda by other means.* Routledge.

Brennan, J. (2017). Propaganda about propaganda. *Critical Review, 29*(1), 34–48. https://doi.org/10.1080/08913811.2017.1290326

Cabot, C. (2017). Climate change and farmer–herder conflicts in West Africa. In H. G. Brauch (Ed.), *Climate change, security risks and conflict reduction in Africa.* Hexagon Series on Human and Environmental Security and Peace, 12. Springer. https://doi.org/10.1007/978-3-642-29237-8_2

Chai, R., Ye, X., & Ma, C. (2019). Greenhouse gas emissions from synthetic nitrogen manufacture and fertilization for main upland crops in China. *Carbon Balance Manage, 14*(20). https://doi.org/10.1186/s13021-019-0133-9

Cheung, R. (2022, August 18). A Chinese city is doing COVID tests on fresh fish. *Vice.* https://www.vice.com/amp/en/article/wxnj3m/chinese-city-tests-fish-for-covid

Chowdhury, A., Kabir, K. H., Abdulai, A.-R., & Alam, M. F. (2023). Systematic review of misinformation in social and online media for the development of an analytical framework for Agri-Food sector. *Sustainability, 15*(6), 4753. https://doi.org/10.3390/su15064753

Cordes, J. (2022). Salmonella outbreak traced to hydroponic farm. https://www.wric.com/news/8news-consumer-alerts/salmonella-outbreak-traced-to-hydroponic-farm/

Daev, E. V., Zabarin, A. V., Barkova, S. M., & Dukel'skaya, A. V. (2016). Distortions of scientific information as a source of the formation of tension in society: The GMO case. *Russian Journal of Genetics: Applied Research, 6*, 633–45.

De Kadt, M. (1983). Energy corporation propaganda: A weapon against public policy. *Review of Radical Political Economics, 15*(3), 35–50.

De Paor, S., & Heravi, B. (2020). Information literacy and fake news: How the field of librarianship can help combat the epidemic of fake news. *Journal of Academic Librarianship, 46*(5), 102218. https://doi.org/10.1016/j.acalib.2020.102218

Dejene, A. (1989). The training and visit agricultural extension in rainfed agriculture: Lessons from Ethiopia. *World Development, 17*(10), 1647–59. https://doi.org/10.1016/0305-750X(89)90034-X

Demestichas, K., Remoundou, K., & Adamopoulou, E. (2020). Food for thought: Fighting fake news and online disinformation. *I.T. Professional, 22*(2), 28–34. https://doi.org/10.1109/MITP.2020.2978043

Dugas, M. J., Hedayati, M., Karavidas, A., Buhr, K., Francis, K., & Phillips, N. A. (2005). Intolerance of uncertainty and information processing: Evidence of biased recall and interpretations. *Cognitive Therapy and Research, 29,* 57–70. https://doi.org/10.1007/s10608-005-1648-9

Food and Agriculture Organization (FAO) (2017). *Inclusive rural communication services: Building evidence, informing policy.* Food and Agriculture Organization of the United Nations. https://www.fao.org/3/i6535e/i6535e.pdf

Food and Agriculture Organization (FAO) (2020). *Extension and advisory services: At the frontline of the response to COVID-19 to ensure food security.* Food and Agriculture Organization of the United Nations. https://doi.org/10.4060/ca8710en

Fought, T. (2013). GMO sugar beets destroyed in 'sabotage' at Oregon plots. *The Spokesman-Review.* https://www.spokesman.com/stories/2013/jun/21/gmo-sugar-beets-destroyed-in-sabotage-at-oregon/

Fuchs, C. (2018). Propaganda 2.0: Herman and Chomsky's propaganda model in the age of the internet, big data and social media. In J. Pedro-Carañana, D. Broudy, & J. Klaehn (Eds.), *Propaganda model today: Filtering perception and awareness: Filtering perception and awareness* (pp. 71–91). University of Westminster Press.

Gillam, C. (2020). Finding and following the facts in an era of fake news. In D. B. Sachsman & J. M. Valenti (Eds.), *Routledge handbook of environmental journalism* (pp. 83–94). Routledge. https://doi.org/10.4324/9781351068406

Guess, A., & Lyons, B. (2020). Misinformation, disinformation, and online propaganda. In N. Persily & J. Tucker (Eds.), *Social media and democracy: The state of the field, prospects for reform* (pp. 10–33). Cambridge University Press.

Gumucio-Dagron, A., & Tufte, T. (Eds.) (2006). *Communication for social change anthology: Historical and contemporary readings.* Communication for Social Change Consortium.

Haas, M. L. (2012). *The clash of ideologies: Middle Eastern politics and American security.* Oxford University Press.

Helfand, D. J. (2016). *A survival guide to the misinformation age: Scientific habits of mind.* Columbia University Press. https://doi.org/10.1080/17447143.2014.982655

International Food Policy Research Institute (IFPRI) (2002). Green revolution: Curse or blessing? [Issue Brief.] https://www.ifpri.org/publication/green-revolution

Jayagopal, V., & Basser, K. K. (2022). Data management and big data analytics: Data management in digital economy. In Information Resources Management Association (Ed.), *Research anthology on Big Data analytics, architectures, and applications* (pp. 1614–33). IGI Global. https://doi.org/10.4018/978-1-6684-3662-2

John, D. A., & Babu, G. R. 2021. Lessons from the aftermaths of green revolution on food system and health. *Frontiers in Sustainable Food Systems, 5,* 644559. https://doi.org/10.3389/fsufs.2021.644559

Jolls, T., & Johnsen, M. (2017). Media literacy: A foundational skill for democracy in the 21st century. *Hastings Law Journal, 69*(5), 1379–408. https://repository.uclawsf.edu/cgi/viewcontent.cgi?article=3826&context=hastings_law_journal

Kapferer, J. N. (1989). A mass poisoning rumor in Europe. *Public Opinion Quarterly, 53*(4), 467–81. https://doi.org/10.1086/269167

Kelly, K. L. (1998). A systems approach to identifying decisive information for sustainable development. *European Journal of Operational Research, 109*(2), 452–64.

Kincaid, J., & Cole, R. L. (2019). Attachments to multiple communities, trust in governments, political polarization, and public attitudes toward immigration in the United States. In J. Jedwab & J. Kincaid (Eds.), *Identities, trust, and cohesion in federal systems: Public perspectives* (pp. 147–80). McGill-Queen's University Press. https://doi.org/10.2307/j. ctvdtpjdq.9

Kochi, S. (2022). Fact check: Cattle in Kansas were killed by high temperatures, humidity. *USAToday.com*. https://www.usatoday.com/story/news/factcheck/2022/06/30/fact-check-cattle-kansas-killed-high-temperatures-humidity-bill-gates/7720674001/

Kolchinsky, E. I., Kutschera, U., Hossfeld, U., & Levit, G. S. (2017). Russia's new Lysenko-ism. *Current Biology, 27*(19): R1042–47. https://doi.org/10.1016/j.cub.2017.07.045

Kuo, R., & Marwick, A. (2021, August 12). Critical disinformation studies: History, power, and politics. Harvard Kennedy School (HKS), *Misinformation Review*. https://doi.org/10.37016/mr-2020-76

Leahy, S. (2021, March 2021). Revealed: Seafood fraud happening on a vast global scale. *The Guardian*. https://www.theguardian.com/environment/2021/mar/15/revealed-seafood-happening-on-a-vast-global-scale

Lee, A. (2020). Online hoaxes, existential threat, and internet shutdown: A case study of securitization dynamics in Indonesia. *Journal of Indonesian Social Sciences and Humanities, 10*(1), 17–34. https://doi.org/10.14203/jissh.v10i1.156

Maygar, J. (2022). Indoor, vertical farming has a big role in sustainable agriculture. *Forbes.com*. https://www.forbes.com/sites/sap/2022/02/02/indoor-vertical-farming-has-a-big-role-in-sustainable-agriculture/?sh=650a499e6013

Mintz, K. (2017). Arguments and actors in recent debates over U.S. genetically modified organisms (GMOs). *Journal of Environmental Studies and Sciences, 7*(1), 1–9. https://doi.org/10.1007/s13412-016-0371-z

Mohammad, B., & Heena, J., Parveez, P., Syed, B., Subha, G., Asif, B., Rajesh, W., & Kausar, Q. (2015). Fraudulent adulteration/substitution of meat: A review. *International Journal of Recent Research and Applied Studies, 2*, 22–33.

Mojeed, A. (2022, July 24). Inside: Slaughterhouses where animal wastes fuel green-house gas emission. *Premium Times*. https://www.premiumtimesng.com/news/headlines/544460-special-report-inside-slaughterhouses-where-animal-wastes-fuel-greenhouse-gas-emission.html

Morozov, E. (2017, January 8). Moral panic over fake news hides the real enemy – The digital giants. *The Guardian*. https://www.theguardian.com/commentisfree/2017/jan/08/blaming-fake-news-not-the-answer-democracy-crisis

Naaum, A. M., Warner, K., Mariani, S., Hanner, R. H., & Carolin, C. D. (2016). Seafood mislabeling incidence and impacts. In A. M. Naaum & R. H. Hanner (Eds.), *Seafood authenticity and traceability* (pp. 3–26). Academic Press. https://doi.org/10.1016/B978-0-12-801592-6.00001-2

Naus, T. (2018, August 29). Is vertical farming really sustainable? *EIT Food*. [Blog.] https://www.eitfood.eu/blog/is-vertical-farming-really-sustainable

New Zealand, Ministry for Primary Industries (2021, September 2). Heads up on tail-docking. *Facebook.com*. https://m.facebook.com/MPIgovtnz/photos/p.4336059366475806/4336059366475806/

Nganchamung, T., Robson, M. G., & Siriwong, W. (2017). Chemical fertilizer use and acute health effects among chili farmers in Ubon Ratchathani province, Thailand. *Journal of Health Research, 31*(6), 427–35. https://doi.org/10.14456/jhr.2017.53

Norwood, F. B., Oltenacu, P. A., Calvo-Lorenzo, M. S., & Lancaster, S. (2015). *Agricultural and food controversies: What everyone needs to know*. Oxford University Press. https://doi.org/10.1093/wentk/9780199368433.001.0001

Nwankwo, C. (2022). Grammar of geopolitics: Geopolitical imaginations of farmer-herder conflicts in Nigeria. *Journal of Geography, 44*, 1–18. https://doi.org/10.26650/JGEOG2022-888146

Nwoye, C. I. (2022). Chinese trawlers with an illegal fishing record have been licensed by Senegal. *Quartz*. https://qz.com/africa/1915624/senegal-okays-chinese-boats-with-illegal-fishing-record-greenpeace/

Ogbette, A. S., Idam, M. O., Kareem, A. O., & Ogbette, D. N. (2019). Fake news in Nigeria: Causes, effects and management. *Information and Knowledge Management, 9*(2), 96–99. https://doi.org/10.7176/IKM/9-2-10

Omoregie, U., & Ryall, K. (2023). *Misinformation matters: Online content and quality analysis*. CRC Press.

Peng, X. (1987). Demographic consequences of the great leap forward in China's provinces. *Population and Development Review, 13*(4), 639–70. https://doi.org/10.2307/1973026

Peters, O. (2018, November 30). Indiscriminate use of chemicals for storage of beans. *Nigerian Tribune*. https://tribuneonlineng.com/indiscriminate-use-of-chemicals-for-storage-of-beans/

Ptak, A. (2022, August 22). Polish minister warns of "fake news" spread in Germany about Oder river pollution. *Notes from Poland.com*. https://notesfrompoland.com/2022/08/22/polish-minister-warns-of-fake-news-spread-in-germany-about-oder-river-pollution/

Quizon, J., Feder, G., & Murgai, R. (2001). Fiscal sustainability of agricultural extension: The case of the farmer field school approach. *Journal of International Agricultural and Extension Education, 8*(1), 13–24. https://doi.org/10.5191/jiaee.2001.08102

Rains, S. A., & Tukachinsky, R. (2015). An examination of the relationships among uncertainty, appraisal, and information-seeking behavior proposed in uncertainty management theory. *Health Communication, 30*(4), 339–49. https://doi.org/10.1080/10410236.2013.858285

Reuters. (2022, August 12). Germany and Poland search for cause of mass fish die-off in river Oder. *Reuters.com*. https://www.reuters.com/world/europe/mass-fish-die-off-german-polish-river-blamed-unknown-toxic-substance-2022-08-12/

Reviglio, U., & Agosti, C. (2020). Thinking outside the black-box: The case for "algorithmic sovereignty" in social media. *Social Media + Society, 6*(2), 2056305120915613. https://doi.org/10.1177/2056305120915613

Reznik, S., & Fet, V. (2019). The destructive role of Trofim Lysenko in Russian science. *European Journal of Human Genetics, 27*, 1324–25. https://doi.org/10.1038/s41431-019-0422-5

Roozenbeek, J., & Van der Linden, S. (2019). Fake news game confers psychological resistance against online misinformation. *Palgrave Communications, 5*(65), 1–10. https://doi.org/10.1057/s41599-019-0279-9

Ryan, C. D., Schaul, A. J., Butner, R., & Swarthout, J. T. (2020). Monetizing disinformation in the attention economy: The case of genetically modified organisms (GMOs). *European Management Journal, 38*(1), 7–18. https://doi.org/10.1016/j.emj.2019.11.002

Salemink, K., Strijker, D., & Bosworth, G. (2017). Rural development in the digital age: A systematic literature review on unequal ICT availability, adoption, and use in rural areas. *Journal of Rural Studies, 54*, 360–71. https://doi.org/10.1016/j.jrurstud.2015.09.001

Seed World Europe Videos. (2020, November 25). Fake news, misinformation and disinformation in agriculture. *Seed World Europe*. https://european-seed.com/2020/11/fake-news-misinformation-and-disinformation-in-agriculture/

Sekabira, H., Tepa-Yotto, G. T., Ahouandjinou, A. R., Thunes, K. H., Pittendrigh, B., Kaweesa, Y., & Tamò, M. (2023). Are digital services the right solution for empowering smallholder farmers? A perspective enlightened by COVID-19 experiences to inform smart IPM. *Frontiers in Sustainable Food Systems, 7*, 983063. https://doi.org/10.3389/fsufs.2023.983063

Selick, K. (2018, March 2). You can now go to prison in Canada for providing raw milk. Seriously. *Financial Post*. https://financialpost.com/opinion/you-can-now-go-to-prison-in-canada-for-providing-raw-milk-seriously

Servaes, J. (2018). *Handbook of communication for development and social change*. Springer.

Servaes, J., & Lie, R. (2014). New challenges for communication for sustainable development and social change: A review essay. *Journal of Multicultural Discourses*, *10*(1), 124–48.

Shepherd, C. J., & Little, D. C. (2014) Aquaculture: are the criticisms justified? II – Aquaculture's environmental impact and use of resources, with specific reference to farming Atlantic salmon. *World Agriculture*, *4*(2), 37–52.

Soyfer, V. (1989). New light on the Lysenko era. *Nature*, *339*, 415–20. https://doi.org/10.1038/339415a0

Treen, K. M., Williams, H. T. P., & O'Neill, S. J. (2020). Online misinformation about climate change. *Wiley Interdisciplinary Reviews: Climate Change*, *11*(5), e665. https://doi.org/10.1002/wcc.665

Tripoli, M., & Schmidhuber, J. (2018). *Emerging opportunities for the application of blockchain in the agri-food industry*. FAO and ICTSD. Licence: CC BY-NC-SA 3, 2018.

Tsiboe, F., & Turner, D. (2022). *Risk in agriculture*. https://www.ers.usda.gov/topics/farm-practices-management/risk-management/risk-in-agriculture/

Uchegbu, N., Nnamocha, T., & Ishiwu, C. (2020). Natural food colourants juxtaposed with synthetic food colourant: A review. *Pakistan Journal of Nutrition*, *19*, 404–19. https://doi.org/10.3923/pjn.2020.404.419

Van der Linden, S. (2023). *Foolproof why misinformation infects our minds and how to build immunity*. W. W. Norton & Company.

Van Mele, P., Wanvoeke, J., & Zossou, E. (2010). Enhancing rural learning, linkages, and institutions: The rice videos in Africa. *Development in Practice*, *20*(3), 414–21. https://www.jstor.org/stable/27806717

Waisbord, S. (2018). Truth is what happens to news: On journalism, fake news, and post-truth. *Journalism Studies*, *19*(13), 1866–78. https://is.muni.cz/el/fss/podzim2019/ZURb1608/um/readings/week_10/Truth_is_What_Happens_to_News.pdf

Wardle, C., & Derakhshan, H. (2017). *Information disorder: Toward an interdisciplinary framework for research and policy making*. Council of Europe report: DGI(2017)09. https://rm.coe.int/information-disorder-toward-an-interdisciplinary-framework-for-researc/168076277c

Yadav, J., Yadav, A., Misra, M., Rana, N. P., & Zhou, J. (2023). Role of social media in technology adoption for sustainable agriculture practices: Evidence from Twitter Analytics. *Communications of the Association for Information Systems*, *52*, 833–51. https://doi.org/10.17705/1CAIS.05240

Zalom, F. G. (1993). Reorganizing to facilitate the development and use of integrated pest management. *Agriculture, Ecosystems & Environment*, *46*(1–4), 245–56. https://doi.org/10.1016/0167-8809(93)90028-N

Zhang, C., Wohlhueter, R., & Zhang, H. (2016). Genetically modified foods: A critical review of their promise and problems. *Food Science and Human Wellness*, *5*(3), 116–23. https://doi.org/10.1016/j.fshw.2016.04.002

Zuboff, S. (2015). Big other: Surveillance capitalism and the prospects of an information civilization. *Journal of Information Technology*, *30*(1), 75–89. https://doi.org/10.1057/jit.2015.5

# 5 Agricultural extension, social media, and the dilemma of path-dependency

*Gordon A. Gow*

## Introduction

The importance of corporate social media (CSM) as a ready-to-use information and communication technologies (ICT) solution for the agriculture sector was highlighted by the COVID-19 pandemic. Particularly in the Global South, CSM tools and platforms have been critical to creative community responses for locating food stockpiles and obtaining vegetables from farmers (GRAIN, 2020). CSM has played an important role in agriculture and rural development, with Facebook, WhatsApp, Telegram, YouTube, and Viber among the most extensively utilized social networking and content sharing platforms (Chander & Rathod, 2020; Ramos et al., 2020).

Facebook's worldwide outages on October 4 and October 9, 2021 (Taylor, 2021), which affected its subsidiary platforms Messenger, Instagram, and WhatsApp, have underscored the reliance of individuals and small businesses on a few centralized communications platforms. Whistleblower Frances Haugen accused Facebook of "prioritizing profit over public safety" with its use of controversial algorithm designs and content moderation policies (Milmo, 2021a). This has further downgraded Facebook's reputation, which was tarnished by the Cambridge Analytica scandal (Confessore, 2018), and the revelation of covert *emotional contagion* experiments on users (Meyer, 2014).

Other CSM platforms have faced similar criticism for failing to effectively address misinformation about COVID-19 vaccinations and treatments (Spring, 2020). These disclosures have had less of an impact on the general population in countries of the Global South (Punit et al., 2021), but they nonetheless continue to raise concerns about the quality of information shared through social media for those involved in agriculture extension and rural development (Klerkx, 2021; Lubell & McRoberts, 2018; Wang & Song, 2020).

Scholars have claimed that these actions are typical of the prevailing business model in a social media ecosystem centered on *surveillance capitalism*, which profits from unfettered access to user-generated content and personal trace data (Srnicek, 2016; Zuboff, 2018). This business model has spawned a *social media logic* (Van Dijck & Poell, 2013) and a *datafication* imperative (Mejias & Couldry, 2019) that encourages users to align their behaviors in ways that are responsive to the commercial interests of platform owners. This logic is pervasive, extending

DOI: 10.4324/9781003282075-7

from the user interface to content moderation practices, algorithm design, and corporate governance (Gehl, 2015; Van Dijck & Poell, 2013; Van Dijck et al., 2018).

The two most recent events involving Facebook, as well as the evidence-based critiques presented by scholars and stakeholders, including one of the founders of Facebook (Hughes, 2019), point to a broad set of issues unlikely to be resolved anytime soon, despite regulatory efforts now underway in some countries, such as the United Kingdom's online safety bill (Milmo, 2021b). As more people become aware of the systemic problems with CSM, agricultural extension and rural development policymakers and practitioners should assess whether they have become overly reliant on these popular platforms and what unintended negative consequences this may have.

This chapter proposes that those involved in agricultural extension and advisory services (EAS) and rural development should think about the path-dependent risks of CSM and take efforts to avoid reliance on these platforms. This chapter will sketch parameters for an ICT diversification approach based on non-commercial, decentralized social media platforms as one option, while acknowledging the continuing value of CSM. It will propose that a long-term diversification strategy is critical to creating inclusive digital ecosystems, especially among smallholders and rural communities in the Global South.

## Agricultural extension and social media

Some of the earliest published views on the potential for social media in agriculture can be traced back to the Web2forDev conference held in Rome, 2007 (Barth & Rambaldi, 2009; Kreutz, 2008). When nascent social networking sites were initially garnering attention, the international development community was introduced to the concept of *the social Web* during this gathering. The availability of free social networking platforms sparked considerable interest in the possibility for peer-based knowledge generation (Benkler, 2011) and post-institutional possibilities for group action (Shirky, 2008). As a result of this upbeat outlook, an *extension 2.0* movement arose within the agricultural sector to consider ways in which social media might serve the various communication roles for EAS services (Ballantyne, 2009, 2010; Ballantyne et al., 2010).

The digital divide between urban and rural communities and its impact on social media use in agriculture extension was an early focus for researchers in the Global North (Cornelisse et al., 2011; Hunt et al., 2012; Kader, 2013). In another early study on social media adoption, Gilbert et al. (2010) looked at the social dynamics of 'friending' and online interactions among rural communities by comparing the behavioral differences among more than 3,000 rural and urban MySpace users in the United States. In Ontario, Canada, Chowdhury and Hambly Odame (2013) conducted a ground-breaking study on social media use among agricultural and rural development stakeholders. Their research looked at how social media could improve "collective learning processes and co-creation of knowledge," concluding that it was still in "an early exploratory phase" with limitations stemming from differing perspectives on its value, inherent risks, and credibility as a communications tool (p. 97).

Others such as Mulyandaril et al. (2012) began to consider social media adoption in the Global South as a branch of *cyber extension*, which took advantage of the growing use of mobile phones and text messaging (Aker, 2011; De Silva et al., 2012). Researchers in extension education, such as Dissanayeke et al. (2012), who examined the role of Twitter among young farmers in Sri Lanka, began to investigate the possibilities of social media for mobile learning. Strong's (2012) study on loan distribution to Mexican farmers revealed that social media could aid in overcoming information asymmetries and improving communications between government agencies and farmers. Social media has also been investigated as an innovative knowledge mobilization strategy for agricultural populations in the Global South, with the potential to help boost production (Chisita, 2012) and encourage youth to remain in farming (Ofunya Afande et al., 2012). During this adoption period, capacity-building programs related to social media in agricultural extension were implemented (Medhi-Thies et al., 2015). For example, the United States Agency for International Development's (USAID) handbook on social media for agricultural development practitioners included training materials and best practices (Andres & Woodard, 2013). A training course on using social media for development was also launched by the United Nations Food and Agriculture Organization (FAO) (FAO, 2013).

With the rise in popularity of social media, the Global Forum for Rural Advisory Services (GFRAS) undertook a global survey on its use in agriculture in 2015. The survey's findings suggested that by strengthening ties between extension and other stakeholders in the agricultural innovation system, social media could play a key role in a *new extensionist* approach to advisory services (Davis, 2015). It cautioned that "choosing a specific platform needs much deliberation on the part of [extension] organizations" (Bhattacharjee & Raj, 2016, p. 27). The survey's findings foreshadowed the large-scale moderation challenges now facing CSM platforms in recognizing that social media were easier to manage with small groups rather than large audiences (Bhattacharjee & Raj, 2016, p. 30). The survey suggested that a new role for extension practitioners would be to participate as facilitators responsible for cultivating online knowledge exchange using social media for small group interactions.

As social media continued to grow, researchers began efforts to measure its impact in the agricultural sector. One significant finding from these studies was that the adoption of social media has taken place mostly through ad hoc individualized efforts rather than organized initiatives led by public extension services (Thakur & Chander, 2018). Research found that farmers are seeking more timely and localized information than what was provided by institutional initiatives, such as the Kisan Call Centers in India (Devanand & Kamala, 2018; Thakur et al., 2017). WhatsApp has emerged as one of the more popular social media platforms for farmer groups and extension services (Devanand & Kamala, 2018; Kamani et al., 2016; Naruka et al., 2017; Thakur & Chander, 2017).

Early skepticism about social media's contribution to EAS has been superseded by increased acceptance (Thakur & Chander, 2018), with practitioners seeing it as a cost-effective communication medium. Hashem et al. (2021) in a multi-country cross-sectional survey of ICT use during COVID-19 provides supporting evidence that social media is currently the most widely utilized form of communications in agriculture. Much of the published research remains focused on information effects with fewer studies looking at the more qualitative aspects of networked learning and behavioral effects associated with social media use (Spielman et al., 2021). Recent studies are examining the social dynamics of online interactions with EAS on these platforms (Aguilar-Gallegos et al., 2021; Munthali et al., 2021). Klerkx (2021, p. 278) has recently stated that while "there is [now] a considerable body of research on social media in relation to extension and advisory services," more research is needed to understand social dynamics including the role of influencers and their role in online interactions and knowledge sharing.

Studies on the unintended negative consequences of CSM for the agriculture sector are not yet evident in the literature. The spread of misinformation and other harmful content has been noted (Kamani et al., 2016; Lubell & McRoberts, 2018; Stroud, 2019), it will draw the attention of EAS researchers and practitioners. Other issues such as digital self-determination and the concentration of power in CSM platforms appear to have received limited attention within the EAS community despite gaining significance elsewhere.

### *Techlash* and demands for responsible innovation

The agricultural sector, particularly in the Global South, will surely continue to benefit from CSM. It is usually the most cost-effective, user-friendly, and feature-rich ICT available to extension practitioners and smallholder farmers. Despite efforts by central governments to impose digital agriculture ecosystems, such as India's controversial Agristack initiative (ASHA Kisan Swaraj, 2021; Internet Freedom Foundation, n.d.; Kapil, 2021; Pal, 2021), CSM may prove to be the preferable alternative for many farmers and small businesses in agriculture.

Despite its widespread popularity, increasing public awareness of CSM's drawbacks has spawned "a more generalized opposition to technology itself" (Atkinson et al., 2019, p. 5), with the term *techlash* appearing in recent years. The large technology companies embroiled in techlash are also increasingly active in developing digital platforms for the agriculture sector. These include initiatives such as Microsoft's FarmBeats and Amazon Web Services partnerships with Bayer and other companies that use its cloud service. Activist groups are concerned that these partnerships may lead to a greater concentration of power in the hands of a few global conglomerates (GRAIN, 2021).

Opposition to India's Agristack program and the broader techlash movement are rooted in a perceived imbalance of interests within emerging digital ecosystems. For instance, activists in India have called for greater participation in the decision-making processes that will shape the country's digital agriculture future (ASHA Kisan Swaraj, 2021). The absence of representation from extension providers, farmers, and small rural enterprises expected to adopt digital technology is a subject of criticism in both the Global North and Global South (Bronson, 2018;

Jakku et al., 2019; Rotz et al., 2019). These stakeholders see themselves as having few opportunities to influence the sector's long-term digital transformation to match their needs and aspirations.

Similar concerns have been expressed about commercial digital ecosystems as policy research organizations push for more inclusive policies and practices (West, 2021):

> improving transparency regarding how algorithms work, taking public concerns seriously, and working with policymakers on guardrails that protect human values likely would strengthen public support for the [tech] sector. Making sure technology works for people is one of the most important things tech executives can do going forward.

Calls for transparency and public accountability in the digital tech sector are closely related to the responsible innovation frameworks proposed for genetically modified crops (Von Schomberg, 2013). In both a responsible innovation pathway is seen to follow a policy-practice nexus established on "precautionary principles" (Ahteensuu, 2010, p. 616), which emphasizes anticipation, inclusiveness, reflexivity, and responsiveness (Macnaghten, 2016). Tworek (2019) argues that social media firms should abandon their 'move fast and break things' mentality in favor of incorporating more precautionary measures into their design-release cycle.

### The control dilemma, path-dependency, and the Choice Framework

Despite concerns that the techlash resistance will stifle innovation efforts more broadly (Atkinson et al., 2019), the movement highlights a central concern within scholarship on responsible innovation and, more broadly, for those who study the social shaping of technology: the so-called *control dilemma*, which was first articulated by Collingridge (1982) and is now a widely cited idea within the field of Science and Technology studies (Macnaghten, 2016, p. 283):

> Collingridge's control dilemma [refers to the situation] where dynamics of control lead technologies as they develop to become increasingly less amenable to societal shaping in the face of dynamics of power, lock-in and incumbent interests.

Collingridge described a two-horned dilemma as a trade-off between "controllability" and "corrigibility," in which "a developed technology is more clear [*sic*] in its implications but more entrenched in the face of efforts to reshape it" (Genus & Stirling, 2018, p. 64). As companies and individual users become more reliant on the innovation, it develops a path-dependency or lock-in that binds it to an entrenched activity system and social practice. The control dilemma has inspired a large corpus of applied policy research rooted in various technology assessment programs that date back to the 1970s (Grunwald, 2014), which laid the foundation for today's responsible "innovation movement" (Genus & Stirling, 2018, p. 62).

The continued dominance of the QWERTY keyboard design is an example of technological path-dependency that affects everyday practice. This layout is still the industry standard for practically all ICT devices today (Noyes, 1998). By spreading out the most frequently used letters on the keyboard, QWERTY was created to prevent mechanical typebars from becoming entangled during fast touch typing. With digital keyboards this problem no longer exists, but QWERTY remains locked-in as a social practice constituted by three interrelated factors of materiality, competence, and meaning of a social practice (Shove et al., 2012). The factors are manifest in economies of scale in the production of physical keyboards and software, users' pre-existing keyboarding skills, and widespread cultural expectations for keyboard layout, which has an impact on marketing and training.

While human factors studies on their merits have not been conclusive (Buzing, 2003), the question as to which keyboard layout is superior is perhaps best answered with 'it depends.' The significant point of the QWERTY example is that when choices become available to users, awareness of those choices is essential to circumventing an unintended or unwanted lock-in situation with a technology. Drawing on Amartya Sen's capabilities approach, Kleine's Choice Framework (Kleine, 2013) can serve a normative role in an ICT diversification strategy by emphasizing 'choice itself' as an essential outcome. The objective of establishing choice for users is organized into four stages of exposure and experimentation with ICTs. In the context of social media platforms, for instance, the first stage of diversification is to make community members aware of the 'existence' of choice. The second stage is to assist community members to develop a 'sense of choice' by demonstrating specific examples of how these alternatives might be integrated into existing communication practices. The third stage is to support the 'use of choice' by providing training and other supports to encourage users to experiment with alternatives. The fourth stage of diversification is to recognize and sustain the *achievement of choice* through systematic evaluation of small-scale pilot deployments. The final stage of evaluation includes understanding points of failure and building on successful efforts and, as with action research, provides the basis for further learning and refinement of practice.

### Digital colonialism and the Global South

Path-dependency raises pressing questions about the role of CSM in an emerging form of *digital colonialism* (Couldry & Mejias, 2021). Avila (2020) views digital colonialism as a process by which large corporations establish hegemonic positions within the digital ecosystem "to get their hands on the missing datasets of the global poor" (p. 47). Such an arrangement has implications for extension practitioners and others relying on corporate social media and for digital agriculture ecosystems established in collaboration with large technology firms as Avila (2020, p. 50):

> Workforces receiving training today will only be trained and prepared to use the technologies produced by the current wave of tech companies, creating circles of dependence. Skills developed by workers will be connected to specific

products and therefore benefit the profitability of the few. It will have a tremendous impact in stifling the development of new cultures of collaboration.

Kwet (2020, p. 5) sees evidence for digital colonialism in the "structural domination" of the global digital ecosystem by a small number of primarily United States-based technology companies. He argues this domination is reproducing North-South colonial dependencies in three areas in which CSM are implicated:

- Circumscribing access and innovation with intellectual property restrictions and digital rights management in software and hardware available to users in the Global South.
- Exerting control over access and distribution of user-generated content through centrally governed moderation and the creation of 'walled gardens' that limit or prevent cross-platform interoperability and portability of user data.
- Extraction of monopoly rents through net outflows of user and trace data from social media platforms that are in turn sold to advertisers as value-added marketing intelligence targeted at the consumers in the Global South.

These forms of control are rooted in and reinforced through deeper layers of technology lock-in extending to the global topology of the Internet infrastructure itself. The siting of Internet Exchange Points (IXPs) outside the borders of countries in the Global South also results in a 'tromboning' effect when local Internet service providers (ISPs) are required to use international links to exchange data. Profits accrue to the large companies that provide these peering and transiting services at the expense of local economic development and networking capabilities (Kende, 2021, p. 120). Tromboning also impacts the affordability and quality of local broadband services, as was determined by the work of United Nations Economic and Social Commission for Asia and the Pacific (UNESCAP), as local ISPs are required to route data outside the country even when it is destined for other local ISPs (2019).

While acknowledging the contributions of CSM in overcoming the digital divide in the Global South, Kwet (2019, p. 16) reminds us that this arrangement comes bundled with forms of domination that extend beyond global flows of capital to include new possibilities for "imperial state surveillance" and "ideological domination" that reinforce and normalize big tech's hegemony within a deterministic narrative about the inevitable advent of Industry 4.0 (Moll, 2021).

Regulatory reform to address unwanted problems associated with CSM has been slow and uncertain. It is likely that these efforts will result in relatively minor changes to the underlying business logic that has tended to pit private interests against public values (Van Dijck et al., 2018) while perpetuating neo-colonial dependencies. It is prudent for policymakers and practitioners in the agriculture sector to consider the structural consequences of relying on CSM and the long-term consequences it may have for digital self-determination within the sector more broadly (Salemink et al., 2017). There are now opportunities to begin this self-examination by considering the availability of open-source and decentralized social media platforms capable of expanding the range of choice available to stakeholders.

## Diversifying digital habitats beyond CSM

CSM represent a collection of tools and platforms that together form a *digital habitat* (Wenger et al., 2009) within which agricultural communities of practice exchange knowledge, carry out social networking, and engage in social learning. Further diversifying the digital habitats that exist today will be difficult. CSM have proven to be cost-effective and popular among extension practitioners and many farmers. They are locked-in by powerful network effects and by offering advanced technical features in a small package that requires a relatively modest investment in a smartphone and data plan. Most of the costs and risks that come with system administration and security are borne by the platforms in exchange for collecting personal and trace data of their users. Many users are willing to accept this bargain and will be reluctant to forego the perceived and genuine benefits of CSM in exchange for less familiar options.

This does not negate the need for EAS services to address the long-term implications of becoming overly reliant on CSM within their digital habitats. As the agricultural sector becomes more digitalized, decisions taken today will either hamper or enable rural community development, smallholders, and their families to achieve digital self-determination (Remolina & Findlay, 2021). EAS services can play a crucial role in an ICT diversification strategy by promoting inclusivity and assisting with equitable, economically, and environmentally sustainable digital transformation (Florey et al., 2020; Klerkx, 2020).

Other choices that can be introduced into the digital habitat are now available. The 'Fediverse,' for example, has been created by the alternative social media (ASM) movement, which is a collection of open-source projects that mirror many of the CSM platforms (Fediverse, 2023; Robinson, 2021). These initiatives foster autonomous but interconnected online communities by using a decentralized ('federated') network topology and transparent governance mechanisms (Gehl, 2017; Zulli et al., 2020). While many of these initiatives face significant technical challenges and uncertain financial support, some projects, such as Mastodon or Diaspora, are well-established enough to serve as a foundation for conducting exploratory research with EAS organizations (Gow, 2020).

Table 5.1 shows a sample of the current ASM projects that can be used to supplement or replace CSM within a digital habitat. Most of these projects are made

*Table 5.1* Selected open-source ASM projects within the Fediverse as of late 2021

| Project | Social media function | Active instances | User base |
|---------|----------------------|------------------|-----------|
| Mastodon | Microblogging | 3,250 | 3,004,436 |
| Pleroma | Social networking | 1,112 | 75,474 |
| PeerTube | Video sharing | 1,100 | 228,879 |
| Pixelfed | Photo sharing | 215 | 63,127 |
| Diaspora | Social networking | 187 | 859,455 |
| Funkwhale | Audio streaming | 108 | 5,352 |

*Source:* the-federation.info.

up of a small number of dedicated users, but their open-source licensing allows any group or organization to acquire the software code and run their own instance.

The availability of these non-commercial ASM provides one route to diversification, but there are considerable technical, resource, digital literacy, and user acceptance barriers to overcome (Gehl, 2017). Although research is underway to examine these challenges and to investigate these issues and identify other non-CSM alternatives (Zuckerman & Rajendra-NicoluccAi, 2020), the most immediate contribution of this movement may be to spark discussion and debate among extension providers and rural development practitioners about their communication needs and aspirations as they relate to CSM and its long-term implications for digital self-determination.

Introducing users to a broader range of digital tools and platforms, for example, can strengthen informational capabilities (Gigler, 2011, 2015) within agricultural communities of practice, enabling members to contribute meaningfully to responsible innovation initiatives for digital agriculture ecosystems (Bronson, 2019; Lajoie-O'Malley et al., 2020).

The ability to choose from a variety of open-source, non-commercial software tools and platforms may have positive consequences for data residency because servers can be set up in-country, decreasing the requirements for 'tromboning,' and the expense and reliance on foreign intermediaries. Maintaining data residency aids the formation of community-owned data trusts in the agriculture sector (Wylie & McDonald, 2018). The expansion of the digital habitat to include ASM opens the door to local data ownership, control, access, and possession—the so-called OCAP® (ownership, control, access, and possession) principles—proposed by Indigenous organizations in the Global North calling for data sovereignty (Scassa, 2018). In countries of the Global South, OCAP® concepts may become widely acknowledged as relevant to their development agendas.

A diversification strategy can assist in the fight against misinformation and help diffuse online conflict by adding social networking tools and platforms that place more focus on peer-based governance and community-led moderations procedures than their CSM counterparts. In their study of ASM microblogging platform Mastodon, for example, Zulli et al. (2020) discovered that its small-scale independent online communities established a form of convivial interest-based sociality that represents "a sharp departure from CSM where users have little corrective authority if they find the tenor of a post inappropriate" (Zulli et al., 2020, p. 1198).

The decentralized character of ASM platforms like Mastodon is critical to encouraging this ethos of cooperative sociality. Rather than creating a single server that connects all users globally, separate online communities ('instances') can be established and then choose to link to other instances for sharing information and building social networks. A regional interest group could, for example, establish a small, separate Mastodon instance for internal social networking and knowledge exchange. The group may choose to connect its instance to other Fediverse-based communities while still maintaining control over local governance and moderation policies.

## Conclusion

This chapter has proposed that policymakers and practitioners involved in agricultural extension and rural development should start thinking about the path-dependent risks of CSM and take efforts to avoid becoming unduly reliant on these platforms, which have systemic problems related to the promulgation of misinformation and harmful content. The chapter then considered one possibility for an ICT diversification strategy incorporating Fediverse-based ASM diversifying the digital habitat for agricultural and rural development communities of practice creates opportunities for strengthening informational capabilities, data residency, and digital self-determination.

A first step in an ICT diversification plan will be to raise awareness of the possibilities for non-commercial apps and platforms to supplement or replace CSM. The next stage will be to test these options with small groups of users in a controlled setting. This is where technical challenges will surface, and it is possible that the uncertainty associated with this phase could deter many potential users from adopting ASM. For example, setting up a stand-alone instance of Mastodon necessitates a system administrator for downloading and installing the software code. Hosting an instance necessitates access to a reliable and secure server with adequate bandwidth, and additional technical support on a day-to-day basis. These supports, notably access to server space and bandwidth, are among the most pressing challenges that EAS organizations and community groups, particularly in the Global South, face. These challenges, however, may not be insurmountable with partnerships for pooling resources and talent to provide shared infrastructure and support.

Expanding digital skills training and gaining user acceptance and uptake of unfamiliar social media tools will be difficult. Although the practical similarities between many ASM and CSM platforms may lower some of these barriers, diversifying a digital habitat necessitates technical competencies combined with adaptive leadership. Despite these challenges, EAS services can take the lead in encouraging discussion of these prospects and participating in quasi-experimental and participatory action research designs involving non-commercial social media initiatives. A small step in this direction is now underway in partnership with post-secondary and in-service technology stewardship training for extension services in the Global South (Gow, 2021), but more could be done with coordinated efforts among extension organizations in both the Global North and Global South reaching out to the open-source software community.

There will be no easy fixes or solutions to the control dilemma created by CSM. However, decisions being made today will shape future opportunities for EAS organizations and smallholders to assume greater role in cultivating their digital habitats and to retain control over the valuable data that will be generated within those habitats. If digital self-determination is to be a long-term development goal for the agriculture sector in the Global South, it will require a concerted and sustained effort among all stakeholders to undertake forward-looking conversations about CSM, digital dependencies, and community-based ICT initiatives within the agriculture sector.

# References

Aguilar-Gallegos, N., Klerkx, L., Romero-García, L. E., Martínez-González, E. G., & Aguilar-Ávila, J. (2021). Social network analysis of spreading and exchanging information on Twitter: The case of an agricultural research and education centre in Mexico. *The Journal of Agricultural Education and Extension, 28*(1), 115–36. https://doi.org/10.1080/13892 24X.2021.1915829

Ahteensuu, M. (2010). Agricultural biotechnology and the precautionary principle. *Sociology Compass, 4*(8), 616–27. https://doi.org/10.1111/j.1751-9020.2010.00308.x

Aker, J. C. (2011). Dial "A" for agriculture: A review of information and communication technologies for agricultural extension in developing countries. *Agricultural Economics, 42*(6), 631–47. https://doi.org/10.1111/j.1574-0862.2011.00545.x

Andres, D., & Woodard, J. (2013, September). *Social media handbook for Agricultural development practitioners* (USAID). USAID/FHI 360. https://thisisjoshwoodard.com/wp-content/uploads/2018/03/SocialMediaforAgHandbook.pdf

ASHA Kisan Swaraj. (2021). Response to consultation paper on IDEA-India Digital Ecosystem of Agriculture. [Email online.] https://kisanswaraj.in/2021/06/30/civil-society-response-on-gois-idea-consultation-paper-on-digitisation-in-indian-agriculture/

Atkinson, R. D., Brake, D., Castro, D., Cunliff, C., Kennedy, J., McLaughlin, M., New, J., & McQuinn, A. (2019, October 28). A policymaker's guide to the "Techlash"—What it is and why it's a threat to growth and progress. *Information and Technology and Innovation Foundation.* https://itif.org/publications/2019/10/28/policymakers-guide-techlash/

Avila, R. (2020). Against digital colonialism. In J. Muldoon & W. Stronge (Eds.), *Platforming equality: Policy challenges for the digital economy* (pp. 47–57). Autonomy Research Ltd. https://autonomy.work/wp-content/uploads/2020/09/platforming-equality-V3.pdf

Ballantyne, P. (2009). Accessing, sharing and communicating agricultural information for development: Emerging trends and issues. *Information Development, 25*(4), 260–71. https://doi.org/10.1177/0266666909351634

Ballantyne, P. (2010). Agricultural information and knowledge sharing: Promising opportunities for agricultural information specialists. *Agricultural Information Worldwide, 3*(1), 4–9. https://hdl.handle.net/10568/1535

Ballantyne, P., Maru, A., & Porcari, E. M. (2010). Information and communication technologies—Opportunities to mobilize agricultural science for development. *Crop Science, 50*, S-63–69. https://doi.org/10.2135/cropsci2009.09.0527

Barth, A., & Rambaldi, G. (2009). The Web2forDev story: Towards a community of practice. *Participatory Learning and Action, 59*, 95–104. https://www.iied.org/sites/default/files/pdfs/migrate/G02846.pdf

Benkler, Y. (2011). *The penguin and the leviathan: The triumph of cooperation over self-interest.* Crown Business.

Bhattacharjee, S., & Raj, S. (2016). Social media: Shaping the future of agricultural extension and advisory services. GFRAS interest group on ICT4RAS discussion paper. GFRAS. https://doi.org/10.13140/RG.2.2.10815.56488

Bronson, K. (2018). Smart farming: Including rights holders for responsible agricultural innovation. *Technology Innovation Management Review, 8*(2), 7–14. https://doi.org/10.22215/timreview/1135

Bronson, K. (2019). Looking through a responsible innovation lens at uneven engagements with digital farming. *NJAS—Wageningen Journal of Life Sciences, 90–91*, 100294. https://doi.org/10.1016/j.njas.2019.03.001

Buzing, P. (2003). Comparing different keyboard layouts: Aspects of QWERTY, DVORAK and alphabetical keyboards. Delft University of Technology Articles. Retrieved from https://www.researchgate.net/publication/252214871_Comparing_Different_Keyboard_ Layouts_Aspects_of_QWERTY_DVORAK_and_alphabetical_keyboards

Chander, M., & Rathod, P. (2020). Reorienting priorities of extension and advisory services in India during and post COVID-19 pandemic: A review. *Indian Journal of Extension Education*, *56*(3), 1–9. https://acspublisher.com/journals/index.php/ijee/article/view/4424

Chisita, C. T. (2012). *Knotting and networking agricultural information services through Web 2.0 to create an informed farming community: A case of Zimbabwe.* [Conference presentation.] World Library and Information Congress: 78th IFLA General Conference and Assembly, Helsinki, Finland. https://www.ifla.org/past-wlic/2012/205-chisita-en.pdf

Chowdhury, A., & Hambly Odame, H. (2013). Social media for enhancing innovation in agri-food and rural development: Current dynamics in Ontario, Canada. *Journal of Rural and Community Development*, *8*(2), 97–19. https://journals.brandonu.ca/jrcd/article/view/1007/232

Collingridge, D. 1982. *The social control of technology.* St. Martin's Press.

Confessore, N. (2018, April 4). Cambridge Analytica and Facebook: The scandal and the fallout so far. *The New York Times.* https://www.nytimes.com/2018/04/04/us/politics/cambridge-analytica-scandal-fallout.html

Cornelisse, S., Hyde, J., Raines, C., Kelley, K., Ollendyke, D., & Remcheck, J. (2011). Entrepreneurial extension conducted via social media. *Journal of Extension*, *49*(6), article 6TOT1. https://archives.joe.org/joe/2011december/tt1.php

Couldry, N., & Mejias, U. A. (2021). The decolonial turn in data and technology research: What is at stake and where is it heading? *Information, Communication & Society*, *26*(4), 786–802. https://doi.org/10.1080/1369118X.2021.1986102

Davis, K. (2015, August 15). The new extensionist: Core competencies for individuals. *GFRAS* Brief no. 3. Global Forum for Rural Advisory Services. http://www.g-fras.org/en/knowledge/gfras-publications.html?download=358:the-new-extensionist-core-competencies-for-individuals

De Silva, L., Goonetillake, J. S., Wikramanayake, G. N., & Ginige, A. (2012). *Towards using ICT to enhance flow of information to aid farmer sustainability in Sri Lanka.* [Conference presentation.] ACIS 2012: Proceedings of the 23rd Australasian Conference on Information Systems 2012, Geelong, Australia.

Devanand, I. I., & Kamala, I. M. (2018). WhatsApp groups: A powerful tool and farming solution for sustainable agricultural development. In R. K. Naresh (Ed.), *Research trends in agricultural sciences* (Vol. 7, pp. 93–100). AkiNik Publishers.

Dissanayeke, U., Hewagamage, K. P., Ramberg, R., & Wikramanayake, G. N. (2012). *Twitter in mLearning: Enhancing agriculture knowledge among a group of young famers in Kandy district, Sri Lanka.* [Conference presentation.] 5th Annual UCSC Research Symposium, University of Colombo, Sri Lanka.

Fediverse. (2023). About Fediverse. https://fediverse.party/en/fediverse

Florey, C., Hellin, J., & J. Balié, J. (2020). Digital agriculture and pathways out of poverty: The need for appropriate design, targeting, and scaling. *Enterprise Development and Microfinance*, *31*(2), 126–40. https://doi.org/10.3362/1755-1986.20-00007

Food and Agriculture Organization of the United Nations (FAO). (2013). Food and agriculture organization of the United Nations E-Learning Centre: Social media for development. http://www.fao.org/elearning/#/elc/en/Course/W2

Gehl, R. (2015). The case for alternative social media. *Social Media + Society*, *1*(2). https://doi.org/10.1177/2056305115604338

Gehl, R. (2017). Alternative social media: From critique to code. In J. Burgess, A. Marwick, & T. Poell (Eds.), The *SAGE handbook of social media* (pp. 330–50). SAGE.

Genus, A., & Stirling, A. (2018). Collingridge and the dilemma of control: Towards responsible and accountable innovation. *Research Policy, 47*(1), 61–69. https://doi.org/10.1016/j.respol.2017.09.012

Gigler, B.-S. (2011). Informational capabilities: The missing link for the impact of ICT on development. http://dx.doi.org/10.2139/ssrn.2191594

Gigler, B.-S. (2015). *Development as freedom in a digital age: Experiences from the rural poor in Bolivia.* The World Bank.

Gilbert, E., Karahalios, K., & Sandvig, C. (2010). The network in the garden: Designing social media for rural life. *American Behavioral Scientist, 53*(9), 1367–88. https://doi.org/10.1177/0002764210361690

Gow, G. A. (2020). Alternative social media for outreach and engagement: Considering technology stewardship as a pathway to adoption. In M. Adna (Ed.), *Handbook of research on using new media for citizen engagement and participation* (pp. 160–80). IGI Global.

Gow, G. A. (2021). Turning to alternative social media. In L. Sloan & A. Quan-Haase (Eds.), The *SAGE handbook of social media research methods* (pp. 160–80), SAGE. https://era.library.ualberta.ca/items/8fb2da29-cb60-4553-a074-f984a66661a1/view/3c0a1677-3bae-4903-88ed-5d4b8838cea2/Gow_Platform%20Challenge_preprint.pdf

GRAIN (2020, May 15). Millions forced to choose between hunger or Covid-19. *GRAIN.* https://grain.org/en/article/6465-millions-forced-to-choose-between-hunger-or-covid-19

GRAIN (2021, January 21). Digital control: How Big Tech moves into food and farming (and what it means). *GRAIN.* https://grain.org/en/article/6595-digital-control-how-big-tech-moves-into-food-and-farming-and-what-it-means

Grunwald, A. (2014). Technology assessment for responsible innovation. In J. Van den Hoven, N. Doorn, T. Swierstra, B.-J. Koops, & H. Romijn (Eds.), *Responsible Innovation 1: Innovative solutions for global issues* (pp. 15–31). Springer Netherlands. https://doi.org/10.1007/978-94-017-8956-1_2

Hashem, N. M., Hassanein, E. M., Hocquette, J. F., Gonzalez-Bulnes, A., Ahmed, F. A., Attia, Y. A., & Asiry, K. A. (2021). Agro-livestock farming system sustainability during the COVID-19 era: A cross-sectional study on the role of information and communication technologies. *Sustainability (Switzerland), 13*(12), 6521. https://doi.org/10.3390/su13126521

Hughes, C. (2019, May 9). It's time to break up Facebook. *The New York Times.* https://www.nytimes.com/2019/05/09/opinion/sunday/chris-hughes-facebook-zuckerberg.html

Hunt, W., Birch, C., Coutts, J., & Vanclay, F. (2012). The many turnings of agricultural extension in Australia. *The Journal of Agricultural Education and Extension, 18*(1), 9–26. https://doi.org/10.1080/1389224X.2012.638780

Internet Freedom Foundation (n.d.). https://internetfreedom.in/

Jakku, E., Taylor, B., Fleming, A., Mason, C., Fielke, S., Sounness, C., & Thorburn, P. (2019). "If they don't tell us what they do with it, why would we trust them?" Trust, transparency and benefit-sharing in Smart Farming. *NJAS—Wageningen Journal of Life Sciences, 90–91*, 100285. https://doi.org/10.1016/j.njas.2018.11.002

Kader, A. A. (2013). Opportunities for international collaboration in postharvest education and extension activities. *Acta Horticulturae*, 1363–70. https://doi.org/10.17660/ActaHortic.2013.1012.184

Kamani, K. C., Ghodasara, Y. R., Soni, N. V., & Parsaniya, P. S. (2016). Empowering Indian agriculture with WhatsApp: A positive step toward digital India. *International Journal of Agriculture Sciences, 8*(13), 1210–12.

Kapil, S. (2021, June 23). Agristack: The new digital push in agriculture raises serious concerns. *DownToEarth.* https://www.downtoearth.org.in/news/agriculture/agristack-the-new-digital-push-in-agriculture-raises-serious-concerns-77613

Kende, M. (2021). *The flip side of free: Understanding the economics of the Internet.* MIT Press.

Kleine, D. (2013). *Technologies of choice? ICTs, development, and the capabilities approach.* MIT Press.

Klerkx, L. (2020). Advisory services and transformation, plurality and disruption of agriculture and food systems: Towards a new research agenda for agricultural education and extension studies. *The Journal of Agricultural Education and Extension, 26*(2), 131–40. https://doi.org/10.1080/1389224X.2020.1738046

Klerkx, L. (2021). Digital and virtual spaces as sites of extension and advisory services research: Social media, gaming, and digitally integrated and augmented advice. *The Journal of Agricultural Education and Extension, 27*(3), 277–86. https://doi.org/10.1080/1389224X.2021.1934998

Kreutz, C. (2008). The participatory web: New potentials of ICT in rural areas. Deutsche Gesellschaft für Technische Zusammenarbeit (GTZ) GmbH. https://wocatpedia.net/images/6/65/GIZ_%282008%29_The_Participatory_Web_-_New_Potentials_of_ICT_in_Rural_Areas.pdf

Kwet, M. (2019, March 13). Digital colonialism is threatening the Global South. *Al Jazeera.* https://www.aljazeera.com/opinions/2019/3/13/digital-colonialism-is-threatening-the-global-south

Kwet, M. (2020). Fixing social media: Toward a democratic digital commons. *Markets, Globalization & Development Review, 5*(1), Article no. 4. https://doi.org/10.23860/MGDR-2020-05-01-04

Lajoie-O'Malley, A., Bronson, K., van der Burg, S., & Klerkx, L. (2020). The future(s) of digital agriculture and sustainable food systems: An analysis of high-level policy documents. *Ecosystem Services, 45*, 101183. https://doi.org/10.1016/j.ecoser.2020.101183

Lubell, M., & McRoberts, N. (2018). Closing the extension gap: Information and communication technology in sustainable agriculture. *California Agriculture, 72*(4), 236–42. https://doi.org/10.3733/ca.2018a0025

Macnaghten, P. (2016). Responsible innovation and the reshaping of existing technological trajectories: The hard case of genetically modified crops. *Journal of Responsible Innovation, 3*(3), 282–89. https://doi.org/10.1080/23299460.2016.1255700

Medhi-Thies, I., Ferreira, P., Gupta, N., O'Neill, J., & Cutrell, E. (2015). *Krishi Pustak: A social networking system for low-literate farmers.* [Conference presentation.] CSCW '15: Proceedings of the 18th ACM Conference on Computer Supported Cooperative Work & Social Computing, February, 1670–81. https://doi.org/10.1145/2675133.2675224

Mejias, U. A., & Couldry, N. (2019). Datafication. *Internet Policy Review, 8*(4). https://doi.org/10.14763/2019.4.1428

Meyer, R. (2014, June 28). Everything we know about Facebook's secret mood-manipulation experiment. *The Atlantic.* https://www.theatlantic.com/technology/archive/2014/06/everything-we-know-about-facebooks-secret-mood-manipulation-experiment/373648/

Milmo, D. (2021a, October 4). Facebook 'tearing our societies apart': Key excerpts from a whistleblower. *The Guardian*. https://www.theguardian.com/technology/2021/oct/04/facebook-tearing-our-societies-apart-key-excerpts-from-a-whistleblower-frances-haugen

Milmo, D. (2021b, November 10). TechScape: What to expect from the online safety bill. *The Guardian*. https://www.theguardian.com/technology/2021/nov/10/techscape-online-safety-bill-ofcom

Moll, I. (2021). The myth of the fourth Industrial Revolution. *Theoria, 68*(167), 1–38. https://doi.org/10.3167/th.2021.6816701

Mulyandaril, R. S. H., Sumaraja, N. K., & Lubis, D. P. (2012). Cyber extension as a communications media for vegetable farmer empowerment. *Journal of Agricultural Extension and Rural Development, 4*(3), 77–84, Article no. 43C25302914. https://doi.org/10.5897/JAERD11.145

Munthali, N., van Paassen, A., Leeuwis, C., Lie, R., van Lammeren, R., Aguilar-Gallegos, N., & Oppong-Mensah, B. (2021). Social media platforms, open communication and problem solving in the back-office of Ghanaian extension: A substantive, structural and relational analysis. *Agricultural Systems, 190*, 103123. https://doi.org/10.1016/j.agsy.2021.103123

Naruka, P. S., Verma, S., Sarangdevot, S. S., Pachauri, C. P., Kerketta, S., & Singh, J. P. (2017). A study on role of WhatsApp in agriculture value chains. *Asian Journal of Agricultural Extension, Economics & Sociology, 20*(1), 1–11. https://doi.org/10.9734/AJAEES/2017/36498

Noyes, J. (1998). QWERTY—The immortal keyboard. *Computing & Control Engineering Journal, 9*(3), 117–22. https://doi.org/10.1049/cce:19980302

Ofunya Afande, F., Maina, W. N., & Maina, P. M. (2012). Youth engagement in agriculture in Kenya: Challenges and prospects. *Journal of Culture, Society and Development, 7*, 4–19. https://www.iiste.org/Journals/index.php/JCSD/article/view/22759

Pal, S. (2021, June 28). Agristack policy aimed at benefiting corporates, dispossessing farmers: Experts. *Newsclick.in*. https://www.newsclick.in/agristack-policy-aimed-benefiting-corporates-dispossessing-farmers-experts#:~:text=Agristack%20Policy%20Aimed%20at%20Benefiting%20Corporates%2C%20Dispossessing%20Farmers%3A%20Experts,-Sumedha%20Pal%20%7C%2028&text=Under%20the%20proposed%20stack%2C%20data,further%20be%20linked%20to%20Aadhaar

Punit, I. S., Adegoke, Y., Hilton, S., Ormerod, A. G., & Christopher, N. (2021, November 9). Why the rest of the world shrugged at the Facebook Papers. *Rest of World*. https://restofworld.org/2021/rest-of-the-world-reaction-facebook-papers/

Ramos, A. K., Duysen, E., Carvajal-Suarez, M., & Trinidad, N. (2020). Virtual outreach: Using social media to reach Spanish-speaking agricultural workers during the COVID-19 pandemic. *Journal of Agromedicine, 25*(4), 353–56. https://doi.org/10.1080/10599 24X.2020.1814919

Remolina, N., & Findlay, M. J. (2021). The paths to digital self-determination—Foundational theoretical framework. SMU Centre for AI & Data Governance Research Paper no. 03/2021. https://ssrn.com/abstract=3831726

Robinson, J. (2021). The Federation: Welcome to the new social web. https://the-federation.info/

Rotz, S., Duncan, E., Small, M., Botschner, J., Dara, R., Mosby, I., Reed, M., & Fraser, E. D. G. (2019). The politics of digital agricultural technologies: A preliminary review. *Sociologia Ruralis, 59*(2), 203–29. https://doi.org/10.1111/soru.12233

Salemink, K., Strijker, D., & Bosworth, G. (2017). Rural development in the digital age: A systematic literature review on unequal ICT availability, adoption, and use in rural areas. *Journal of Rural Studies, 54*, 360–71. https://doi.org/10.1016/j.jrurstud.2015.09.001

Scassa, T. (2018, September 4). Data ownership. CIGI Papers no. 187, Ottawa Faculty of Law Working Paper no. 2018–26. http://dx.doi.org/10.2139/ssrn.3251542

Shirky, C. (2008). *Here comes everybody: The power of organizing without organizations.* Penguin Books.

Shove, E., Pantzar, M., & Watson, M. (2012). *The dynamics of social practice: Everyday life and how it changes.* Sage.

Spielman, D., Lecoutere, E., Makhija, S., & Van Campenhout, B. (2021). Information and communications technology (ICT) and agricultural extension in developing countries. *Annual Review of Resource Economics*, *13*(1), 177–201. https://doi.org/10.1146/annurev-resource-101520-080657

Spring, M. (2020, June 4). Social media firms fail to act on Covid-19 fake news. *BBC News.* https://www.bbc.com/news/technology-52903680

Srnicek, N. (2016). *Platform capitalism.* Polity.

Strong, R. (2012). Improving loan distribution to farmers: Informational needs of Mexican banks. *Journal of International Agricultural and Extension Education*, *19*(3), 63–74. https://doi.org/10:5191/jiaee.2012.19306

Stroud, J. L. (2019). Tackling misinformation in agriculture. https://doi.org/10.1101/2019.12.27.889279

Taylor, J. (2021, October 5). Facebook outage: What went wrong and why did it take so long to fix after social platform went down? *The Guardian.* https://www.theguardian.com/technology/2021/oct/05/facebook-outage-what-went-wrong-and-why-did-it-take-so-long-to-fix

Thakur, D., & Chander, M. (2017). Use of social media for livestock advisory services: The case of WhatsApp in Himachal Pradesh, India. *Indian Journal of Animal Sciences*, *87*(8), 1034–37. https://doi.org/10.56093/tjans.v87i8.73531

Thakur, D., & Chander, M. (2018). Use of social media in agricultural extension: Some evidences from India. *International Journal of Science, Environment and Technology*, *7*(4), 1334–46.

Thakur, D., Chander, M., & Sinha, S. (2017). WhatsApp for farmers: Enhancing the scope and coverage of traditional agricultural extension. *International Journal of Science, Environment and Technology*, *6*(4), 2190–201.

Tworek, H. (2019, November 19). How platforms could benefit from the precautionary principle. Centre for International Governance Innovation. https://www.cigionline.org/articles/how-platforms-could-benefit-precautionary-principle/

United Nations Economic and Social Commission for Asia and the Pacific (UNESCAP) (2019). Estimating the effects of Internet exchange points on fixed-broadband speed and latency. *Asia-Pacific Information Superhighway (AP-IS) Working Paper Series.* https://repository.unescap.org/handle/20.500.12870/1312

Van Dijck, J., & Poell, T. (2013). Understanding social media logic. *Media and Communication*, *1*(1), 2–14. https://doi.org/10.17645/mac.v1i1.70

Van Dijck, J., Poell, T., & de Waal, M. (2018). *The platform society: Public values in a connective world.* Oxford University Press.

Von Schomberg, R. (2013). A vision of responsible research and innovation. In R. Owen, J. Bessant, & M. Heintz (Eds.), *Responsible innovation* (pp. 51–74). Wiley.

Wang, X., & Song, Y. (2020). Viral misinformation and echo chambers: The diffusion of rumors about genetically modified organisms on social media. *Internet Research*, *30*(5), 1547–64. https://doi.org/10.1108/INTR-11-2019-0491

Wenger, E., White, N., & Smith, J. D. (2009). *Digital habitats: Stewarding technology for communities*. CPSquare.

West, D. M. (2021, April 2). Techlash continues to batter technology sector. *Techtank*. Brookings Institution. https://www.brookings.edu/blog/techtank/2021/04/02/techlash-continues-to-batter-technology-sector/

Wylie, B., & McDonald, S. (2018, October 9). What is a data trust? *CIGI online: Big data, platform governance*. Centre for International Governance Innovation. https://www.cigionline.org/articles/what-data-trust/

Zuboff, S. (2018). *The age of surveillance capitalism: The fight for the future at the new frontier of power*. Profile Books.

Zuckerman, E., & Rajendra-NicoluccAi, C. (2020, October 8). Beyond Facebook logic: Help us map alternative social media! *Medium*. https://knightcolumbia.org/blog/beyond-facebook-logic-help-us-map-alternative-social-media

Zulli, D., Liu, M., & Gehl, R. (2020). Rethinking the "social" in "social media": Insights into topology, abstraction, and scale on the Mastodon social network. *New Media & Society*, *22*(7), 1188–205. https://doi.org/10.1177/1461444820912533

**Part 3**

# Lesson for a post-COVID era

Practices, experiences, cases, and tools

# 6 DigitalNWT—adapting digital tools to support remotely managed digital literacy research, education, and communications in Northern Canada

*Rob McMahon, Michael B. McNally,
Samantha Blais, Murat Akcayir, Kyle Napier,
Leanne Goose and Kevin Zhu*

## Introduction

Digital information and communication technologies (ICTs) offer researchers, educators, and communicators many opportunities to connect and collaborate with people located in geographically rural / remote and Indigenous communities. The COVID-19 pandemic illustrates the importance of adapting these tools to support productive and reciprocal relationships over long distances. Many projects, especially in the field of Social Sciences and Humanities, face significant structural, ethical, and methodological barriers to the critical adaptation of digital information and communication technology (ICTs). It is essential that the voices and perspectives of people based in involved communities are substantively included in discussions assessing the efficacy and impact of digitally enabled remote research, education, and communication activities that affect them.

This chapter examines these issues in the context of a participatory digital inclusion project undertaken with rural / remote Indigenous communities in the Northwest Territories (NWT), Canada. For this project the chapter defines all NWT communities as rural and remote as compared to communities across southern Canada. The chapter also draws upon Statistics Canada's (2018, n.p.) definition of a "population centre" as an area with a population of at least 1,000 and a density of 400 or more people per square kilometer.

Drawing on our own experiences, as well as from semi-structured interviews with eight local researchers and digital innovators, the chapter discusses the ways digital ICTs and participatory processes were mobilized to address limitations imposed by COVID-19 and support community-led digital inclusion. Our observations focus on DigitalNWT, a three-year project (2019–2022) that worked to build a foundation for appropriate digital literacy in Nunatsiaq / Denendeh (the NWT) and particularly in the territory's 29 rural / remote communities. Distinct from regional hubs, such as Yellowknife, Inuvik, Hay River, and Fort Smith, travel to and from these small population and primarily Indigenous communities is costly and intermittent, with access to Internet connectivity and digital devices being limited, expensive, and

DOI: 10.4324/9781003282075-9

unreliable (Environics, 2019). At the same time, the people living in these places are engaged in many productive relationships with southern, urban-based researchers (Beaton et al., 2017). This work is guided by community-engaged research protocols and practices established by Northern Indigenous organizations and community members to protect locally held knowledge and provide tangible benefits such as local employment (Dieter et al., 2018; Gwich'in Tribal Council, 2011). The term *Northern* is used to represent described populations within the Northwest Territories.

The DigitalNWT project involved a partnership between university-based researchers, educators, and communicators, both Indigenous and non-Indigenous, and people living and working in rural / remote Indigenous communities. The project was developed in partnership with the Inuvialuit Regional Corporation, Gwich'in Tribal Council, Sahtú Renewable Resources Board, Tłı̨chǫ Government, University of Alberta, Aurora College, Smart Communities NWT / Computers for Schools NWT, and Hands On Media Education. It received administrative support from the MakeWay Shared Platform and was funded by Innovation, Science and Economic Development Canada's Digital Literacy Exchange Program (2019–2022). University-based team members worked remotely alongside Northern residents to research, co-create, and deliver free, openly licensed digital literacy courses through a Train-the-Trainer approach. This involved several interconnected activities, including advocating for evidence-based policy change; researching digital barriers, hopes, and concerns of Northerners; building communities of practice; celebrating NWT-based digital innovators; and sharing capacity around digital inclusion.

DigitalNWT project activities are strongly informed by literature and practices from communications for development, participatory action research, community engagement, and Indigenous research methodologies. The chapter focuses on how the project's activities adapted to remote engagement and collaboration with NWT residents after COVID-19 necessitated a cancellation of all project travel and in-person activities in the spring of 2020. The project's research question asks: How can geographically dispersed teams adopt and adapt digital ICTs and participatory processes in ways that share capacity for remotely managed research, education, and communication activities?

**Literature review**

Remote, digitally enabled research, education, and communication activities present numerous challenges. Obstacles range from ethical considerations and the importance of engaging respectfully in diverse cultural and social contexts to privacy concerns and distrust of outsiders (Bigbee & Lind, 2007; Correa & Pavez, 2016; Pierce & Scherra, 2004; Wilkes, 1999). In Northern and Indigenous contexts, barriers of distance, trust, and connection are compounded by historical and ongoing experiences of settler colonialism (Wolfe, 2006) and extractivism (Greer, 2019) including in the conduct of research and education (Henri et al., 2020). Indigenous communities have been subject to expropriation of land, non-recognition of Indigenous title to land, child abduction, schooling, and assimilation, all of which negatively affect local economies and ways of being and knowing (Coulthard, 2014; Irlbacher-Fox, 2014). In the context of our increasingly digital societies, this can include the problematic extraction of Indigenous knowledges, and most recently, digital data (Duarte, 2017; Kukutai & Taylor, 2016; Stiegman & Castleden, 2015; Tuhiwai Smith, 1999). These conditions have resulted in distrust of external researchers, educators, and communicators, who may engage in activities that are against the desires of communities, such as state surveillance, bureaucratic control, and commercial gain. COVID-19 pandemic restrictions have posed additional barriers, as participant recruitment has become more challenging, and face-to-face interactions have been discouraged (Mabasa & Themane, 2021; Melis et al., 2021). Significant tensions persist between university-based and community-based epistemologies and ethics (Darnell, 2018; Stiegman & Castleden, 2015).

Due to these challenges, special attention, and greater effort are required in university projects involving people located in rural / remote and Indigenous communities (Edwards et al., 2014). Benoit et al. (2005) highlighted the need for community-academic cooperation to build trust among teams operating in rural / remote areas (see Bigbee & Lind, 2007). Henri et al. (2020, pp. 91–92) noted the importance of effective information sharing and early, extensive consultation with local organizations and residents to understand their priorities and communications practices, and ensure relationships reflect mutual learning and respect.

With regard to educational initiatives, special emphasis should be placed on community and land-based instructional approaches given the historically extractive nature of post-secondary institutions' relations with Northern Indigenous communities (Black, 2015). This orientation is also reflected in communications for development in Canada. In Northern Canada, Dalseg and Abele (2015) discuss how Indigenous communities adapted communications technologies to participate in consultations about environmental stewardship and resource extraction.

These and other media, communication, or telecommunications initiatives are applied through community-based participatory action research (CBPR), which advocates for mutually respectful partnerships that position community members as co-researchers and educators (Leavy, 2017; Norman et al., 1998). It is action-oriented, meaning that along with traditional academic goals, research outcomes are designed to support the desired goals of involved communities, such as policy change or curriculum development (Menzies, 2004). This orientation entails collaboration with people affected by the issue being studied, to redress the underrepresentation of disadvantaged or marginalized communities (Green & Mercer, 2001; Macaulay et al., 1999). It requires partnerships, openness, and trust among communities, organizational representatives, and academics who are involved in all aspects of a project (Quarry & Ramirez, 2009; Rhodes et al., 2010). For researchers, community members and policymakers alike, the CBPR approach can contribute to more rigorous research methods and outcomes (Sheikh, 2006; Winter, 2020).

In recent years, CBPR projects have adopted and adapted digital tools (Beaton et al., 2017). These include multi-site video conferencing to conduct focus groups (Beaton et al., 2016; Gratton & O'Donnell, 2011); field trips to build relationships through activities such as a community ICT film festival (Gibson et al., 2012); and the collaborative development of community-based digital libraries (Rathi et al., 2017). Similarly, communications activities typically draw upon digitally mediated forms of connection and collaboration, which extend from social media platforms such as Facebook, Twitter, YouTube, and TikTok (Lalancette & Raynauld, 2020; Loyer, 2020; Wachowich & Scobie, 2010; Young, 2019), to older but still popular tools like community radio (Manatch, 2019; Roth, 2005; Szwarc, 2018). Henri et al. (2020) discuss how research communication strategies should incorporate both traditional media and new media, and should reflect approaches "locally tailored to existing networks, information-sharing pathways, and technological capabilities" (p. 91). While there are examples of technology-related CBPR educational initiatives in the NWT (Arsenault et al., 2019), the success of the Dechinta Centre for Research and Learning underscores the importance of community-oriented, in-person, land-based education (Ballantyne, 2014). The project builds on this work and reflect on our experiences using digitally enabled tools in the context of COVID-19 CBPR. It also incorporates the perspectives of community members the project worked with over the past three years.

## DigitalNWT's adaptation of digital tools for remotely conducted project activities

### Curriculum and course delivery

DigitalNWT curriculum and course delivery consisted of three courses along with workshops on topics including file management, data collection and research, and community networks. Each course included both community-based classes and Train-the-Trainer sessions for instructors. Instructors were primarily Community Adult Educators working at Aurora College who participated in Train-the-Trainer sessions. Both the Train-the-Trainer sessions and community-based classes were originally designed for in-person instruction.

Train-the-Trainer sessions guided participants through course material that they would subsequently teach by providing background information, curriculum, and answering questions. However, pandemic restrictions reduced both the quantity (in terms of time) and quality of the training sessions as they transitioned from in-person to virtual (online video conferencing) gatherings. Training sessions were facilitated by curriculum developers from the University of Alberta (UofA) and Hands On Media Education with the aim of involving Community Adult Educators as collaborative co-creators of the curriculum. The first in-person Train-the-Trainer session in Inuvik (2019) was particularly effective in this regard, with Adult Educators providing substantive feedback on both content (e.g., terms to avoid for Northern learners) and pedagogy (switching from a series of core and optional modules to model that allowed Adult Educators to pick content that was most useful for learners in their own community). By the final online Train-the-Trainer session in 2022 for course three (which substantively involved Northern residents as co-creators), the level of constructive discussion on curricular improvement from Adult Educators was minimal.

*Figure 6.1* Examples of in-person DigitalNWT courses held in NWT communities. Photo credits: Daniel Dokunmu.

After the Train-the-Trainer sessions, trainers delivered courses in their home communities. The DigitalNWT project supported these trainers through providing curriculum, mentorship, funding for professional development opportunities, and digital devices like iPads and laptops. The free courses were offered to any community member who wanted to improve their digital literacy skills. Each attendee received a refurbished laptop as both an incentive and a device with which to practice their new digital literacy skills. This was also a means to disseminate digital devices among NWT residents. Originally, trained Adult Educators taught the community courses with other project partners (e.g., UofA) providing support, training, and materials. Throughout the pandemic, and always in accordance with territorial COVID-19 health restrictions, limited in-person course offerings ran (see Figure 6.1). In some cases, the in-person approach was adapted to include remotely (co-)facilitated workshops hosted by partners from the UofA and Hands On Media Education, in addition to those led by Adult Educators. The DigitalNWT team explored other means to distribute course materials to Northern learners from mailing hard copy materials and Universal Serial Bus (USB) sticks loaded with course content to hosting a weekly radio show, Digital Trapline, on CKLB radio station.

### Research

DigitalNWT (n.d.) employed several methods of digitally enabled remote research, including conducting household surveys using an offline data collection app, gathering Internet performance data through a customized Internet Performance Test (IPT) dashboard, and conducting text message (short message service [SMS]) surveys. Qualitative methods, which are discussed below in the section on communications, included co-creating digital stories with Northern digital innovators. This research contributed to curriculum development as well as academic publications related to access, adoption, and use of digital technologies and Internet connectivity in NWT communities. The images (Figure 6.2) illustrate research activities undertaken by the project.

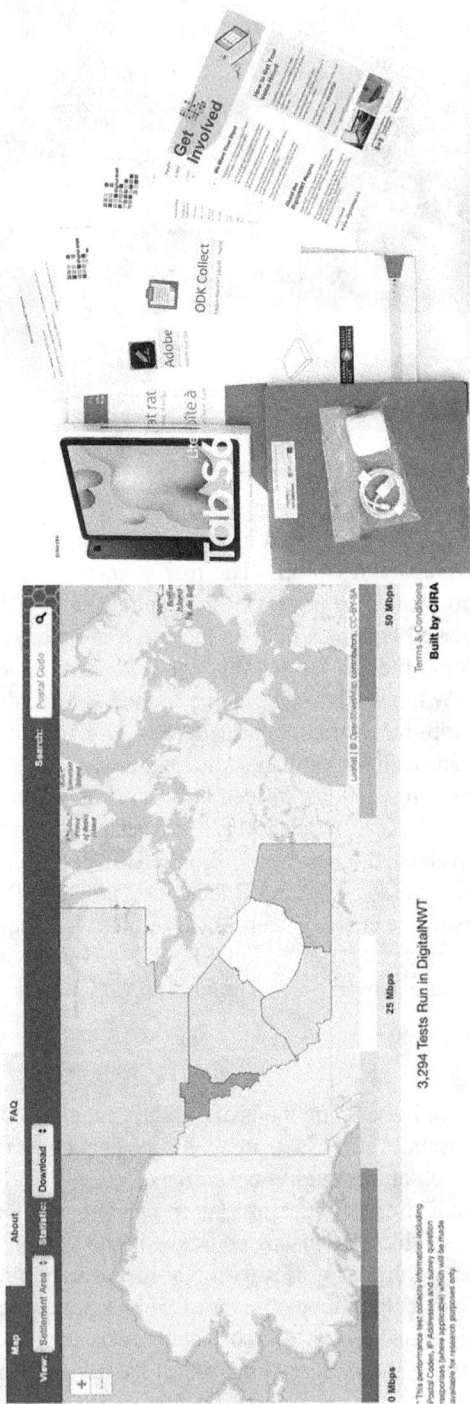

*Figure 6.2* Examples of remote data collection methods used by DigitalNWT researchers. Left: CIRA Internet performance test landing page developed by DigitalNWT and Canadian Internet Registration Authority (Screen capture). Right: Survey kits mailed to local researchers. Photo credit: Murat Akcayir.

After COVID-19 prevented us from traveling to the NWT, the project utilized digital tools and methods to remotely train and hire local researchers to conduct household surveys in their home communities, where available empirical data on digital access, adoption, and use was limited (Morgan et al., 2005). This involved customizing an offline digital data collection app hosted on a tablet, and setting up an automated workflow for local researchers to enter data. The Open Data Kit (ODK), a free and open-source data collection app that can operate offline on an Android-powered tablet, was adopted. Local researchers conducted face-to-face or phone surveys and then entered data into the app, which was set up to automatically upload completed surveys to a data management system at the university once they connected to a WiFi (wireless fidelity) access point.

To orient local researchers, university-based researchers delivered online video conferencing workshops in low-bandwidth environments, that local participants could join by phone and follow through emailed or print-copy slides. The training covered survey methods and ethical and privacy requirements. Discussions during training provided important information about the conditions in local communities and helped refine our methodology. Following the training, university-based researchers prepared and mailed survey kits that included a tablet pre-loaded with the survey app and questions, hard copy training materials, and an iPad to use in a prize draw (see Figure 6.2). Local researchers received payment for their work, and signed a confidentiality agreement to protect the privacy of respondents. After promoting the surveys on community radio and social media, and posters put up locally, they began door-to-door surveys. To address COVID-19 health guidelines, university-based researchers amended the data collection process (e.g., following social distancing guidelines, wearing a mask, wiping down tablets after conducting surveys, conducting surveys via phone). This process involved significant coordination among the project's partner organizations and local communities.

Two rounds of household surveys resulted in relatively high responses across 15 primarily remote / fly-in communities, most with populations of less than 1,000 people. Collaborating with this cohort of trained local researchers not only generated data, but it also helped the university-based researchers establish rapport with our Northern partners. For their part, communities benefited through skill building and short-term employment. Community members learned how to use digital tools that can support data collection and management in sectors, such as health and housing.

Another digitally enabled data collection method involved programmable SMS. SMS services were chosen rather than an online or social media survey due to higher penetration (and usage) rates for 2G services in the North, and the compatibility of SMS services with older devices. Instead of sending an SMS message to every number in a phone number listing, the project created an outreach campaign and invited respondents to text a specific phone number to participate. By making the survey voluntary, each respondent had to take the step of choosing to respond. This approach was chosen to build trust. A chatbot was created to ask survey questions. This approach mimicked a conversation, sending questions one at a time through text, waiting for a response, and then sending the next question accordingly providing a less formal, conversation-like structure.

The SMS survey information was promoted through Facebook / social media, on the project's website, through regular local radio programming, and through community posters. Respondents sent an SMS short code (e.g., DNWT1) to the survey's phone number, and received instructions in response. These include a shortened Informed Consent SMS message. Respondents could access the full informed consent information by dialing the SMS phone number to hear an audio recording or visiting a website (https://www.digitalnwt.ca/surveys). Survey data were transmitted via a third-party platform (https://www.twilio.com/), after which the research team performed sorting and analysis. The SMS surveys increased the overall number of survey respondents, and reached a wider audience within the NWT. The project reached several NWT communities in which local researchers were not yet present.

A third remote available was a digitally enabled research methodology that involved collecting Internet performance data via a platform established by the Canadian Internet Registration Authority (CIRA). This easy-to-use data collection and mapping platform is designed to encourage the crowd sourcing of geospatial data to illustrate digital access divides by engaging community members in monitoring their home Internet services (GlobeNewswire, 2021). Researchers have pointed to the importance of mapping digital divides, and of moving beyond supply-side mapping that typically provides information on existing or planned infrastructure and services in a region or community, and that illustrates advertised packages noting speeds, costs, etc., such as that employed by the Canada's National Broadband Internet Availability Map (see Hambly & Rajabiun, 2021). Problems existed with these approaches including missing or limited data, lack of granularity, and discrepancies between supply-side and demand-side data (Kahan, 2019; Office of Auditor General of Canada, 2018). The IPT used in this project measures demand-side experiences of Internet access regarding how end-users connect. It also reflected capacity-sharing among local community members who contributed data about their services, and university researchers who utilized digital tools and participatory methods to engage them in the monitoring and mapping process. We recognize that supply-side measures face reliability issues. For example, high levels of users sharing a household connection can affect Internet speed and quality, or old customer equipment (e.g., home routers) can impact results (Goldberg, 2020). However, in communities where alternatives are otherwise unavailable, the IPT provides important data on user experience, and data to complement the results of other tests.

We partnered with CIRA to set up an NWT-specific dashboard to integrate geographic information system (GIS) map overlays on top of IPT data. This compared Internet upload and download speeds across the territory, and in relation to individual municipalities, with a specific focus on representing data from Indigenous communities and regions. To encourage participation and share capacity, we conducted online outreach activities, including the #NWTDigitalDivide contest. The project also developed a curriculum module and online workshop that demonstrates how to record household Internet data (pricing, speed, reliability) and reported it to the national telecommunications regulator. As of June 1, 2022, a total of 2,870 tests had been conducted through the project's IPT portal since we launched the campaign in 2020 Fall.

### Communications and community engagement

The communications and community engagement component of DigitalNWT focused on building and sustaining an online community around NWT approaches to digital literacy, showcasing the successes of NWT-based digital innovators, and building rapport, familiarity, and trust. DigitalNWT's social media presence on Facebook, Twitter, YouTube, and Instagram included producing and sharing videos as well as promoting education opportunities and research activities. These included three campaigns (1) digital stories of NWT innovators, (2) short 'how-to' videos and photos and music from NWT artists and musicians, and (3) promoting research activities and opportunities to participate. The images in Figure 6.3 illustrate some of the people featured by DigitalNWT.

The #NWTDigitalDivide campaign supported our research activities. We promoted the IPT and SMS surveys through social media and popular news outlets (see Cohen & Stassen, 2020; Sibley, 2020). To engage communities, the project evolved from offering tablets and gift cards as prizes in the first two years of the project to hiring a local Indigenous artist to create customized moosehide bags. With respect to SMS surveys, our campaign helped us learn about the efficacy of (digitally mediated) word of mouth. This was illustrated when a prominent NWT digital innovator shared information about our first SMS survey on their social media pages, driving dozens of SMS participants in two days. This experience illustrated the effect of trust and reputation in supporting digitally enabled remote research methods particularly in the context of lean forms of communication like SMS.

*Figure 6.3* Two digital innovators from the NWT featured in the #NWTDigitalInnovators campaign. Left: Leanne Goose, Dene Nation, Northern Arts and Cultural Centre (Screen capture). Right: Jacey Firth-Hagen, #SpeakGwichinToMe (Screen capture).

The DigitalNWT project team also co-created digital stories that profile digital innovators from communities across the NWT. Digital storytelling involves participants creating a narrative through an edited combination of digital photos and audio; it reflected a low-cost but engaging form of participatory filmmaking that prioritized process over product (Wheeler, 2012). Researchers acted as facilitators guiding participants through steps in the co-design process from planning out a story to assembling components of video, audio, and text, to sharing the final product. Digital storytelling empowers participants to shape their own responses to questions posed by a researcher / facilitator, and has been used in disciplines such as community development and international development (Marshall et al., 2021). Digital storytelling has been used by Indigenous communities in projects to re-story and reclaim narratives (Iseke & Moore, 2011; Poitras Pratt, 2020; Whiskeyjack, n.d.). In generating digital stories, the project drew from Indigenous scholars with respect to issues including Indigenous intellectual property, sovereignty over media development, and stewardship of digital records (First Nations Information Governance Centre, 2014; Wemigwans, 2018).

Digital storytelling is typically employed via in-person workshop settings, through organizations such as the StoryCenter and Hands On Media Education—one of DigitalNWT's partner organizations. Due to geographic distances and COVID-19 health protocols, we adapted this approach to use remote methods supported by digital tools such as video conferencing and email. Participants had multiple opportunities to review and approve their stories, and could request the removal of their video at any time. In their digital stories, 17 Dene, Inuit, Métis, and non-Indigenous innovators shared their work and identified challenges related to digital technologies in the NWT. Inspired by the Fogo Process, digital storytelling in this case was employed as a tool of social change to represent the experiences and expertise of participants to audiences (including policymakers) while also positioning the act of media co-creation as collective action to support community development, which in this case was digital inclusion (Crocker, 2008; Newhook, 2009; Snowden, 1998).

## Perspectives from community partners

Northerners contributed to these digitally enabled remote research, education, and communication activities in multiple ways. It is essential that their voices are featured to ensure that our claims to be an ethical, reciprocal, and engaged project that reflects the perspectives of Northern residents.

DigitalNWT employed Indigenous Peoples from Northern communities as graduate research assistants, local surveyors, and curriculum reviewers and instructors. These team members were involved in co-designing research, outreach, data aggregation, and publishing methods. They were particularly included in community surveys, digital storytelling, CIRA Internet performance speed test outreach, and in representing DigitalNWT publicly in social media and the news. Indigenous team members co-created and shared authorship for publications and presentations.

Of the 24 authors and contributors, 10 were Indigenous Peoples from the NWT and two were non-Indigenous residents of the territory. The project also worked to highlight Northern digital innovators, musicians, and artists, such as through featuring Northern Indigenous music and visual art across project materials.

For DigitalNWT to build trust with communities, the project needed to create and sustain authentic relationships. Community trust is essential when gathering fulsome, representational data; Indigenous scholars point to this approach as supported by methods like visiting or conversation (Gaudet, 2019; Kovach, 2009). This work required visiting communities and meeting in-person. The ability to travel and conduct face-to-face research was restricted because of COVID-19. To build relations among Northern participants and to raise awareness of our project, we monitored our social media pages to ensure consistency and promptly responded to inquiries. We asked recognized Indigenous innovators, also working as graduate research assistants, to hold public draws using digital tools (e.g., Facebook Live). We developed a contact list of interested stakeholders, including Indigenous and municipal governments as well as news organizations, to inform them of upcoming community activities. Indigenous scholars ensured DigitalNWT's most public-facing activities and communications represented tone, messaging, and vernacular of the Northwest Territories. Northern partners acted as "social animators" (Crocker, 2008, p. 73) and helped organize community-based research activities, such as household surveys and digital storytelling.

Not all Indigenous team members were included at the beginning of the research and curriculum design processes. Once included, they provided invaluable input. Indigenous scholars contributed to the development of the video digital storytelling process, to the planning and conduct of research interviews, to the co-development of curriculum materials, to the preparation of academic publications, and to the project's consideration of intellectual property and copyright. Critically, their involvement represented the relationship to the communities and lands addressed through this project. We learned that Northern gift cards are not a strong incentive for residents of smaller rural communities, since many do not have a Northern store. Instead, we adopted prizes created by Northern artists, such as custom-made moosehide bags. We observed increased engagement with our social media platforms after introducing these more culturally relevant prizes. The wording of survey and interview questions was revised several times. Suggestions from Northern and Indigenous researchers improved the data collection protocols through the inclusion of more culturally appropriate and sensitive wording. With regard to curriculum development, it was not until the final course where Northern perspectives were fully integrated. This was done by using an analogy for the process of digital storytelling as setting up a canvas tent, and including short videos from notable Northerners discussing each step in the process. The curriculum also included custom-designed illustrations from an Inuvik-based illustrator. This experience provided valuable advice to inform future practice.

To gain deeper insight into remotely managed data collection activities, the project presented perspectives from four local researchers involved in the community surveys, as well as four digital innovators. Semi-structured interviews were conducted with these researchers and innovators via online video conferencing and telephone. Transcripts were analyzed using an inductive qualitative data approach supported by the Taguette open access data analysis tool (www.Taguette.org).

During the surveying process, collaboration with local surveyors required considerable coordination, including numerous telephone calls, text messages, emails, and online video conferencing meetings. Sometimes local surveyors could not receive project updates because of electrical or communications outages. In one community a COVID-19 outbreak delayed data collection and the local surveyor contacted respondents by telephone and entered their responses using a tablet. Remote, digitally enabled data collection may take longer than planned.

All interviewed local researchers indicated they felt supported through the remote research process. To prepare local researchers, online workshops were organized on a variety of topics, including survey methods as well as ethical and privacy requirements. Reflecting on this training, one interviewee stated it was a real learning experience and an opportunity to learn about new technology and data management. Another noted: "[the] best experience working with DigitalNWT is the workshops ... they're like really good at explaining things for me" (Participant 7). Others requested more face-to-face interaction with university-based researchers during training. One researcher who joined the training by telephone said: "it was really hard to ... understand where people are coming from, although we hear it," and requested more face-to-face interaction with peers (Participant 5). Another local researcher (Participant 7) said that "people understand better in-person rather than virtually, but everyone's different. I would prefer virtually." These suggestions pointed to a combination of face-to-face and online engagement in future training.

The interviews also reflected significant benefits for university-based researchers. As noted by Morgan et al. (2005), local data collectors know the local context and appropriate methods to obtain a higher response rate. Local researchers provided suggestions to improve incentives for participants. For example, rather than holding a prize draw then providing each respondent with a small gift certificate. Local researchers also helped communicate the benefits of surveys to respondents, such as allowing communities to compare and learn current conditions of Internet services (price, speed, etc.), increasing levels of interaction among people in the community, creating job opportunities, and giving a voice to communities. Particularly in communities of small populations, local researchers know community members and this helped establish rapport and engagement. As one researcher stated:

> I did a lot of communication with people, young to old, so I really had that, um, that relationship with the community, and I know everybody in the community. And it's, how to work with Elders and stuff as well, like communicate with them, so I already had that experience.
>
> (Participant 5)

Local researchers also provided invaluable expertise in translating and explaining survey questions, including in the local Indigenous language. As Participant 6 commented:

> [Y]ou'll get better insight as to … how you should phrase like a certain sentence or how you should approach the subject or how you should … structure the sentences and like the questions and, and, and I think that's really important …. I would … go see the Elders. And then I go have coffee or bannock with them, just sit there. And then, then they'll talk about like themselves and stuff like that and they're more relaxed.

Regarding survey adaptations in response to the COVID-19 pandemic, two sub-themes emerged from our interviews: (1) changes in survey practices, and (2) health and social distancing practices. Changes in survey practices included informing respondents when they could expect the researchers to arrive. For example, one interviewed researcher said that they phoned or messaged people on Facebook before visiting their home, which helped with response rates. Health and social distancing practices required surveyors to interact with participants through the phone or online (e.g., FaceTime) rather than conducting surveys in-person. Several local researchers found it difficult to conduct surveys over the phone, which felt impersonal for them. As Participant 7 stated:

> My experience working as a local surveyor during the pandemic … [was] it was hard to interact with others because of COVID. And at that time there was like an outbreak here. So I had to figure out how could I do this survey while being home. So, instead I reached out to people by phone or they call me and I use social media to get the word out there.

These insights point to the importance of in-person, on-the-ground local researchers, and social animators. As Henri et al. (2020) pointed out, although they can help mitigate some logistical challenges, "Internet-enabled communications technologies cannot, in and of themselves, lead to better interpersonal communication or research partnerships" (p. 92). We suggest that while interactions between university-based researchers and local surveyors can reflect a blend of intensive in-person training paired with ongoing remote, digitally enabled support, community surveys should primarily be conducted through face-to-face methods if possible.

With respect to educational activities, COVID-19 significantly impacted the delivery of courses. While limited in-person courses ran at different times in different communities, the Train-the-Trainer approach was most prominently impacted. Due to time and connectivity constraints, the collaborative, three-day in-person Train-the-Trainer event held in 2019 (pre-pandemic) could not be replicated online. Shortened online training sessions were less amenable to constructive discussion of the curriculum and evolved from mutual exchanges between the curriculum developers and community instructors, with the latter providing significant feedback on materials, to more unidirectional discussions of how to teach the material. However, the need for remote or co-facilitation by the curriculum development team enabled better insights into how learners received the materials. The diminution of the Train-the-Trainer sessions was also offset by enriched and more explicit instructional materials, specifically extensive facilitator guides and development of training videos.

With respect to communications, we focus here on the digital stories, which helped the project establish trust and reputation. The digital innovators appreciated the process and opportunity to showcase their projects and ideas in ways that felt relevant and useful for their communities. This helps reaffirm the capacity for digital media content creation in Indigenous communities, despite limitations experienced by the digital divide.

We interviewed four digital innovators to obtain a deeper understanding of their experiences and opinions regarding the digital stories. During the interviews, we asked about their motivations to participate and record a digital story. The main reasons include raising awareness of Northern issues (e.g., connectivity), their perceived importance of the topic (digital literacy and connection), their desire to revitalize and promote Indigenous languages, and appreciating the project's objectives. Participant 8 said,

> the other reason [I participated] is so that I could try to improve access to the Internet in my own community. And it's a slow process; but it's still a process, and at least I will be able to make a change.

Participant 2 stated, "I liked the project. It seemed like it was run by a good team. I really respect the team and the work that's being done." These and related statements highlight the importance of reciprocity and trust for community-based team members.

Regarding possible future improvements, some digital innovators have suggested involving more project members based in the communities. Others suggested more community visits from the university-based research team. As Participant 8 noted:

> So, I guess, number one [suggestion] would be—and I know it's COVID—but I'm pretty sure that communities would appreciate meeting the people, not just the surveyors, but I'm pretty sure that the communities would like visits from DigitalNWT.

These suggestions provided ideas about how the team could continue to strengthen their connections with communities. They also connected to practices identified and described in past communications for development initiatives such as the Fogo Process.

## Conclusion

In rural / remote areas, physical distances between university-based researchers, educators, and communicators, and team members based in communities, presented significant challenges. COVID-19 compounded these challenges as we were required to rapidly adapt digitally enabled forms of engagement due to our inability to travel to Northern communities. In many ways, our activities reflected practices established through the Fogo Process. As Newhook (2009) wrote,

> some of the most innovative elements of what would become the Fogo Process were not part of a careful plan, but were responses to circumstances which arose during the production and post-production processes; both production partners were familiar and comfortable with adjusting their approaches as events required.
>
> (p. 172)

In the context of COVID-19, this involved the DigitalNWT team and we adopted a variety of digital tools and participatory processes.

With respect to the research, we hired local surveyors who knew the area and lived in the participating communities. We worked closely with Indigenous team members to refine data collection and recruit participants through methods, such as SMS surveys and CIRA's IPT. This approach provided mutual benefits for all involved. For example, the local researchers, who spoke local Indigenous languages, could interpret and explain survey questions to local respondents. At the same time, they gained experience with new technologies and data management. Educational activities benefited from the insights of local team members as instructors and curriculum developers, who in turn gained professional development opportunities to adapt the openly licensed curriculum. The success of communication and community engagement further illustrated how digitally mediated activities required a blend of expertise from partners based in participating communities and in universities (Henri et al., 2020; Quarry & Ramirez, 2009).

Our experiences with DigitalNWT pointed to areas of improvement for future activities. We noted that in all cases remote, digitally mediated engagements were most effective when accompanied by opportunities for team members to meet and interact in-person. Our interviews with Northern team members suggested a blend of intensive in-person interaction, during training and initial contact, with regular periods of remote engagement, such as through email, text messaging, telephone calls, and video conferencing. In our future work, we intend to build upon these insights.

Our project illustrated that amid unforeseen and difficult circumstances such as the COVID-19 pandemic, digitally mediated research, education, and communication processes can be adapted—if relations of trust and reciprocity are in place. For future activities, we recommend that university-based co-design strategies adopt blended methods that incorporate interpersonal meetings with digitally mediated engagements in ways that benefit all involved parties. As communication for development initiatives such as the Fogo Process made clear, such activities require authentic and substantial engagement between academic and community partners. While this approach takes considerable effort to set up particularly in sensitive contexts that reflect historic and ongoing effects of settler colonialism and extractivism. We suggest that it provides concrete ways to build reciprocal projects that reflect the desires and goals of involved participants.

## Acknowledgments

The authors acknowledge the DigitalNWT project's funding through a Digital Literacy Exchange Program grant from Innovation, Science and Economic Development, Canada (grant number: 511811). We also thank everyone living and working in the communities that we worked with, learned from, and interviewed for this project. The following people agreed to have their names listed here; to those who wish to remain anonymous, thank you for all your contributions. Mahsi to: Belinda Pea'a, Les Baton, Shawna Nerysoo, and Violet Kikoak.

## References

Arsenault, R., Bourassa, B., Diver, S., McGregor, D., & Witham, A. (2019). Including indigenous knowledge systems in environmental assessments: Restructuring the process. *Global Environmental Politics, 19*(3), 120–32. https://doi.org/10.1162/glep_a_00519

Ballantyne, E. F. (2014). Dechinta Bush University: Mobilizing a knowledge economy of reciprocity, resurgence and decolonization. *Decolonization: Indigeneity, Education & Society, 3*(3), 67–85.

Beaton, B., McMahon, R., O'Donnell, S., Hudson, H., Whiteduck, T., & Williams, D. (2016). *Digital technology adoption in Northern and remote Indigenous communities.* Prepared for Innovation, Science and Economic Development Canada. First Mile Connectivity Consortium. http://firstmile.ca/wp-content/uploads/2016-ISED-FMCC.pdf

Beaton, B., Perley, D., George, C., & O'Donnell, S. (2017). Engaging remote Indigenous communities using appropriate online research methods. In N. Fielding, R. M. Lee, & G. Blank (Eds.), *The Sage handbook of online research methods* (2nd ed., pp. 563–77). Sage.

Benoit, C., Jansson, M., Millar, A., & Phillips, R. (2005). Community-academic research on hard-to-reach populations: Benefits and challenges. *Qualitative Health Research, 15*(2), 263–82. https://doi.org/10.1177/1049732304267752

Bigbee, J. L., & Lind, B. (2007). Methodological challenges in rural and frontier nursing research. *Applied Nursing Research, 20*(2), 104–06. https://doi.org/10.1016/j.apnr.2007.01.001

Black, K. (2015). Extracting northern knowledge: Tracing the history of post-secondary education in the Northwest Territories and Nunavut. *The Northern Review, 40*, 35–61. https://thenorthernreview.ca/nr/index.php/nr/article/download/462/512

Cohen, S., & Stassen, J. (September 26, 2020). *Is your internet painfully slow? This N.W.T. organization wants to hear about it.* CBC North. https://www.cbc.ca/news/canada/north/nwt-digital-divide-campaign-1.5732772

Correa, T. & Pavez, I. (2016). Digital Inclusion in Rural Areas: A Qualitative Exploration of Challenges Faced by People From Isolated Communities. *Journal of Computer-Mediated Communication, 21*(3): 247–263. https://doi.org/10.1111/jcc4.12154

Coulthard, G. S. (2014). *Red skin, white masks: Rejecting the colonial politics of recognition.* University of Minnesota Press.

Crocker, S. (2008). Filmmaking and the politics of remoteness: The genesis of the Fogo Process on Fogo Island, Newfoundland. *Shima: The International Journal of Research into Island Cultures, 1*(2), 59–75. https://shimajournal.org/issues/v2n1/g.-Crocker-Shima-v2n1.pdf

Dalseg, S. K., & Abele, F. (2015). Language, distance, democracy: Development decision making and Northern communications. *The Northern Review, 41,* 207–40. https://doi. org/10.22584/nr41.2015.009

Darnell, R. (2018). Reconciliation, resurgence, and revitalization: Collaborative research protocols with contemporary First Nations communities. In M. Asch, J. Borrows, & J. Tully (Eds.), *Resurgence and reconciliation: Indigenous-settler relations and earth teachings* (pp. 229–44). University of Toronto Press.

Dieter, J., McKim, L. T., Tickell, J., Bourassa, C. A., & Lavallee, J. (2018). The path of creating co-researchers in the File Hills Qu'Appelle Tribal Council. *The International Indigenous Policy Journal, 9*(4), 1. https://doi.org/10.18584/iipj.2018.9.4.1

DigitalNWT. (n.d.). *Digital literacy and innovation in the NWT.* https://www.digitalnwt.ca

Duarte, M. E. (2017). *Network sovereignty: Building the Internet across Indian country.* University of Washington Press.

Edwards, K. M., Mattingly, M. J., Dixon, K. J., & Banyard, V. L. (2014). Community matters: Intimate partner violence among rural young adults. *American Journal of Community Psychology, 53*(1), 198–207. https://doi.org/10.1007/s10464-014-9633-7

Environics. (2019). *Research on telecommunications services in Northern Canada—Final report.* https://epe.lac-bac.gc.ca/003/008/099/003008-disclaimer.html?orig=/100/200/301/ pwgsc-tpsgc/por-ef/crtc/2021/023-20-e/POR023-20-Final-Report.html

First Nations Information Governance Centre. (2014). *Ownership, control, access and possession (OCAP™): The path to First Nations information governance.* https://achh.ca/ wp-content/uploads/2018/07/OCAP_FNIGC.pdf

Gaudet, J. C. (2019). Keeoukaywin: The visiting way—Fostering an Indigenous research methodology. *Aboriginal Policy Studies, 7*(2), 47–64. https://doi.org/10.5663/aps. v7i2.29336

Gibson, K., Thomas, L., O'Donnell, S., Lockhart, E., & Beaton, B. (2012). Cocreating community narratives: How researchers are engaging First Nation community members to cowrite publications. [Paper presentation.] *Qualitative Analysis Conference,* St. John's, NL, Canada. http://susanodo.ca/wp-content/uploads/2017/10/2012-Qualitatives.pdf

GlobeNewswire. (2021, May 19). CIRA surpasses 1 million Internet performance tests. [News release.] https://www.globenewswire.com/news-release/2021/05/19/2232686/0/ en/CIRA-surpasses-1-million-Internet-Performance-Tests.html

Goldberg, M. (2020). CIRA fails its performance test. http://mhgoldberg.com/blog/?p=14098

Gratton, M. F., & O'Donnell, S. (2011). Communication technologies for focus groups with remote communities: A case study of research with First Nations in Canada. *Qualitative Research, 11*(2), 159–75. https://nrc-publications.canada.ca/eng/view/accepted/?id= ab607b11-e299-442c-9871-c4d6f7745507

Green, L. W., & Mercer, S. L. (2001). Can public health researchers and agencies reconcile the push from funding bodies and the pull from communities? *American Journal of Public Health, 91*(12), 1926–29. https://doi.org/10.2105/AJPH.91.12.1926

Greer, A. (2019). Settler colonialism and beyond. *Journal of the Canadian Historical Association/Revue de la Société historique du Canada, 30*(1), 61–86. https://doi.org/10.7202/ 1070631ar

Gwich'in Tribal Council (2011). *Conducting traditional knowledge research in the Gwich'in settlement area: A guide for researchers.* https://nwtresearch.com/sites/default/files/ gwich-in-social-and-cultural-institute_0.pdf

Hambly, H., & Rajabiun, R. (2021). Rural broadband: Gaps, maps and challenges. *Telematics and Informatics, 60,* 101565. https://doi.org/10.1016/j.tele.2021.101565

Henri, D. A., Brunet, N. D., Dort, H. E., Hambly Odame, H., Shirley, J., & Gilchrist, H. G. (2020). What is effective research communication? Towards cooperative inquiry with Nunavut communities. *ARCTIC, 73*(1), 81–98. https://doi.org/10.14430/arctic70000

Irlbacher-Fox, S. (2014). Traditional knowledge, co-existence and co-resistance. *Decolonization: Indigeneity, Education & Society, 3*(3), 145–58.

Iseke, J., & Moore, S. (2011). Community-based Indigenous digital storytelling with Elders and youth. *American Indian Culture and Research Journal, 35*(4), 19–38.

Kahan, J. (2019, April 8). It's time for a new approach for mapping broadband data to better serve Americans. Microsoft on the Issues. [Blog.] https://blogs.microsoft.com/on-the-issues/2019/04/08/its-time-for-a-new-approach-for-mapping-broadband-data-to-better-serve-americans/

Kovach, M. (2009). *Indigenous methodologies: Characteristics, conversations, and contexts.* University of Toronto Press.

Kukutai, T., & Taylor, J. (2016). *Indigenous data sovereignty: Toward an agenda.* ANU Press.https://doi.org/10.22459/CAEPR38.11.2016

Lalancette, M., & Raynauld, V. (2020). Online mobilization: Tweeting truth to power in an era of revised patterns of mobilization in Canada. In T. Small & H. Jansen (Eds.), *Digital politics in Canada: Promises and realities* (pp. 223–44). University of Toronto Press.

Leavy, P. (2017). *Research design: Quantitative, qualitative, mixed methods, arts-based, and community-based participatory research approaches.* The Guilford Press.

Loyer, J. (2020, April 23). Indigenous TikTok is transforming cultural knowledge. *Canadian Art.* https://canadianart.ca/essays/indigenous-tiktok-is-transforming-cultural-knowledge/

Mabasa, L. T., & Themane, M. J. (2021). Experiences of educational researchers in the use of the case study design during the Covid-19 pandemic: Lessons from a South African rural setting. *International Journal of Qualitative Methods, 20*, 16094069211022567. https://doi.org/10.1177/16094069211022567

Macaulay, A. C., Commanda, L. E., Freeman, W. L., Gibson, N., McCabe, M. L., Robbins, C. M., & Twohig, P. L. (1999). Participatory research maximises community and lay involvement. *BMJ, 319*(7212), 774. https://doi.org/10.1136/bmj.319.7212.774

Manatch, M. (2019). *Spoken from the heart: Indigenous radio in Canada.* https://en.ccunesco.ca/-/media/Files/Unesco/Resources/2019/08/SpokenFromTheHeartIndigenousRadioInCanada.pdf

Marshall, C., Rossman, G. B., & Blanco, G. L. (2021). *Designing qualitative research* (7th ed.). Sage.

Melis, G., Sala, E., & Zaccaria, D. (2021). Remote recruiting and video-interviewing older people: A research note on a qualitative case study carried out in the first Covid-19 Red Zone in Europe. *International Journal of Social Research Methodology, 25*(4), 477–82. https://doi.org/10.1080/13645579.2021.1913921

Menzies, C. R. (2004). Putting words into action: Negotiating collaborative research in Gitxaala. *Canadian Journal of Native Education, 28*(1), 15–32.

Morgan, L. L., Fahs, P. S., & Klesh, J. (2005). Barriers to research participation identified by rural people. *Journal of Agricultural Safety and Health, 11*(4), 407–14. https://doi.org/10.13031/2013.19719

Newhook, S. (2009). The godfathers of Fogo: Donald Snowden, Fred Earle and the roots of the Fogo Island Films, 1964–1967. *Newfoundland and Labrador Studies, 24*(2), 171–97. https://www.erudit.org/en/journals/nflds/2009-v24-n2-nflds24_2/nflds24_2art01/

Norman, D. W., Bloomquist, L. E., Freyenberger, S. G., Regehr, D. L., Schurle, B. W., & Janke, R. R. (1998). Farmers attitudes concerning on-farm research: Kansas survey results. *Journal of Natural Resources and Life Sciences Education, 27*(1), 35–41. https://doi.org/10.2134/jnrlse.1998.0035

Office of Auditor General of Canada (OAG) (2018). Report 1—Connectivity in rural and remote areas. https://www.oag-bvg.gc.ca/internet/English/att__e_43221.html

Pierce, C., & Scherra, E. (2004). The challenges of data collection in rural dwelling samples. *Online Journal of Rural Nursing and Health Care, 4*(2), 25–30. https://doi.org/10.14574/ojrnhc.v4i2.197

Poitras Pratt, Y. (2020). *Digital storytelling in Indigenous education: A decolonizing journey for a Métis community.* Routledge.

Quarry, W., & Ramirez, R. (2009). *Communication for another development: Listening before telling.* Zed Books.

Rathi, D., Shiri, A., & Cockney, C. (2017). Environmental scan: A methodological framework to initiate digital library development for communities in Canada's North. *Aslib Journal of Information Management, 69*(1), 76–94. https://doi.org/10.1108/AJIM-06-2016-0082

Rhodes, S. D., Malow, R. M., & Jolly, C. (2010). Community-based participatory research: A new and not-so-new approach to HIV/AIDS prevention, care, and treatment. *AIDS Education and Prevention: Official Publication of the International Society for AIDS Education, 22*(3), 173–83. https://doi.org/10.1521/aeap.2010.22.3.173

Roth, L. (2005). *Something new in the air: The story of First Peoples television broadcasting in Canada.* McGill-Queen's University Press.

Sheikh, A. (2006). Why are ethnic minorities under-represented in US research studies? *PLoS Medicine, 3*(2), 166–67. https://doi.org/10.1371/journal.pmed.0030049

Sibley, S. (2020, September 15). Campaign seeks to highlight 'agonizing' NWT internet access. Yellowknife, NT: Cabin Radio. https://cabinradio.ca/45193/news/dehcho/campaign-seeks-to-highlight-agonizing-nwt-internet-access/

Snowden, D. (1998). Eyes see; ears hear. In D. Richardson & L. Paisley (Eds.), *The first mile of connectivity: Advancing telecommunications for rural development through a participatory communication approach* (pp. 60–73). Communication for Development. Food and Agriculture Organization of the United Nations (FAO). https://www.fao.org/3/x0295e/x0295e.pdf

Statistics Canada. (2018). Population centre (POPCTR): Detailed definition. https://www150.statcan.gc.ca/n1/pub/92-195-x/2011001/geo/pop/def-eng.htm

Stiegman, M. L., & Castleden, H. (2015). Leashes and lies: Navigating the colonial tensions of institutional ethics of research involving Indigenous Peoples in Canada. *International Indigenous Policy Journal, 6*(3). https://doi.org/10.18584/iipj.2015.6.3.2

Szwarc, J. (2018). Indigenous broadcasting and the CRTC: Lessons from the licensing of Native Type B radio. https://crtc.gc.ca/eng/acrtc/prx/2018szwarc.htm

Tuhiwai Smith, T. L. (1999). *Decolonizing methodologies: Research and Indigenous Peoples.* University of Otago Press; Zed Books. https://nycstandswithstandingrock.files.wordpress.com/2016/10/linda-tuhiwai-smith-decolonizing-methodologies-research-and-indigenous-peoples.pdf

Wachowich, N., & Scobie, W. (2010). Uploading selves: Inuit digital storytelling on YouTube. *Études/Inuit/Studies, 34*(2), 81–05. https://doi.org/10.7202/1003966ar

Wemigwans, J. (2018). *A digital bundle: Protecting and promoting Indigenous knowledge online.* University of Regina Press.

Wheeler, J. (2012). Using participatory video to engage in policy processes: Representation, power, and knowledge in public screenings. In E-J. Milne, C. Mitchell, & N. D. Lange (Eds.), *Handbook of participatory video* (pp. 365–82). Rowman & Littlefield.

Whiskeyjack, L. (n.d.). *Sage and sweetgrass.* [YouTube channel.] https://www.youtube.com/channel/UCmmOvY9kutX1Vat1ktUoRtA

Wilkes, L. (1999). Metropolitan researchers undertaking rural research: Benefits and pitfalls. *Australian Journal of Rural Health, 7*(3), 181–85. https://doi.org/10.1046/j.1440-1584.1999.00210.x

Winter, L. A. (2020). Assessing JoinLite—A recruitment tool for all of us Pennsylvania. [Unpublished Master's thesis, University of Pittsburgh.]

Wolfe, P. (2006). Settler colonialism and the elimination of the native. *Journal of Genocide Research, 8*(4), 387–409. https://doi.org/10.1080/14623520601056240

Young, J. C. (2019). Rural digital geographies and new landscapes of social resilience. *Journal of Rural Studies, 70*, 66–74. https://doi.org/10.1016/j.jrurstud.2019.07.001

# 7 The paradox of digital participation and misinformation

## Lessons from online agri-food communities of practice in Trinidad and Tobago

*Jeet Ramjattan, Ataharul Chowdhury,*
*Khondokar Humayun Kabir and Wayne Ganpat*

### Introduction

With the increased use of Information and Communication Technology (ICT) in agriculture, farmers can access a wealth of information that can improve their yields and profitability (World Bank, 2017). Yet, with the increased digital participation of farmers on various ICT-based platforms they are also exposed to an increasing amount of misinformation that can have severe effects (Luo et al., 2021). According to research, the prevalence of false or misleading information on digital platforms (DP) such as social media (SM), farming forums, and online marketplaces can lead to incorrect decisions and practices, which can impact farmers' livelihoods, resulting in financial losses (Lynas et al., 2022; Watts, 2018). Some types of misinformation that farmers may face on DP include false claims about efficacy of certain products or practices, exaggerated benefits of particular technologies or inputs, and inadequate or erroneous information regarding the risks of particular methods (Chowdhury et al., 2023; Menon, 2020; Spencer-Jolliffe, 2023).

Moreover, the ease of sharing and disseminating information on online platforms without appropriate fact-checking or verification can expedite the spread of misinformation (Cook et al., 2017). Hence, farmers must be cautious and examine the information they obtain on DP cautiously, as well as seek trustworthy information sources (Ivanchuk, 2023; Rust et al., 2021; Singh & Verma, 2023). In addition, agricultural organizations, researchers, and politicians must establish measures to reduce the spread of misinformation and increase the dissemination of correct and trustworthy information to farmers (Dilleen et al., 2023; van der Linden et al., 2017; West, 2017; Witze, 2021).

Technological advancements have raised expectations that different agricultural actors, including farmers and advisory service providers, will tap its potential (Steinke et al., 2020). The use of various Internet-based communication technologies by members of the agri-food community has revolutionized the manner in which rural people communicate. Over the past two decades, researchers have focused on farmers' digital participation to help them cope with new technologies, climate change, and market uncertainty. "Digital Participation" refers to farmers'

DOI: 10.4324/9781003282075-10

ability to participate in educational and networking activities via the Internet, SM, mobile technology, and forms of ICTs (Gilman & Peixoto, 2019). Research shows that despite investments in ICTs, projected returns have not always been achieved due to a lack of capacity among farmers and extension workers to properly utilize their digital participation (Birke et al., 2018; Kabir et al., 2022).

DPs can also provide fertile ground for spreading misinformation and disinformation, highlighting needs of farmers and other agri-food actors' ability to ingest and respond to information. With technological breakthroughs such as SM and smartphones, knowledge and information exchanges happen exponentially. However, not all information transmitted in this manner is reliable, raising concerns about how misinformation appears as facts spread, deceiving people. The spread of speculative, unsubstantiated, ambiguous, or conflicting statements or portrayals, whether intended to deceive others or not, is known as misinformation (Bode & Vraga, 2017; Tan et al., 2015, p. 675). Misinformation, generally, refers to any false, fraudulent information or claim that circulates on SM. 'Misinformation' is a term that encompasses disinformation, fake news, rumor, spam, trolls, urban legends, and all have the false message as a common characteristic that, if unchecked, can cause concern or negative impacts via SM (Nguyen et al., 2023). Misinformation arises without deliberate intent to deceive. For example, if new information about developments becomes available, the first explanation of what is happening may not contain all necessary information; clarifications and updates may be offered later. Nonetheless, misinformation can still be shared even if discrepancies in the information shared are corrected in future statements (Tan et al., 2015, p. 676). Disinformation is a subcategory of misinformation as it is purposefully crafted to be deceptive and misleading (Wardle & Derakhshan, 2017). Regardless of whether the sharer is aware of the error, both types of information can be extensively disseminated (Gebel, 2021).

The agricultural sector in Trinidad and Tobago (T&T) is undergoing a transformative shift propelled by the influential role of social media. Farmers and extension specialists actively participate in online Communities of Practice (OCoPs) through SM platforms (SMPs). These OCoPs enable regular interaction among individuals sharing a common concern or passion, fostering collective learning within a specific domain. SMPs such as Facebook and WhatsApp effectively facilitate the formation of OCoPs, aligning with key characteristics of the CoP framework established by Wenger and Wenger-Trayner (2015). Agricultural extension agents utilize these platforms to expand their professional networks and connect with like-minded individuals. Consequently, these platforms facilitate the establishment of agricultural associations, uniting individuals within the agricultural community and promoting a shared sense of purpose. CoPs are recognized as social learning networks, defined by Wenger et al. (2011). OCoPs, operating within this context, utilize SMPs to connect individuals sharing common concerns or passions (Johnson, 2001). OCoPs serve as platforms for knowledge sharing and collective expertise development within the community.

Due to the expanding controversy around agricultural processes, anyone in a OCoP is able to distribute misleading information. Several controversial topics, including genetically modified organisms (GMOs), organic agriculture, artificial intelligence, pesticide resistance, acceptance of synthetic meats, and water scarcity, are polarizing the online social climate (Pfeiffer et al., 2022). Farmers' skepticism, rejection, and contrarianism, primarily form due to their digital participation in different platforms, have made it harder for agricultural extension and advisory services (AEAS) to communicate complex science (Treen et al., 2020, p. 12). Globally, AEAS use various ICTs and encourage rural farmers to participate in various DPs, connecting them to a network of actors. According to Treen et al. (2020), different actors are involved in funding, generating, and amplifying false information in an organized manner that rural farmers are unable to distinguish between accurate information, advice, and claims. When someone posts a viral message or misleading information in OCoP, others run the risk of resharing it without verifying its legitimacy (Bellows & Moore, 2013). To "enlighten" their followers, some users disseminate false material under the impression that it is accurate. Additionally, individuals may transmit knowledge because it is intriguing, notable, or has the potential to boost their social standing or prominence in some way (Valenzuela et al., 2019). According to extension service providers, agricultural information is available in numerous formats and from numerous sources, including the government, research institutions, and global and regional agricultural associations. The proliferation of SM is reshaping agricultural and rural communities in T&T. Despite this the lack of digital literacy makes it difficult for farmers to acquire and locate authoritative, relevant, and reliable information that can assist them in making informed decisions about their livelihoods (Renwick, 2010). Farmers and extension specialists are establishing OCoPs on SM in order to share their knowledge and develop networks to discuss ideas, acquire new skills, and exchange information. These networks are enabling the provision of more effective agricultural advisory and extension services by allowing agents to respond in real time (Ramjattan et al., 2020). The Ministry of Agriculture of Trinidad and Tobago has deployed ICT-based extension options during the COVID-19 pandemic, such as apps and online learning and information exchange platforms, to improve agricultural extension services (AES) (Purdue University, 2020).

SMPs like Instagram, Facebook, WhatsApp, and YouTube have become vital for sharing agricultural information, enabling farmers and extension workers to access and exchange knowledge on best practices, innovative methods, and new technologies (Colussi et al., 2022; Ghosh et al., 2021; Kesherwani et al., 2022; Md Nordin et al., 2021; Thakur & Chander, 2018). These platforms facilitate the sharing of textual, visual, audio, and video agricultural content, fostering co-learning and network development among farmers and stakeholders (Baker, 2020). SM provides access to manufacturing, environmental, and marketing expertise, benefiting local food systems and sustainable agriculture. Facebook is commonly used by rural populations seeking agricultural extension and technical support, while WhatsApp is popular for quick and cost-effective information sharing (Kumar, 2023; Suchiradipta & Saravanan, 2016).

If agents and administrators of platforms are trained in knowledge management and effective use of ICTs and SM for educational purposes, it is anticipated ICTs will increase food system participation and education (Ramjattan et al., 2020). However, farmers encounter different challenges with their digital participation, but information credibility and bias are the most pervasive. Many agricultural development initiatives are still centered on helping farmers improve their ability to utilize various DPs as their information and knowledge sources (Birke et al., 2018; Kabir et al., 2022). In addition, research studies have not yet considered the possibility that farmers could be exposed to misinformation as a result of their digital participation in a variety of ICT-based platforms.

## Purpose of the study

The chapter aims to investigate the use of SM by members (e.g., farmers, extension agents) of OCoP, and the perception of agri-food misinformation as well as its consequences for digital participation. This study addresses:

1  What are OCoP agricultural sector members' SM behaviors, including time spent and preference for media use?
2  How do members of agricultural OCoP, including extension and advisory service providers, perceive misinformation, its diffusion, consequences for digital participation, and countermeasures?

## Methodology

The study employed a mixed-methods approach and consider Facebook and WhatsApp as online OCoPs.

### *Identifying and selecting agricultural online communities of practice*

#### *Facebook*

An online search using relevant search terms was conducted to identify individuals and organizations from T&T's agricultural sector who use the platform to access agriculturally related information. Facebook-suggested related pages were also considered. Data collected were coded into categories using an Excel spreadsheet and coding guide, which included specific variables pertaining to CoPs such as discussed topics, shared interests, and collaborative activities.

#### *WhatsApp*

WhatsApp groups were identified and selected using publicly available online links and a Facebook search. Initially, 15 groups were identified. For those groups, an electronic database was created. Groups were classified according to their level of dedication to specific agricultural issues and topics discussed. Following the initial identification of available WhatsApp groups, the researcher requested and joined those that were actively discussing agricultural-related subjects ($n = 10$). After contacting the group's administrators and members, the purpose of the exploratory study was conveyed. Group members were invited to participate anonymously in the research study.

### Data collection

The study utilized a survey research design to gather data about SM use, understanding misinformation-related concerns of members of agricultural OCoPs and combating strategies among OCoP members and administrators. Data were collected through a structured, online administered, and pilot-tested questionnaire completed by OCoP members using Facebook and WhatsApp SMPs. The study's questionnaire used close-ended questions (Likert-type scale, ranking, choice option scale) to obtain data from both Facebook and WhatsApp group members as these were identified as platforms most used by farmers and extension agents. Initially, 71 Facebook pages and 10 WhatsApp chat groups were identified. The Facebook pages were formed from 2008 to 2021 and the WhatsApp groups were formed from 2016 to 2021. Respondents were selected from these groups, and a total of 165 individuals responded from both platforms out of 258 contacted through the snowball sampling technique. The survey was conducted from March to May 2021 using Google Forms' Web-based survey administration software and administered as an online link via email. Through the Facebook messenger application and WhatsApp messaging, respondents were contacted and requested to take the survey and assist in further recruitment of respondents using the snowball sampling technique of soliciting referrals of potential respondents.

The survey instrument contained three sections: (1) demographic and socioeconomic profile; (2) use of social media; and (3) misinformation-related questions, such as understanding, concerns, beliefs, dissemination and combating strategies. Participants were asked to read the questionnaire and follow instructions. It was up to participants to take part in the study. They were all made aware of its objective, as well as the study's objective.

### Content validity assessment

Content validity assessment of the instrument was done prior to the survey to assess how scales represented the concepts used. A panel of seven extension experts reviewed the questionnaire, which comprised three Extension Directors, three university lecturers with more than 20 years of experience in the T&T extension system, and one ICT subject matter specialist. Three items that did not adequately depict the concepts were eliminated and a few other items required minor modifications. A pre-test was conducted to determine the effectiveness of the question's wording, format, skip patterns, timing, and order. The pre-test was conducted with five agricultural extension agents and five farmers to test the practicality and understanding of the questionnaire before conducting the survey.

*Data analysis*

The survey was administered online using Google Forms and distributed to OCoP members through Facebook and WhatsApp. A total of 165 questionnaires were completed, representing a response rate of 65%. The collected data were organized and coded in Excel before being imported into SPSS, version 22, for analysis. Descriptive statistics, such as frequencies and percentages, were used to analyze data (Table 7.1).

*Table 7.1* Instrument sections and measurements

| Questionnaire section and information | Variables | Measurement scale |
|---|---|---|
| **Section 1** Demographic and socio-economic | Gender, age, education, and organization affiliation | **Gender** Male = 1, Female = 2 **Age** <20 = 1; 21–30 = 2; 31–40 = 3; 41–50 = 4; >50 = 5 **Education** Primary = 1, Secondary = 2; Technical vocational = 3; Undergraduate = 4; Post-graduate = 5 **Organization affiliation** Public sector = 1; Private sector = 2; Farmers / producers = 3; Non-profits = 4; Other = 5 |
| **Section 2** SM use behavior | Frequency of use | 4 = Most often; 3 = Sometimes; 2 = Least often; 1 = Do not use |
| **Section 3** Misinformation-related questions | Defining / understanding misinformation (10 items) | 3.1 Four items to obtain respondents' definition of misinformation with option to pick more than one. 3.2 Three items about the credibility of information with an option to pick more than one. 3.3 Three items about a person's likelihood to believe in agricultural scientists. |
| | Misinformation concerns (13 items) Misinformation beliefs (11 items) | 1 = Strongly disagree; 2 = Disagree; 3 = Agree; 4 = Strongly agree 1 = Yes; 0 = No |
| | Misinformation circulation (six items) | (1) Which medium do you consider to be the most prevalent source of misinformation? i. Newspapers; ii. Face to face; iii. Social media / Internet; iv. Television; v. Radio (2) Strategies to deal with misinformation and rank the topics (write down the items number here) Which medium do you consider to be the most prevalent source of misinformation? i. Newspapers; ii. Face to face; iii. Social media / Internet; iv. Television; v. Radio |

*(Continued)*

*Table 7.1* (Continued)

| Questionnaire section and information | Variables | Measurement scale |
|---|---|---|
| | | (3) Which of the following social media platform is most responsible for spreading misinformation?<br>i. Facebook; ii. WhatsApp; iii. Twitter; iv. Instagram; v. Pinterest; vi. Blogs; vii. Other<br>(4) How often do you encounter groups or individuals who spread misinformation on your social media networks?<br>i. Not often<br>ii. Often<br>iii. Very often |
| | Misinformation strategies | (5) How do you deal with misinformation? Please write your response. |
| | Misinformation topics | (6) What do you consider to be the most common agriculturally related misinformation topics being circulated?<br>i. Genetically Modified Organisms (GMOs)<br>ii. Climate change<br>iii. Chemical pesticide and environmental protection<br>iv. Food and nutrition, safety, and security<br>v. Other |

# Results

## Demographic and socio-economic characteristics

This section presents the descriptive findings from the survey on misinformation, including frequencies, percentages, and means. Table 7.2 shows the demographic and socio-economic profile of OCoPs. The online survey of agricultural extension OCoPs members on Facebook and WhatsApp received 165 replies. Respondents included males (49%) and females (51%).

Most respondents (33%) belonged to the mid-age grouping (31–40), 2% of respondents were <20 years old, followed by 13% in the 21–30 age group categories, 31% in the 41–45 age group categories, and 21% were >50 years. It was found that none of the respondents attained primary school education alone ; 13% completed secondary school, 13% had technical vocational training, 33% had an undergraduate degree, and 41% had a postgraduate degree. Also, 32% of respondents worked in the private sector, 49% were employed in the public sector, 10% were farmers or producers, and 9% were affiliated with non-profits or other groups.

*Table 7.2* Demographic and socio-economic profile of agricultural extension CoPs' social media (*n* = 165)

| Parameters | Categories | % |
|---|---|---|
| Gender | Female | 51 |
| | Male | 49 |
| Age (years) | <20 | 2 |
| | 21–30 | 13 |
| | 31–40 | 33 |
| | 41–50 | 31 |
| | >50 | 21 |
| Education Level | Primary | 0 |
| | Secondary | 13 |
| | Diploma, Certificate | 13 |
| | Undergraduate | 33 |
| | Postgraduate | 41 |
| Type of organization affiliation / working with | Public sector | 49 |
| | Private sector | 32 |
| | Farmers / Producer | 10 |
| | Non-profits and other | 9 |

### Social media use behavior

#### Social media use

Seven SMPs were mentioned in the survey questions. WhatsApp was the most widely used platform, with only 2% of respondents claiming they didn't use it. However, of the 98% who used WhatsApp, 71% said they used it frequently while 25% said they used it occasionally, and 2% said they used it infrequently. Ten percent of respondents said they did not use the Facebook platform, while 43% said they used it most often, 38% said they used it occasionally, and 8% said they used it the least. The YouTube platform showed a similar response to that of Facebook as 11% did not use YouTube. The majority of respondents, comprising 78%, reported not utilizing any additional social networking platforms. Following this, 72% mentioned non-usage of Twitter, while 39% did not engage with Instagram. Additionally, 26% abstained from other professional networking platforms, and 21% did not employ MS Teams, as illustrated in Table 7.3.

#### Time spent on social media

As illustrated in Figure 7.1, most respondents (49%) spent >3 but <4 hours per day on social media, followed by 28% who spent >2 but <3 hours per day, and 14% who spent >1 hour but <2 per day. According to respondents, the least time spent on SM was "Few times a week." This option was chosen by 2% of the respondents, indicating that at the absolute least, individuals spent at least a small amount of time every week, and 7% spent less than an hour each day.

*Table 7.3* Social media use (*n* = 165)

| Social media | 1<br>Do not<br>use (%) | 2<br>Least<br>often (%) | 3<br>Sometimes<br>(%) | 4<br>Most<br>often (%) |
|---|---|---|---|---|
| Facebook | 10 | 8 | 38 | 43 |
| WhatsApp | 2 | 2 | 25 | 71 |
| Instagram | 39 | 17 | 25 | 19 |
| YouTube | 11 | 8 | 44 | 39 |
| Professional networking<br>  (e.g., Research Gate, LinkedIn) | 26 | 27 | 39 | 9 |
| Twitter | 72 | 18 | 9 | 1 |
| MS Teams, Zoom, Skype | 21 | 19 | 38 | 22 |
| Other | 78 | 7 | 14 | 1 |

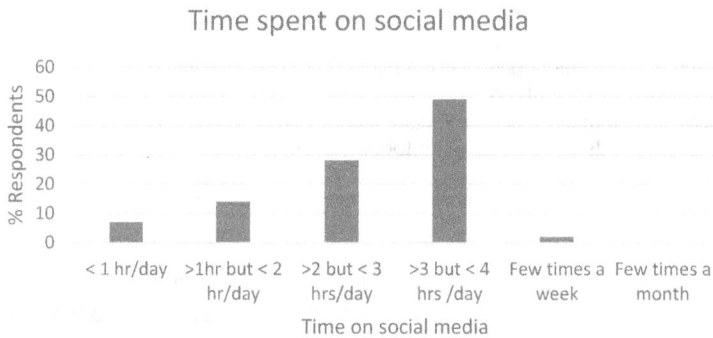

*Figure 7.1* Time spent on social media (*n*=165).

## Misinformation and its related issues

*Understanding of misinformation*

Respondents were asked to select one or more definitions from the following four statements that best describe misinformation:

1  =Misinformation is false information.
2  =Misinformation is information that is false information deliberately shared to harm people.
3  =People may share false information, not knowing it is false.
4  =People may share partial information, which is not the whole truth. Hence it is misinformation.

The results (Table 7.4) showed that one-third of the respondents (30%) chose all the choices as appropriate definitions of misinformation, and 9% of respondents

*Table 7.4* Understanding of misinformation (*n* = 165)

| Statement and number(s) | *f* | % |
| --- | --- | --- |
| False information (1) | 15 | 9 |
| Deliberately shared to harm people (2) | 5 | 3 |
| Not knowing it is false (3) | 13 | 8 |
| Share partial information (4) | 10 | 6 |
| False information + deliberately shared to harm people + not knowing it is false + share partial information (1, 2, 3, 4) | 50 | 30 |
| False information + deliberately shared to harm people (1, 2) | 5 | 3 |
| False information + deliberately shared to harm people + not knowing it is false (1, 2, 3) | 6 | 4 |
| False information + deliberately shared to harm people + share partial information (1, 2, 4) | 3 | 2 |
| False information + not knowing it is false (1, 3) | 8 | 5 |
| False information + not knowing it is false + share partial information (1, 3, 4) | 19 | 11 |
| False information + share partial information (1, 4) | 2 | 1 |
| Deliberately shared to harm people + not knowing it is false (2, 3) | 6 | 4 |
| Deliberately shared to harm people + not knowing it is false + share partial information(2, 3, 4) | 7 | 4 |
| Deliberately shared to harm people + share partial information (2, 4) | 1 | 1 |
| Not knowing it is false + share partial information (3, 4) | 15 | 9 |
| Total | 165 | 100 |

chose item statement 1, "Misinformation is false information." While 3% chose item 2, "Misinformation is false information intentionally shared to harm people," 8% chose item 3, "People may share false information, not knowing it is false," and 6% chose option 4 as the most appropriate definition of misinformation, "People may share partial information, which is not the entire truth." This implies that 30% of respondents felt that "misinformation" should include both deliberate and unintentional awareness of misinformation and incomplete information being disseminated. These are some of the possible reasons why individuals spread misinformation under a variety of circumstances.

Why do you think people share anything without examining its credibility?

Respondents were asked why they believe individuals share anything without verifying their credibility. With the option of selecting more than one answer from the three statements listed below:

1 People share something that supports their ideology or belief system.
2 People share something if they see them shared by their friends and family.
3 People share something because they want to be part of a group.

It was found (Table 7.5) that 33% of respondents chose all three options while 23% selected both options 1 and 2, i.e., shared because of similarity to ideology and friend and family, 22% of the respondents selected option 1 "People share things that support their ideology or belief system." Of the respondents, 5% chose

*Table 7.5* Sharing behavior (*n* = 165)

| Statement and number(s) | *f* | % |
|---|---|---|
| Support their ideology (1) | 36 | 22 |
| Shared by their friends and family (2) | 8 | 5 |
| want to be part of a group (3) | 5 | 3 |
| Support their ideology + shared by their friends and family (1, 2) | 38 | 23 |
| Support their ideology + shared by their friends and family + want to be part of a group (1, 2, 3) | 54 | 33 |
| Support their ideology + want to be part of a group (1, 3) | 19 | 11 |
| Shared by their friends and family + want to be part of a group (2, 3) | 5 | 3 |
| Total | 165 | 100 |

*Table 7.6* Users trust in scientists (*n* = 165)

| Statement and number(s) | *f* | % |
|---|---|---|
| Companies want them to say (1) | 68 | 41 |
| Do not understand farmer's problems (2) | 39 | 23 |
| Part of the government (3) | 11 | 7 |
| Companies want them to say + do not understand farmer's problems (1, 2) | 24 | 14 |
| Companies want them to say + do not understand farmer's problems + part of the government (1, 2, 3) | 5 | 3 |
| Companies want them to say + part of the government (1, 3) | 14 | 8 |
| Do not understand farmer's problems + part of the government (2, 3) | 4 | 2 |
| Total | 165 | 100 |

option 2, i.e., shared things following friends and family, and 3% indicated that becoming part of a group as a reason for sharing unverified information.

Why do you think people are less likely to believe in agricultural scientists?

Respondents were asked why people are less likely to believe in agricultural scientists. They had the option to provide more than one answer below:

1  Because they say what the companies want them to say.
2  Scientists do not understand farmers' problems.
3  Scientists are part of the government.

It was found that option 1 (Table 7.6), "Because they say what the companies want them to say," was the most selected response (41%) to the question, followed by option 2, "Scientists do not understand farmer's problems," and 14% responded option 1 and 2. In comparison, 7% chose option 3 on its own.

### *Misinformation concerns*

To gain a more comprehensive understanding of the distribution of concerns related to misinformation, the frequencies were converted into percentages. Summarized results can be found in Table 7.7. The findings from the section on misinformation

*Table 7.7* Misinformation concern (*n* = 165)

| Item | Statements | SA | A | D | SD |
|---|---|---|---|---|---|
| 6 | I am concerned that social media can help spread misinformation rapidly. | 71 | 25 | 2 | 2 |
| 1 | I am concerned that misinformation can put peoples' livelihoods at risk. | 64 | 34 | 1 | 1 |
| 4 | I am concerned that misinformation may provide false solutions causing more harm than good. | 48 | 47 | 3 | 2 |
| 8 | I am concerned that misinformation can affect networking and relationships among members on social media. | 40 | 52 | 7 | 1 |
| 3 | I am concerned that online content on sensitive topics makes it difficult to distinguish facts from fake news. | 45 | 42 | 12 | 1 |
| 12 | I am concerned that misinformation can appear as crowdsourcing (obtaining services and ideas by soliciting contributions). | 26 | 61 | 12 | 1 |
| 2 | I am concerned that false information may change my understanding of the facts. | 34 | 48 | 11 | 7 |
| 11 | I am concerned that professionals can contribute to the misinformation spread. | 29 | 54 | 13 | 4 |
| 10 | I am concerned that members may not share similar values of integrity and honesty. | 19 | 56 | 24 | 1 |
| 7 | I am concerned that my group / media platform shares misinformation. | 24 | 41 | 31 | 4 |
| 9 | I am concerned about trusting other members for advice. | 18 | 47 | 32 | 3 |
| 5 | I am concerned that I do not have the skills and competence to deal with the misinformation spread. | 15 | 25 | 44 | 16 |

concerns indicated that most participants agreed or strongly agreed with statements 6, 1, 4, and 8, which address concerns about the rapid spread of misinformation, the potential impact on livelihoods, and the effects on networking and relationships within social media groups (SMGs). However, statements 5, 7, and 9 received a higher number of disagreement responses, suggesting varying levels of concern regarding the influence of misinformation spread, trust issues, and personal skills and competence in dealing with misinformation within their social circles. These results underscore the widespread recognition of misinformation as a significant issue and emphasize the perceived need for action.

Table 7.7 presents the percentages of responses for each statement, providing a breakdown of participant agreement and disagreement. For instance, statement item number 5 reveals that 44% of participants disagreed with the statement "I am concerned that I do not have the skills and competence to deal with misinformation spread," indicating that respondents did not express concern about their ability to handle misinformation. Conversely, item 8, "I am concerned that misinformation can affect networking and relationships among members on social media," received relatively high levels of concern suggesting that participants expressed worry about the impact of misinformation within SMGs. Furthermore, item 6, "I am concerned that social media can help spread misinformation rapidly," indicated that respondents were concerned about the rapid dissemination of misinformation through social media channels (SMCs). Lastly, statement 1, "I am concerned that misinformation can put people's livelihoods at risk," garnered the second-highest percentage of agreements, indicating that participants were concerned about the potential impact of misinformation on their livelihoods.

### Misinformation beliefs

Data regarding respondents' beliefs about misinformation were gathered using binary choice option questions, with respondents being asked 11 items in this section. Percentages were calculated based on their responses, and the results are presented in Table 7.8, arranged in descending order of percentage. Among the item statements, the highest percentage of affirmative responses (98%) was observed for statement 6, which inquired about individuals' perspectives on fact verification and information validation before sharing.

This suggests that respondents believed in the importance of verifying and fact-checking information to ensure the reliability and accuracy of sources and claims prior to sharing them. Similarly, statement 2 received a substantial percentage of "yes" responses (97%), indicating that most respondents believed that experts should provide explanations for promotions and claims to address any doubts that may arise. Additionally, statement 3 garnered a high percentage of "yes" responses (96%) as respondents expressed their belief in the effectiveness of experts' actions against misinformation. This highlights the significant reliance of CoP members on the expertise of professionals in countering misperceptions.

*Table 7.8* Misinformation beliefs ($n = 165$)

| | Statements | Yes | No |
|---|---|---|---|
| 1 | I believe responding with corrective scientific information can remedy the effect of misinformation. | 87 | 13 |
| 2 | I believe people should seek clarifications from the experts before believing promotions and claims. | 97 | 3 |
| 3 | I believe that researchers and agri-experts should do more to counteract misinformation. | 96 | 4 |
| 4 | I believe I can distinguish misinformation from facts. | 72 | 28 |
| 5 | I believe information to be correct when it comes from influential persons within my network. | 42 | 58 |
| 6 | I believe people should check the facts and authenticate the information received before sharing it. | 98 | 2 |
| 7 | Although I know the risks involved, I believe sharing information instantaneously creates awareness among my contacts. | 47 | 53 |
| 8 | I believe I can share information just as I receive it because I do not have time to validate it. | 15 | 85 |
| 9 | I believe it is important to raise awareness of fake news when I receive it. | 90 | 10 |
| 10 | I believe fake news senders should be alerted to cross-check the facts first before sharing. | 94 | 6 |
| 11 | I believe fake news senders should be blocked and reported. | 72 | 28 |

In contrast, statement 8 did not align with most respondents' beliefs, with 85% of them responding negatively ("no"). This indicates that they would refrain from sharing information that they hadn't had sufficient time to validate.

Furthermore, for statement 5, the responses indicated that the belief expressed was not widely held, as 58% of respondents believed that influential individuals could also propagate misinformation within their networks. Moreover, more than half of the respondents (53%) believed that the potential harm of sharing false information outweighed the awareness generated by sharing it. Overall, these findings provide insights into respondents' beliefs about misinformation and underscore the varying perspectives regarding fact verification, expert involvement, content sharing, and the potential risks associated with misinformation dissemination.

### Misinformation circulation

#### Most prevalent source of misinformation

SM and the Internet were considered the most prevalent sources of misinformation (89%), while radio was selected as the least prevalent source of misinformation circulation. These results (Figure 7.2) suggested that conventional media outlets were not considered major sources of misinformation. The diversified traditional forms indicated a low selection as a source of misinformation circulation, while SM and the Internet were recognized as the most prevalent sources of misinformation circulation.

#### Platforms most responsible for spreading misinformation

Figure 7.3 showed that respondents selected Facebook (62%) followed by WhatsApp (28%) as the sites mainly responsible for misinformation dissemination, while Twitter, Instagram, and others were extremely low (2%, 2%, and 8%, respectively).

#### Frequency of encountering misinformation on social media

The results in Figure 7.4 showed that 31% of respondents selected "not often" option for their encounters with groups or individuals spreading disinformation on their SM networks, while 69% of the respondents selected the often and very often categories.

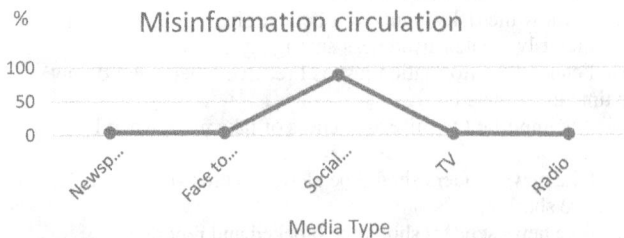

*Figure 7.2* Misinformation circulation (*n*=165).

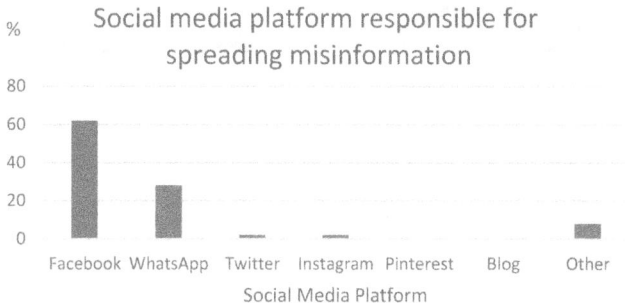

*Figure 7.3* Misinformation spread by SMPs (*n*=165).

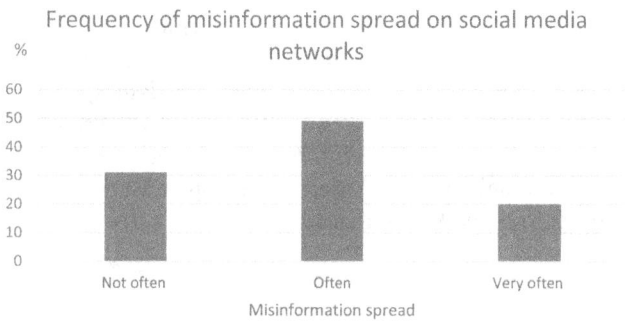

*Figure 7.4* Misinformation frequency (*n*=165).

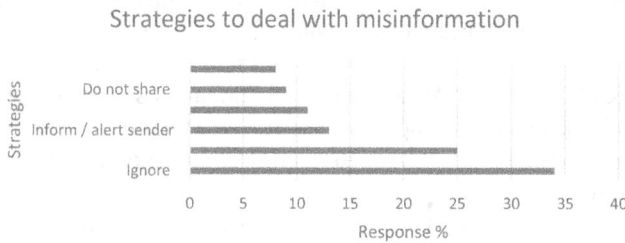

*Figure 7.5* Strategies to deal with misinformation (*n*=165).

*Strategies to combat misinformation and the most prevalent misinformation themes related to agriculture*

According to Figure 7.5, more than a third of respondents said they ignore misinformation without taking steps to limit its spread. In contrast, 66% of those polled took the following actions in response to misinformation: 25% dealt with misinformation by providing accurate and credible information to the sender. Eleven percent tried to verify and confirm the information by double-checking the facts.

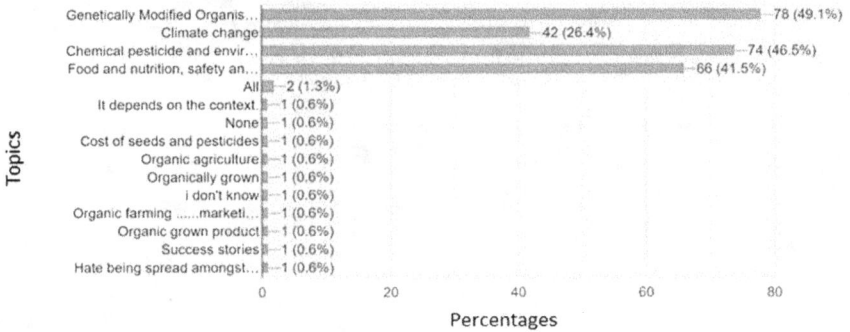

*Figure 7.6* Misinformation topics (*n*=165).

According to Figure 7.6, 49% of the respondents reported that genetically modified organisms were the most common misinformation topic being circulated. This was followed by chemical pesticides and environmental issues, 46.5%, the second most common topic circulated. Food and nutrition topics were 41.5%, the third most common discussed topics, and climate change issues were 26.4%, ranked as the fourth most commonly discussed topic. Other topics assessed, such as organic farming, success stories, marketing-related information, and input costs, were minimally discussed.

## Discussion

This study found that agricultural stakeholders, including farmers and extension officers, utilized SM at fairly high levels. WhatsApp, Facebook, and YouTube were the most popular SMPs. Twitter and Instagram, on the other hand, were less utilized as 90% of respondents indicated that misinformation circulated through Facebook and WhatsApp. These findings have significant implications for agricultural CoPs since they highly use these SMPs as popular information-sharing channels. SMPs, such as Facebook and WhatsApp, were considered the most popular platforms for spreading misinformation compared to traditional media outlets, including radio, television, and newspapers.

Due to the pervasive use and popularity of SMPs (Facebook, WhatsApp) as the primary medium for agricultural communication within the community it serves, it is imperative to adopt proactive measures for identification and analysis of potentially detrimental content. Such measures are essential to empower users with the ability to implement appropriate mitigation strategies aimed at curbing circulation of misinformation. Literature review reveals that these platforms not only function as gateways for information dissemination and connectivity among CoPs but also play a crucial role in evaluating users' activities pertaining to information circulation and credibility assessment. However, it is important to acknowledge that SMPs can also serve as prominent hubs for transmission of misinformation, thus perpetuating its wide-scale distribution.

The study's findings indicated that implementation of regulatory mechanisms, such as editorial, peer review processes, and fact-checking, on SMPs could potentially mitigate dissemination of misinformation. By adopting established frameworks of quality control and fact-checking commonly employed by traditional information sources, the circulation of misinformation through SMCs, which often reaches a wide audience before inaccuracies can be effectively addressed, may be curtailed.

The study also revealed that participants held the view that responding with corrective scientific information can serve as a viable remedy to counter the effects of misinformation. Furthermore, participants expressed the belief that seeking clarifications from experts before accepting promotions and claims can help ensure accuracy and reliability of information. Researchers and agricultural experts were perceived to have a crucial role in countering misinformation, and participants expressed a desire for them to engage more actively in this regard.

Moreover, the participants demonstrated confidence in their ability to distinguish misinformation from information. They emphasized the importance of individuals checking and verifying facts before sharing information, highlighting the need to authenticate the received information for accuracy.

Combating misinformation on SM is unlikely to be handled via individual attempts to correct or prevent the spread of misinformation; instead, individuals must work together to detect and refute misinformation (Chen et al., 2021; Sharma et al., 2021). This emphasizes the need for information creators and disseminators to take responsibility for mitigating the impact of misinformation by verifying and confirming material before dissemination. Furthermore, information-sharing behavior should be approached with caution and due diligence as indicated by the respondents' definition of misinformation, which encompasses purposeful and unintentional sharing of information with the knowledge of incomplete facts and not necessarily the whole truth. Additionally, almost half of respondents (49%) rated their misinformation interactions as "often," and 20% rated their misinformation encounters as "very often" within their SM networks.

The majority of respondents expressed concern regarding potential impact of misinformation on networking and relationships among members within OCoPs. Participants shared strong agreement regarding their concerns about detrimental effects of misinformation. They expressed deep concerns that misinformation has the potential to jeopardize people's livelihoods, recognizing the grave risks associated with relying on false or inaccurate information within their agricultural practices. The rapid spread of misinformation through SMPs was a significant concern, as participants acknowledged the platform's ability to disseminate misinformation quickly and widely, potentially amplifying its impact and reaching a broader audience. Moreover, participants voiced their anxieties about the misleading nature of misinformation, highlighting the potential for false solutions that may ultimately exacerbate problems rather than offering genuine benefits. Recognizing the importance of making informed decisions and implementing effective strategies, participants emphasized crucial role of accurate and reliable information in ensuring positive outcomes and minimizing harm.

These shared concerns underscored the participants' understanding of potential consequences and risks associated with misinformation within their agricultural communities. Their collective agreement on these concerns reinforces the urgent need for interventions and strategies to address and mitigate harmful effects of misinformation, safeguarding livelihoods, and promoting informed decision-making for the betterment of agricultural practices.

GMOs proved to be the most common misinformation topic being discussed by OCoP members. Additionally, respondents were concerned that professionals could also contribute to misinformation circulation regarding the truth behind specific agricultural innovations and mistrust in government and experts. The most common misinformation topic being circulated otherwise were chemical and environmental problems, food and nutrition, and climate change. SMPs were positioned as an effective means of disseminating unofficial information, resulting in widespread reports about these topics, making it harder for members to obtain scientific knowledge and make sound choices about them. Organic farming, success stories, marketing-related information, and input costs were also cited as misinformation topics, although to a lesser extent. Furthermore, high levels of a lack of confidence in scientists and government agents reflected most respondents' belief that influential individuals might spread misinformation in their networks and that scientists do not understand farmers. The findings remind us recent studies highlighted that EAS professionals must play important roles in moderating controversial topics (Leal et al., 2020), proactively communicating about these topics, and building trust to prevent communities from misinformation related to GMOs, climate change, use of chemicals, etc. (Gibson et al., 2022).

Over a third of respondents acknowledged that they disregarded misleading information without trying to prevent it from being disseminated further. The majority, however, dealt with the situation by using different strategies, such as informing the sender and providing correct information so the validated form could be recirculated. Another strategy used by some members was to get the facts and validate the information. Others decided that it was more advisable to delete the content and prevent further spreading of misinformation. This practice of avoiding resharing was also used as a chosen strategy to keep circulation under effective control. A substantial portion of the solution was centered on the users' ability to discern the difference between opinion and fact. The actual outcome of the increased interest in misinformation has been a greater concern about the reality that most people are not equipped to distinguish between correct and misleading information. Although most people are confident in their ability to detect misinformation, it may be more challenging than expected (Bickham et al., 2018; Stecula, 2017).

The respondents' beliefs and concerns identified in this study provide valuable insights into misinformation dissemination among agricultural CoPs in T&T. This underscores the importance of developing mitigation methods to address issues collectively rather than making decisions based on unsubstantiated information, which may result in failure. This has serious implications for agricultural CoPs as agriculture has been identified as one of the most critical areas of development with potential to make a meaningful difference in terms of growth and expansion, and poverty reduction. However, agricultural productivity is restricted by various issues, including a lack of access to improved technology, finance, high input costs, pest and disease, and farmers' skills and entrepreneurship (Food and Agriculture Organization [FAO] & Caribbean Development Bank [CDB], 2019). Agriculture is critical to society, economy, ecology, and physical well-being. Therefore, the public must have correct perceptions of agriculture, particularly about crops, food production, animals, and food safety (Terry & Lawver, 1995). Due to the delicate nature of the sector, misinformation may have a major disruptive influence on the attainment of agriculture's social and economic goals. Therefore, it is critical to monitor and regulate the degree of misinformation circulation. Individuals may be more equipped to assess the quality of information they get if they adopt a direct approach, such as raising their degree of skepticism and incredulity. Attempting to combat or prevent misinformation spread on SM as a unilateral effort is unlikely to be effective. As a result, agricultural CoP members, experts, professionals, extension agents, and farmers should identify and correct misinformation jointly. Unless proper measures are implemented, the issue will worsen as more people subscribe to information online.

## Conclusion

In T&T, agricultural CoPs are using SMPs like Facebook and WhatsApp to enhance knowledge sharing and improve their livelihoods. However, research findings underscore concerns regarding prevalence of misinformation within these online communities. Participants demonstrated a clear understanding of misinformation, including its definition, classification, and interpretation. They highlighted strategies to counteract misinformation, such as seeking verification from experts within their networks to validate false claims. The participants recognized the potential impact that shared knowledge could have on their perspectives and decision-making processes.

To mitigate against misinformation, CoP members emphasized the importance of verifying accuracy of information before sharing it. They stressed the need for information providers and disseminators to validate data they share, aiming to prevent both intentional and unintentional spread of inaccurate or insufficient information. Recognizing vital role of the agriculture sector, it becomes crucial to monitor the information shared, as dissemination of inaccurate information can impede the achievement of goals. Individuals who adopt a more skeptical approach may possess better skills in assessing the authenticity of information.

The study underscores the significance of developing digital literacy, particularly among agricultural knowledge intermediaries such as extension and advisory service professionals. These intermediaries can play a crucial role in combating the circulation of misinformation and assisting users in making informed decisions regarding their agricultural practices. As SMPs like Facebook and WhatsApp serve as gateways for connecting, generating, and spreading information within CoPs, it becomes imperative to evaluate the credibility of information, its origins, and the circulation process. Credibility evaluations should be actively monitored by relevant authorities, professionals, and experts such as editors and peer reviewers.

Engagement of OCoP administrators is vital in combating misinformation within agricultural communities. By offering expert advice, verifying the accuracy of information, and monitoring its credibility, these administrators can establish strong mechanisms to mitigate the negative effects of misinformation. Through their proactive involvement, agricultural CoPs can ensure dissemination of reliable knowledge, leading to informed decision-making, enhanced agricultural practices, and the agricultural sector's sustainable development.

## References

Baker, E. (2020). ICTs and social media: Tools for agricultural extension in the Caribbean region. *ICTs and social media: Tools for agricultural extension in the Caribbean region.* Ontario Agricultural College. https://www.uoguelph.ca/oac/icts

Bellows, L., & Moore, R. (2013, September 28). *Nutrition misinformation: How to identify fraud and misleading claims – 9.350.* Extension. https://extension.colostate.edu/topic-areas/nutrition-food-safety-health/nutrition-misinformation-how-to-identify-fraud-and-misleading-claims-9-350/

Bickham, A., Howard, C., & Simmons, S. L. (2018). *Overview of fake news: For public organizations.* STARS. https://stars.library.ucf.edu/publicsectormedialiteracy/7

Birke, F. M., Lemma, M., & Knierim, A. (2018). Perceptions towards information communication technologies and their use in agricultural extension: Case study from South Wollo, Ethiopia. *The Journal of Agricultural Education and Extension, 25*(1), 47–62. https://doi.org/10.1080/1389224x.2018.1524773

Bode, L., & Vraga, E. K. (2017). See something, say something: Correction of global health misinformation on social media. *Health Communication, 33*(9), 1131–40. https://doi.org/10.1080/10410236.2017.1331312

Chen, K., Luo, Y., Hu, A., Zhao, J., & Zhang, L. (2021). Characteristics of misinformation spreading on social media during the COVID-19 outbreak in China: A descriptive analysis. *Risk Management and Healthcare Policy, 14,* 1869–79. https://doi.org/10.2147/rmhp.s312327

Chowdhury, A., Kabir, K. H., Abdulai, A.-R., & Alam, M. F. (2023). Systematic review of misinformation in social and online media for the development of an analytical framework for Agri-Food sector. *Sustainability, 15*(6), 4753. https://doi.org/10.3390/su15064753

Colussi, J., Morgan, E. L., Schnitkey, G. D., & Padula, A. D. (2022). How communication affects the adoption of digital technologies in soybean production: A survey in Brazil. *Agriculture, 12*(5), 611. https://doi.org/10.3390/agriculture12050611

Cook, J., Lewandowsky, S., & Ecker, U. K. (2017). Neutralizing misinformation through inoculation: Exposing misleading argumentation techniques reduces their influence. *PLoS One, 12*(5). https://doi.org/10.1371/journal.pone.0175799

Dilleen, G., Claffey, E., Foley, A., & Doolin, K. (2023). Investigating knowledge dissemination and social media use in the farming network to build trust in Smart Farming Technology adoption. *Journal of Business & Industrial Marketing, 38*(8), 1754–65. https://doi.org/10.1108/jbim-01-2022-0060

Food and Agriculture Organization (FAO) & Caribbean Development Bank (CDB). (2019). *Study on the state of agriculture in the Caribbean.* FAO.Org. http://www.fao.org/3/ca4726en/ca4726en.pdf?eloutlink=imf2fao

Gebel, M. (2021). Misinformation vs. disinformation: What to know about each form of false information, and how to spot them online. *Business Insider*. https://www.businessinsider.com/misinformation-vs-disinformation

Ghosh, M., Hasan, S. S., Maria, U., Akon, S., Ali, H., Moheuddin, M., & Al Noman, A. (2021). Social media in agricultural extension services: Farmers and Extension Agents Perspective. *European Journal of Humanities and Social Sciences*, *1*(5), 36–43. https://doi.org/10.24018/ejsocial.2021.1.5.143

Gibson, J., Greig, J., Rampold, S., Nelson, H., & Stripling, C. (2022). Can you cite that? Describing Tennessee consumers' use of GMO information channels and sources. *Advancements in Agricultural Development*, *3*(2), 1–16. https://doi.org/10.37433/aad.v3i2.181

Gilman, H. R., & Peixoto, T. C. (2019). Digital participation. In S. Elstub & O. Escobar (Eds.), *Handbook of democratic innovation and governance* (pp. 105–19). Edward Elgar Publishing.

Ivanchuk, N. (2023). Agriculture weather: Precise data for profitable farming. *EOS Data Analytics*. [Blog.] https://eos.com/blog/weather-in-agriculture/

Johnson, C. M. (2001). A survey of current research on online communities of practice. *The Internet and Higher Education*, *4*(1), 45–60. https://doi.org/10.1016/s1096-7516(01)00047-1

Kabir, K. H., Hassan, F., Mukta, Most. Z., Roy, D., Darr, D., Leggette, H., & Ullah, S. M. A. (2022). Application of the technology acceptance model to assess the use and preferences of ICTs among field-level extension officers in Bangladesh. *Digital Geography and Society*, *3*, 100027. https://doi.org/10.1016/j.diggeo.2022.100027

Kesherwani, B., Rout, S., Padhy, D., & Ravichandran, S. (2022). Social media help farmers: For improving agriculture practices. *International Journal of Biology, Pharmacy and Allied Sciences*, *11*(1). https://doi.org/10.31032/ijbpas/2022/11.1.1051

Kumar, R. (2023). Farmers' use of the mobile phone for accessing agricultural information in Haryana: An analytical study. *Open Information Science*, *7*(1). https://doi.org/10.1515/opis-2022-0145

Leal, A., Rumble, J. N., Lamm, A. J., & Gay, K. D. (2020). Discussing extension agents' role in moderating contentious issue conversations. *Journal of Human Sciences and Extension*, *8*(2). https://doi.org/10.54718/nysf5815

Luo, H., Cai, M., & Cui, Y. (2021). Spread of misinformation in social networks: Analysis based on Weibo Tweets. *Security and Communication Networks*, *2021*, 1–23. https://doi.org/10.1155/2021/7999760

Lynas, M., Adams, J., & Conrow, J. (2022). Misinformation in the media: Global coverage of GMOs 2019–2021. *GM Crops & Food*, 1–10. https://doi.org/10.1080/21645698.2022.2140568

Md Nordin, S., Ahmad Rizal, A. R., & Zolkepli, I. A. (2021). Innovation diffusion: The influence of social media affordances on complexity reduction for decision making. *Frontiers in Psychology*, *12*. https://doi.org/10.3389/fpsyg.2021.705245

Menon, S. (2020). Coronavirus: The human cost of fake news in India. *BBC News*. https://www.bbc.com/news/world-asia-india-53165436

Nguyen, H., Ogbadu-Oladapo, L., Ali, I., Chen, H., & Chen, J. (2023). Fighting misinformation: Where are we and where to go? *Lecture Notes in Computer Science*, 371–94. https://doi.org/10.1007/978-3-031-28035-1_27

Pfeiffer, L. J., Knobloch, N. A., Tucker, M. A., & Hovey, M. (2022). Issues-360™: An analysis of transformational learning in a controversial issues engagement initiative. *Journal of Agricultural Education and Extension*, *28*(4), 439–58. https://doi.org/10.1080/1389224X.2021.1942090

Purdue University (2020). Agricultural extension through technology (ICT). https://ag.purdue.edu/department/ipia/_docs/f2f-assignments-pdfs/tt37-extension-sow---agricultural-extension-through-technology-ict.pdf

Ramjattan, J., Chowdhury, A., Ganpat, W. G., & Kathiravan, G. (2020). *Use of social media by the Agricultural Extension Community of Practices in Trinidad & Tobago*. Association for International Agricultural and Extension Education.

Renwick, S. (2010). Current trends in agricultural information services for farmers in Trinidad and Tobago/Caribbean. In *World Library and Information Congress: 76th IFLA General Conference and Assembly*, Gothenburg, Sweden.

Rust, N. A., Stankovics, P., Jarvis, R. M., Morris-Trainor, Z., de Vries, J. R., Ingram, J., Mills, J., Glikman, J. A., Parkinson, J., Toth, Z., Hansda, R., McMorran, R., Glass, J., & Reed, M. S. (2021). Have farmers had enough of experts? *Environmental Management*, *69*(1), 31–44. https://doi.org/10.1007/s00267-021-01546-y

Sharma, S. K., Singh, J. B., & Chandwani, R. (2021). Rumors vs fake news: How to address misinformation in crisis? ET Government. *ETGovernment.com*. https://government.economictimes.indiatimes.com/news/digital-india/rumors-vs-fake-news-how-to-address-misinformation-in-crisis/76421449

Singh, D., & Verma, B. K. (2023). The utilisation of social media for accessing farming information by progressive farmers. *Bhartiya Krishi Anusandhan Patrika*, *38*(3), 290–95. https://doi.org/10.18805/bkap642

Spencer-Jolliffe, N. (2023). Food misinformation and fake news? Scientific and sustainable communication efforts expand to end misreporting. *Food Navigator Europe*. https://www.foodnavigator.com/Article/2023/11/09/Food-misinformation-and-fake-news-Scientific-and-sustainable-communication-efforts-expand-to-end-misreporting

Stecula, D. (2017, July 26). The real consequences of fake news. *The Conversation*. https://theconversation.com/the-real-consequences-of-fake-news-81179

Steinke, J., van Etten, J., Müller, A., Ortiz-Crespo, B., van de Gevel, J., Silvestri, S., & Priebe, J. (2020). Tapping the full potential of the Digital Revolution for agricultural extension: An emerging innovation agenda. *International Journal of Agricultural Sustainability*, *19*(5–6), 549–65. https://doi.org/10.1080/14735903.2020.1738754

Suchiradipta, B., & Saravanan, R. (2016). Social media: Shaping the future of agricultural extension and advisory services. GFRAS interest group on ICT4RAS discussion paper. GFRAS.

Tan, A. S., Lee, C., & Chae, J. (2015). Exposure to health (mis)information: Lagged effects on young adults' health behaviors and potential pathways. *Journal of Communication*, *65*(4), 674–98. https://doi.org/10.1111/jcom.12163

Terry, R., & Lawver, D. E. (1995). University students' perceptions of issues related to agriculture. *Journal of Agricultural Education*, *36*(4), 64–71. https://doi.org/10.5032/jae.1995.04064

Thakur, D., & Chander, M. (2018). Use of social media in agricultural extension: Some evidence from India. *International Journal of Science, Environment and Technology*, *7*(4), 1334–46.

Treen, K. M., Williams, H. T., & O'Neill, S. J. (2020). Online misinformation about climate change. *WIREs Climate Change*, *11*(5). https://doi.org/10.1002/wcc.665

Valenzuela, S., Halpern, D., Katz, J. E., & Miranda, J. P. (2019). The paradox of participation versus misinformation: Social media, political engagement, and the spread of misinformation. *Digital Journalism*, *7*(6), 802–23. https://doi.org/10.1080/21670811.2019.1623701

van der Linden, S., Leiserowitz, A., Rosenthal, S., & Maibach, E. (2017). Inoculating the public against misinformation about climate change. *Global Challenges, 1*(2). https://doi.org/10.1002/gch2.201600008

Wardle, C., & Derakhshan, H. (2017). *Information disorder: Toward an interdisciplinary framework for research and policymaking, 27*. Council of Europe.

Watts, B. (2018). The influence of social media in agriculture. *Challenge Advisory*. https://www.challenge.org/knowledgeitems/the-influence-of-social-media-in-agriculture/

Wenger, E., Trayner, B., & de Laat, M. (2011). *Promoting and assessing value creation in communities and networks: A conceptual framework*. Heerlen.

Wenger, E., & Wenger-Trayner, B. (2015, June 4). *Introduction to communities of practice: A brief overview of the concept and its uses*. https://wenger-trayner.com/introduction-to-communities-of-practice/

West, D. M. (2017). How to combat fake news and disinformation. *Brookings*. https://www.brookings.edu/articles/how-to-combat-fake-news-and-disinformation/

Witze, A. (2021, May 6). How scientists are fighting fake news and misinformation. *Science News*. https://www.sciencenews.org/article/fake-news-misinformation-covid-vaccines-conspiracy

World Bank. (2017). *ICT in Agriculture: Connecting smallholders to knowledge, networks, and institutions (Updated edition)*. World Bank. https://doi.org/10.1596/978-1-4648-1002-2

# 8 Inclusive digital pathways for agricultural extension

## Exploring micro-level innovation with technology stewardship training in Sri Lanka and Trinidad

*Gordon A. Gow, Ataharul Chowdhury, Uvasara Dissanayeke, Jeet Ramjattan and Wayne Ganpat*

### Introduction

Digital transformation of the agriculture sector countries promises to play an essential role in contributing to Sustainable Development Goals (SDGs) and improving the livelihoods of farmers and their families. However, there are concerns that the agricultural innovation system may be excluding voices of smallholders and other marginalized groups, perpetuating systemic inequalities and creating new power asymmetries between farmers and large-scale agribusiness interests leading to negative long-term social justice and environmental consequences.

Several authorities in the field have identified an emerging and essential role for agricultural extension and advisory services to support inclusive digital transformation pathways (DTPs) that will better serve the interests of all stakeholders, including smallholders in the Global South. While this role may be gaining increasing recognition within the academic community, it is not yet clear how extension practitioners will need to prepare and in what capacity they can be most effective in supporting inclusive digital transition processes and practices. There are essential considerations for those involved in public extension services concerning the normative values behind digital transformation initiatives and policies, particularly concerning local innovation and digital-resistant practices among smallholder communities and other critical stakeholders in the agricultural value chain.

This chapter reports on the assessment of a technology stewardship training program (TSTP) that has been designed to support inclusive digital transformation pathways (DTPs) within agricultural communities of practice. We begin with a conceptual discussion to examine how reframing of "digital agriculture" (DA) as "digitalization in agriculture" (DIA) helps re-articulate policy and practice related to innovation systems and the ability to imagine a plurality of transition pathways. Next the discussion considers the role of public extension and advisory services in supporting an inclusive digital transition, establishing a theoretical framework that integrates situated learning theory from the communities of practice literature with the normative perspective of the Capabilities Approach. This framework informs

DOI: 10.4324/9781003282075-11

the design of a joint research-education initiative that has been introduced and assessed with extension practitioners in Sri Lanka as well as Trinidad and Tobago.

**Digital transition pathways and public extension services**

The agriculture sector is one of many undergoing profound changes because of digital transformation or digitalization (Matos et al., 2020). Digitalization is integral to a broader trend in platform capitalism, prompting new business practices and socio-economic relationships based on a reorganization of markets around creating and exploiting large datasets (Mansell & Steinmueller, 2020; Srnicek, 2016; Zuboff, 2018). Within the agricultural sector, digitalization has been underway for some time, involving high- and low-tech innovations across the value chain, ranging from capital intensive smart farming techniques linked to artificial intelligence and machine learning systems to more frugal strategies that involve the use of social media for knowledge sharing and marketing (Cline, 2019; Daum, 2019). The apotheosis of this vision, as characterized, for example, in Shepherd et al. (2020), is sometimes referred to as "Agriculture 4.0," a term that describes a system-of-systems integration of digital technologies that will enhance productivity through efficiency gains, thereby providing greater food security and improving livelihoods of farmers and their families, while promising to be responsive to environmental concerns.

Social scientists are studying DA in the Global North (Bronson & Knezevic, 2019; Phillips et al., 2019), but more work on "a critical approach toward the pervasive application of digital technologies in developing and emerging country agriculture is much needed" (Klerkx et al., 2019, p. 12). Research can be directed to understand and anticipate the potential impact of digitalization within the smallholder sector, which is responsible for more than 70% of the food produced in the Global South (Fanzo, 2017; FAO, 2020). Smallholders are usually family operated with modest-sized plots of land located in rural areas and traditionally low-tech operations. Smallholders play a crucial role in maintaining the genetic diversity of the food supply and contribute to food security for many cities worldwide thereby playing a vital role in the United Nation's (UN) Sustainable Development Goals (SDGs) (Lowder et al., 2016). An expanding industrialized DA sector creates an uncertain future for the viability of this important group (Nature, 2020; Pereira et al., 2018).

While the long-term impact of the SARS-CoV-2 pandemic on the smallholder sector remains uncertain, it has already revealed significant systemic inequalities even as it accelerates efforts to promote DIA (Ceballos et al., 2020; Mohapatra, 2020). Pre-pandemic concerns were being raised from a human rights perspective that digitalization would reinforce or even strengthen existing inequalities within society (Achiume, 2020), including power asymmetries between large commercial entities and smallholder agricultural operations (Bronson, 2019; Fraser, 2020; Rotz et al., 2019b; Wiseman et al., 2019). There are apprehensions about unintended negative consequences of rapid digitalization on the socio-economic factors within agriculture communities, especially as it changes the nature of work for women,

men, and youth (Hellin & Fisher, 2019; Rotz et al., 2019a). Academics have called for the introduction of inclusive innovation processes that engage a greater diversity of rights holders, paying closer attention to the socio-ethical, political-economic, and gendered impacts of digitalization in the agriculture sector (Fielke et al., 2020; Kalkanci et al., 2019).

Evidence suggests that digitalization may be able to help sustain the independence and improve the livelihoods of smallholders, provided they can be empowered to participate in the innovation process with support from agricultural extension and advisory services (Dlamini & Worth, 2019; Shilomboleni et al., 2020; Steinke et al., 2020). Klerkx states, "what is crucial to acknowledge is that there is *a plurality of transition pathways which co-exist, intersect, collaborate, or compete* [emphasis added]" and that extension and advisory services will play a crucial role in guiding those transitions (Klerkx, 2020, p. 132). There needs to be more research to understand what a "plurality of transition pathways" means for the digitalization of agriculture in practice. Additionally, there is a need to understand what opportunities digitalization presents for smallholders and what role public extension services can play in identifying and acting on those opportunities while contributing to a more inclusive agricultural innovation system.

### Transition pathways: digital agriculture versus digitalization in agriculture

Although the terms "DA" and "DIA" are sometimes used synonymously, they imply different assumptions. Examining those assumptions is an essential step in helping stakeholders and policymakers "steer toward or away from transition pathways that are more or less desirable" while planning for possible future disruptions to individuals and organizations (Fielke et al., 2020, p. 2). Frame Analysis Theory has demonstrated how the promotion of particular vocabularies can serve as a "socially shared organizing principle that works symbolically to shape democratic discourse and influence public opinion" (Winslow, 2017, p. 584), establishing a context for normative-prescriptive policies and programs (Dekker, 2017). Social scientists using frame analysis methods to examine texts, published by the World Bank, the Food and Agriculture Organization (FAO), and the Organisation for Economic Co-operation and Development (OECD), and have concluded that DA tends to serve a dominant narrative in which:

> Priority is given … to the argument that peasant farmers become part of existing global food systems through the digital agricultural transition by gaining access to digital tools and infrastructure (e.g., Internet) and producing more food. This priority fits into the neo-Malthusian narrative and may contribute to technological lock-in within dominant industrial agriculture and food system models.
>
> (Lajoie-O'Malley et al., 2020, p. 9)

One concern with this kind of dominant narrative is its focus on an innovation pathway that posits advanced digital systems as the predominant strategy to achieve food security and improved livelihoods (Hall & Dijkman, 2019). It may marginalize other approaches to agricultural sustainability, such as permaculture or agroecology, while underplaying power struggles and systemic inequities embedded in conventional agricultural innovation systems thinking (Pigford et al., 2018, p. 118).

Viewed through the lens of "activity theory" (Karanasios, 2014, p. 4), this dominant narrative positions DA as a prescriptive unit of analysis, framing it as a pre-constituted *activity system*. An activity system is a performative assemblage comprising individuals and groups (subjects) using tools to achieve specific goals, governed by rules and norms within a particular division of labor and situated within a wider set of established community relations. Positioning DA as a pre-constituted activity system serves an important role by providing policymakers and practitioners with a coherent model for planning purposes. Whether explicit or not, such framing also enacts a form of rhetorical closure (Pinch & Bijker, 1984) that projects an imaginary *innovation outcome* to guide policy and investment decisions. In some cases, this framing leads to the technologically deterministic suggestion that countries in the Global South will be able to "leapfrog" social and economic barriers to development by adopting advanced, integrated digital systems in the agriculture sector (UNCTAD, 2018).

Conversely, Lajoie-O'Malley et al. suggest there is value in establishing counternarratives of the future that posit transition pathways offering "a greater diversity of descriptions of the role of emergent technologies in enabling industrial and alternative food system futures as a first step toward evaluating the potential impact of digital agriculture on ecosystem services" (2020, p. 10). DIA can be employed deliberately to reframe the project as an open-ended and multi-faceted *innovation process* circulating within and shaping existing activity systems. A focus on the innovation process begins by articulating normative goals and principles among rights holders and then proceeds incrementally and is reflexively modelled on a multilevel, agile approach to development. This is consistent with Bijker's concept of "interpretive flexibility" (Doherty et al., 2006, p. 571), which suggests an innovation context in which a plurality of transition pathways can be imagined and explored through a series of micro-level innovations that are "progressively embodied into higher structuration levels" (Brunori et al., 2020, p. 171). Micro-level innovation can be a low-risk strategy accommodating many actors, especially those typically marginalized from processes focused on macro-level innovation outcomes (Hayes & Westrup, 2012). As a form of inclusive innovation, it accords with Heeks's view that a strategy should entail "a gradual deepening and/or broadening of the extent of inclusion of the excluded group in relation to innovation," ultimately embodying a process that takes place "within a frame of knowledge and discourse that it in itself inclusive" (Heeks et al., 2014, p. 178).

The frame established by DIA is better suited to imagining a plurality of transition pathways characterized by different priorities and political-economic arrangements. It can reorient the discourse of policy and practice in a way that values situated learning, invites consideration of local and indigenous innovation, and is respectful of digital-resistant practices within and among smallholder communities. This (re)framing is also consistent with constructivist theories of technology. This is found, in particular, in Feenberg's critical theory of technology in which "subordinate groups may challenge the technical code with impact on design as technologies evolve" (Feenberg, 2005, p. 47). It is an approach that emphasizes situated learning through micro-level innovation, or "innovation niches," and creates opportunities for engendering inclusivity and a diversity of views on what constitutes locally appropriate and sustainable farming practices, the desired features of social life within rural communities, and appropriate roles for and limits on digital technology (Bronson, 2019; Pigford et al., 2018). However, as both Saint Ville et al. (2016) and Renken and Heeks (2013) have observed, effective engagement with smallholders will depend partly on the strength of community-based social networks supported by individuals who are motivated and capable of leading micro-level innovation efforts at the local level.

### Public extension services and digital transition pathways

The term "extension" dates to nineteenth-century England to describe adult education programs initiated by Cambridge and Oxford Universities that *extended* the work of these universities beyond the campus and into the wider community. In the United States, the Land-grant University system took up the term to refer more generally to "a process of obtaining useful information for people and then assisting these people to use the necessary knowledge, skills, and information" (Ganpat, 2013, p. 22). The origins of agricultural extension and advisory services in the Global South were part of efforts by the colonial governments to improve crop yields in agricultural products for export. Those initiatives trace back to the establishment of a botanical garden in Sri Lanka, followed by the establishment of government departments and research institutions, such as the St. Augustine campus of the University of the West Indies.

With the advent of the "Green Revolution" in the 1950s and 1960s, governments in the Global South were encouraged to introduce national agricultural advisory services, focusing on transferring knowledge and skills from research institutions to farmers and farming communities as a strategy to apply modern science to crop production. This approach adopted a "linear model" of technology transfer from expert to farmer, with the extension officer acting as an agent of the state. The linear model has since been superseded by an agricultural innovation systems (AIS) framework "focusing more broadly on the factors that stimulate innovative behavior and stress[ing] linkages and partnerships with a wide range of actors along agricultural value chains, including the agribusiness sector" (Anderson, 2008, p. 9). In combination with the shift toward an AIS framework, the past two decades had a sharp decline in funding of public sector extension and advisory services accompanied by other structural changes that have given rise to a plurality of alternative providers that include the private sector, non-profit, and producer organizations in this domain (Benson & Jafry, 2013; Blum et al., 2020). There are ongoing efforts to re-imagine public sector extensions to complement and enhance the agricultural innovation system in combination with other providers operating through decentralized delivery structures. One vital role for public sector services remains in serving smallholder communities that often have a portfolio of needs, a limited ability to pay for private services and that could benefit from a diversification of livelihood strategies to better support vulnerable and marginal groups, such as women and the poor (Benson & Jafry, 2013, p. 389). In meeting this need, public sector extension could be well-positioned to promote greater inclusivity in digital transition efforts by working with and representing smallholders and marginalized groups within the agricultural innovation system.

Insofar as public extension services can take up this role in promoting inclusivity, it is essential to understand how they are implicated in and will be affected by digital transition (Klerkx, 2020, p. 133). Fielke suggests that a digitalization vision entails a "cultural shift in public agricultural research, development and extension organizations to embrace socio-technical interaction through co-innovation and design principles" (2020, p. 8). Steinke et al. (2020, p. 12) suggest that extension services should introduce information and communication technologies (ICT) applications "as part of routine activities" to "generate further evidence on realistic opportunities and challenges ahead," thereby establishing "digital feedback loops" with smallholders participating in micro-level innovation activities. Another suggestion includes creating a role for trained community members with access to ICT and connectivity as "para-extensionists" who can mediate between extension services and the farming community (p. 12). They also advocate for an approach that prioritizes digital capacity building and agile design and evaluation processes over costly upfront investment projects, concluding that "continued local improvisation and adaptation, constant modification and maintenance of digital agro-advisory services are indispensable" to establish and sustain practical digital innovation in the long run (p. 13).

What does this role entail for public extension practitioners? First, it is consistent with the trend toward recognizing extension services as vital intermediaries within the agricultural innovation system rather than acting solely as agents of the state. It coincides with a rethinking of communication for development as it moves away from linear technology transfer models to consider participatory approaches that value local knowledge and practices while explicitly considering social justice and environmental sustainability as normative priorities (Lie & Servaes, 2015). As observed by Sulaiman et al. (2012, p. 334), "the role of extension has shifted from initially that of a disseminator of information to subsequently that of a facilitator of interaction and ... broker ... playing a wider range of intermediation tasks" within "networks of stakeholders operating in different societal spheres." This re-imagining of the role for public extension gains expression in a Global Forum for Rural Advisory Services (GFRAS) "New Extensionist" position paper (Sulaiman & Davis, 2012). It was first published in 2012 and later developed into a "New Extensionist Learning Kit" to support training and skills development. Within this framework, public extension services are well-positioned to ensure the agricultural innovation system serves smallholders, women, and other disadvantaged segments in the value chain. The New Extensionist framework, as discussed by Davis (2015), emphasizes experimentation and learning as a process of incremental adaptation with competencies founded on identifying and using appropriate ICTs to foster change through facilitation and community engagement (CE).

Even within the New Extensionist framework, it is likely that public sector extension organizations will be tasked by national governments to promote DA while simultaneously encountering a plurality of grassroots digitalization efforts involving entrepreneurs, app developers, and local micro-innovators (Krishnadas & Renganathan, 2021). Public sector extension will then be in a strategic position within the agricultural innovation system to promote transition pathways that engender inclusivity and respond to smallholder needs while remaining accountable to national agricultural policy objectives. Despite their significance for the future of agriculture in these countries, public extension services in the Global South will face difficult challenges in fulfilling this role. Presently, these challenges include limited access to advanced digital technology, gaps in digital skills training, and few opportunities to introduce and evaluate low-risk digitalization pilot projects with the involvement of smallholder communities. In response, there have been calls for training of extension officers to expand ICT adoption within the ranks and with farmers (Davis, 2015; Ganpat et al., 2016; Narine & Harder, 2019; Norton & Alwang, 2020; Wanigasundera & Atapattu, 2019), although there has been less emphasis on the need for capacity building for micro-level innovation in collaboration with other stakeholders in the agricultural innovation system.

If digital transition pathways are to evolve in alignment with principles of inclusivity and social equity for smallholders, public extension practitioners will need training in participatory methods responsive to the interconnected social, economic, and cultural impacts within local contexts as noted by Florey et al. (2020, p. 237):

> A clearer understanding of usage by farmers in target geographies is critical to design more appropriate interventions. Such analyses could result in more appropriate, targeted, and inclusive interventions that could be digital-first, hybrid digital and analogue, or only analogue with the potential for integrating digital at a more appropriate point.

An important point is that understanding local context is essential to formulate and explore a plurality of approaches involving a digitalization of practices. In some cases, a digital-first pathway might be appropriate and desirable. In other cases, a mixed approach involving targeted digital solutions combined with updated or traditional analogue practices may present the way forward. Public extension and advisory services can play a central role in articulating and developing these alternative pathways through micro-level innovation digitalization efforts inclusive of smallholders and other marginal groups.

## Technology stewardship: communities of practice and the capabilities approach

A Joint Education and Training Initiative (JETI) developed and piloted a training course in "technology stewardship" (Wenger et al., 2009), in partnership with the University of Alberta, University of Guelph, University of Peradeniya (Sri Lanka), and University of the West Indies. A total of four pilot courses were conducted between 2016 and 2019, involving some 80 extension officers from across Sri Lanka and the Caribbean representing various public sector and smallholder agricultural organizations. We anticipate that the courses will continue to be offered as in-service training delivered by the Postgraduate Institute of Agriculture at the University of Peradeniya and the Faculty of Food and Agriculture at the University of the West Indies.

The project draws on the communities of practice literature and its emphasis on situated learning as integral to professional development. Wenger et al. (2002, p. 4) defined communities of practice (CoP) as "groups of people who share a concern, a set of problems, a passion about a topic, and who deepen their knowledge and expertise by interacting on an ongoing basis" often informally. The concept applies in numerous fields, such as health, education, and any domain where social learning through group interaction is valued. Extension officers are typically members of one or more CoP, working closely with farmers and their families and other essential stakeholders in the agricultural innovation system. A CoP approach can account for informal networks that comprise a mix of stakeholders, including smallholders, which provide an opportunity for extension officers to give voice to multiple and sometimes competing perspectives on digitalization efforts.

In *Digital habitats: Stewarding technology for communities*, Wenger et al. (2009) introduced the term "technology stewardship" to describe a role for individuals who encourage and support the adoption and use of digital technologies within a CoP. Technology stewardship is essentially a leadership role for cultivating the "digital habitat" with the members of a CoP. A digital habitat describes the collection of digital tools and resources available to the community to carry out communications, knowledge sharing, and social learning activities. Digital habitat shares features with the concept of communicative ecologies, i.e., "complex media environment that is socially and culturally framed" comprised of mutually interacting technical, social, and discursive layers (Hearn & Foth, 2007; Tacchi, 2015).

The technology steward, as noted by Wenger et al. (2009, p. 25), is a community member who pays attention to the life and social practices of the CoP and encourages innovation through experimentation:

> Technology stewards are people with enough experience of the working of a community to understand its technology needs and enough experience with or interest in technology to take leadership in addressing those needs. Stewarding typically includes selecting and configuring the technology and supporting its use in the practice of the community.

Technology stewardship is not an information technology (IT) support function, but instead a multi-faceted role that requires intimate knowledge of the technology-related social practices of community members. It demands skills to engage with members to create visions for the future in alignment with community aspirations, be aware of developments and opportunities in the technology landscape, and encourage and support innovative technology practices in fulfilling community choice.

Technology stewardship is a type of innovation intermediary role that shares features with, for example, "ICT champions" (Renken, 2019), but with an emphasis on facilitating social learning in technological innovation (Stewart & Hyysalo, 2008). Research points to the intermediary as influential in promoting innovative technology practices, particularly with women and other marginalized groups (Ayre et al., 2019; Oreglia & Srinivasan, 2016; Walsham, 2020). It is sometimes problematically enacted as a top-down and supply-side driven role (Wahyunengseh et al., 2020), which continues to perpetuate a linear technology transfer model.

By contrast, technology stewardship lends itself to a transformational approach committed to grassroots development, emphasizing conscientization, conciliation, and collaboration (Ibrahim, 2017). Whereas existing digital training programs developed for the agriculture sector tend to focus on technology implementation goals (FAO, 2013; Raj & Bhattacharjee, 2017), technology stewardship aspires to embrace a "whole community" approach (O'Donnell & Beaton, 2018) led by local needs, available supports, and community-defined goals for development. In some cases, this could entail respecting digital-resistant practices and resorting to non-digital methods that community members feel are better suited to address an immediate need or longer-term concern.

Our program is designed as participatory action research (PAR) guided by a technology-augmented Capability Approach "TaCA" (Haenssgen & Ariana, 2018), incorporating Kleine's (Kleine, 2013) "Choice Framework," and Gigler's "informational capabilities" (ICs) (Gigler, 2015). The model incorporates a Technology Stewardship Field School as the foundation for conducting participatory action research (PAR) with extension officers and their CoPs (Figure 8.1).

Haenssgen and Ariana developed their model to account for the role of technology within the broader domain of human development inspired by Amartya Sen's Capability Approach (Robeyns, 2017). An essential contribution of the TaCA model is its conceptualization of technology in a dual role as a resource ("input") as well as a "conversion factor" to emphasize its generative and transformative dimensions in human development. In the generative dimension, technology serves as a multipurpose resource with specific characteristics that circumscribe but do not determine its use. For example, a smartphone provides the possibility of voice, text, or Internet-enabled communication. How users will integrate the device into the digital habitat of a CoP will vary. This means that technological resources, and digital technology in particular, are the means of serving a variety of practices, so their uptake and use across a wide range of contexts must also be considered as noted by Haenssgen and Ariana (2018, p. 103):

> In [the] transformative dimension, technical objects fulfill functions that are otherwise the domain of conversion factors, namely moderating the translation of other inputs into valued capabilities. ... If we accept that technical items have a dual nature that generates characteristics and modifies the characteristics of other inputs, it leaves open the question how an object acquires transformative qualities. We maintain that *these* qualities are not intrinsic to the object but assigned to it within the socio-technological context.
>
> [Emphasis added]

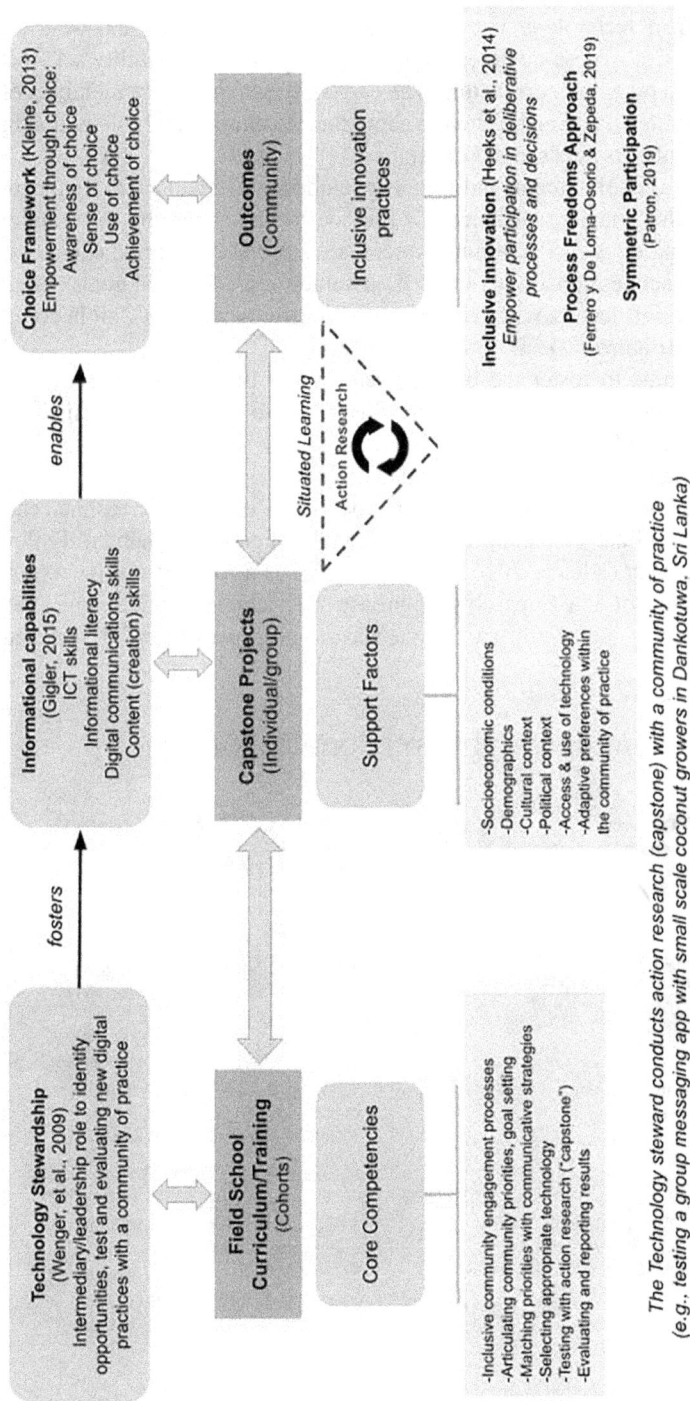

**Technology Stewardship** (Wenger, et al., 2009)
Intermediary/leadership role to identify opportunities, test and evaluating new digital practices with a community of practice

*fosters*

**Informational capabilities** (Gigler, 2015)
ICT skills
Informational literacy
Digital communications skills
Content (creation) skills

*enables*

**Choice Framework** (Kleine, 2013)
Empowerment through choice:
Awareness of choice
Sense of choice
Use of choice
Achievement of choice

**Outcomes** (Community)

Inclusive innovation practices

**Inclusive innovation** (Heeks et al., 2014)
*Empower participation in deliberative processes and decisions*

**Process Freedoms Approach** (Ferrero y De Loma-Osorio & Zepeda, 2019)

**Symmetric Participation** (Patron, 2019)

*Situated Learning*
Action Research

**Capstone Projects** (Individual/group)

**Support Factors**

-Socioeconomic conditions
-Demographics
-Cultural context
-Political context
-Access & use of technology
-Adaptive preferences within the community of practice

**Field School Curriculum/Training** (Cohorts)

**Core Competencies**

-Inclusive community engagement processes
-Articulating community priorities, goal setting
-Matching priorities with communicative strategies
-Selecting appropriate technology
-Testing with action research ("capstone")
-Evaluating and reporting results

*The Technology steward conducts action research (capstone) with a community of practice (e.g., testing a group messaging app with small scale coconut growers in Dankotuwa, Sri Lanka)*

*Figure 8.1* Technology stewardship program (TSP) approach to inclusive digital transition pathways.

In our model, technology stewardship is essential to realizing the transformative qualities of technology within the socio-technological context—the digital habitat—of a CoP. These qualities are manifest through a "capability set" that includes ICs and other support factors. The capability set empowers members of the CoP to deliberate on and make choices about the constitution and role of the digital habitat in serving its needs and aspirations.

IC include a set of outcome indicators developed by Gigler (2015). In the model presented in this chapter, enhanced ICs provide for inclusivity by fostering experiential learning by which community members can make informed choices about technology practices concerning desired outcomes or aspirational goals. Nonetheless, other support factors will also influence inclusivity within a CoP in combination with ICs (Figure 8.2).

The TSP aims to foster the development of ICs through a series of situated learning activities led by an extension officer. These include CE, assisting CoP members to articulate and prioritize aspirational goals as well as designing and evaluating micro-level innovation campaigns.

ICs are a pre-condition that enables members of a CoP to make reasoned choices about uptake and use of digital technology within their digital habitat. Following Kleine's approach (Kleine, 2010, 2013), "choice" is a form of agency expressed in the language of the Capability Approach as "achieved functioning," which when considering inclusivity can be viewed as the actualization of symmetrical

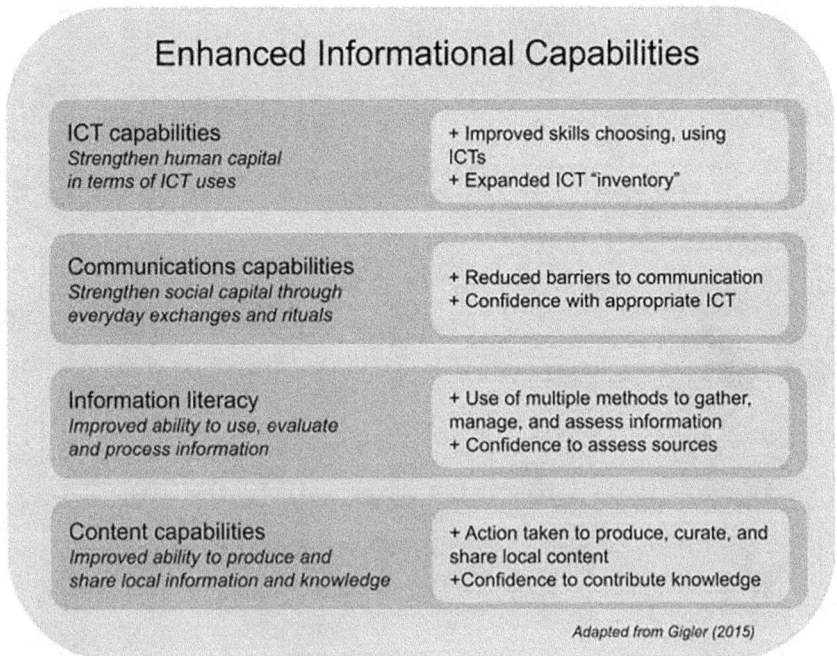

## Enhanced Informational Capabilities

| | |
|---|---|
| **ICT capabilities** *Strengthen human capital in terms of ICT uses* | + Improved skills choosing, using ICTs + Expanded ICT "inventory" |
| **Communications capabilities** *Strengthen social capital through everyday exchanges and rituals* | + Reduced barriers to communication + Confidence with appropriate ICT |
| **Information literacy** *Improved ability to use, evaluate and process information* | + Use of multiple methods to gather, manage, and assess information + Confidence to assess sources |
| **Content capabilities** *Improved ability to produce and share local information and knowledge* | + Action taken to produce, curate, and share local content +Confidence to contribute knowledge |

*Adapted from Gigler (2015)*

*Figure 8.2* ICs and outcome indicators. Adapted from Gigler (2015).

participation (Patrón, 2019) in processes of deliberation on the constitution of the digital habitat of a CoP.

Following Kleine's Choice Framework, technology stewardship directs its effort toward fostering information capabilities with the overall goal of enabling choice, which is considered "a primary development outcome" (2013, p. 45). Choice represents varying degrees of empowerment for community members, and technology stewards are trained to consider their role as enablers within each of four dimensions:

- Make the community aware of the existence of choice (i.e., that digital ways of doing things are possible).
- Help the community to develop a clear sense of choice (i.e., how they might take advantage of digital choices available to them).
- Facilitate and support the use of choice (i.e., assist with trying a new digital practice or deploying unfamiliar digital tools).
- Recognize and sustain the achievement of choice (i.e., report on the outcome of new deployments, analyze and understand points of failure, and acquire resources to build on success).

A technology steward's primary role is, therefore, to facilitate conditions supportive of a "process freedoms approach," emphasizing the involvement of smallholders as "agents and subjects of their own development rather than mere beneficiaries" (Ferrero y De Loma-Osorio & Zepeda, 2019, p. 316). The technology stewardship aims to foster ICs through situated learning activities with micro-level innovation campaigns. It is important to note that enhanced ICs provide a foundation, but do not ensure inclusivity of community members as other support factors must be considered. As indicated in the model, an iterative, action research design provides opportunities for evaluation of inclusive participation and for technology stewards and other members of the CoP to explore various engagement strategies as further learning takes place with subsequent campaigns.

## Technology stewardship program

The technology stewardship program (TSP) provides the vehicle for enacting the stewardship model through PAR. Invited participants receive basic training and then design and lead a capping project with a CoP to which they belong. The capping project component was inspired by an ethnographic action research approach developed by Tacchi et al. (2003) and involves participants in the co-creation of knowledge as both subject and co-researcher (Chevalier & Buckles, 2019, p. 24). The training program serves a dual role. First, as a professional development opportunity with authentic skills training for practitioners. Second, to build capacity to participate in collaborative research on technology stewardship and inclusive innovation.

We have based the curriculum on the "Action Notebook" for technology stewardship from Wenger et al. (2009, p. 147), which sets out "a practitioner-oriented" collection of activities that provide a detailed guide to enacting the role. We have made adaptations for a sector-specific audience. In this case, agricultural extension officers and advisors in Sri Lanka and the Caribbean are the primary focus for this project, although educators and researchers could adapt it for communities of practice in other locations. Our version includes sector-relevant language and localized case study guides while also adding activities in CE, action research methods, and evaluation. The technology stewardship Notebook is available as an open educational resource (Gow et al., 2020c).

### Campaign goal

The training material revolves around a set of activities intended to establish a "campaign goal" that serves as the centerpiece for action research through micro-level innovation efforts. The technology steward's role first involves a consultation with community members, out of which they together develop a campaign objective. Participants are guided through a process that results in a campaign goal structured into three parts:

- States a specific outcome.
- Specific set of ICs.
- Clearly defined CoP.

For example, the following campaign goal statement comes from one created in the classroom by a group of participants from the 2018 cohort in Sri Lanka:

> The goal is to improve awareness and attendance at training events by using digital technology to help organize and schedule meetings with small-scale coconut growers in Dankotuwa ASC Division.

Participants are urged to avoid indicating a specific digital tool or platform in their campaign goal statement. Instead, they identify a specific outcome ("to improve attendance at training sessions") concerning a set of ICs ("organize and schedule meetings") for a defined CoP ("small-scale coconut growers in Dankotuwa ASC Division").

Having articulated a campaign goal statement, the technology steward then facilitates further discussion with the CoP to identify and assess suitable digital tools or platforms before choosing one to integrate into a campaign with the community. The campaign is intended as a micro-level innovation project that may last several weeks to a few months. Stewards are trained to create a campaign evaluation plan that includes formative and summative considerations, emphasizing a holistic, process-oriented approach that reflects on experiential learning within the CoP rather than exclusively on outcomes with the specific digital tool or platform used in the campaign.

The TSP provides a foundation for research based on a multiple, embedded case study design (Yin, 2003). Individual participants in each course receive training as part of a single cohort. For the capping project, each cohort member chooses a focus based on the needs and interests of their CoP. Thus, the study includes multiple units of analysis, namely: (1) the experience and impact of the training program at the cohort level, (2) the experience and impact of training at the individual participant level, and (3) the impact on the CoP to which the technology steward belongs.

Data are collected at various pre-, mid-, and post-training stages using questionnaires, participant observation, and semi-structured interviews. Participants submit an individual action plan (IAP) after the course that indicates if they intend to carry out a capping project and, if so, what activities from the course they intend to apply for that project. In addition, participants who submit a capping project report with some form of documented evidence of completion receive a certificate of advanced standing. We initially asked participants to submit their capping project reports within eight weeks of completing the course, but the timeline for completion tends to be longer in practice.

At the cohort level of analysis, Kirkpatrick's evaluation framework (Kirkpatrick, 1994) was used to assess the training and its suitability in creating capacity for participatory action research. This framework is widely used for evaluating organizational, community, and ICT-related training programs, considering results at four levels of evaluation: (1) participant reaction, (2) learning objectives, (3) behavioral change, and (4) impact. Details on our assessment methodology and findings are available elsewhere (Gow et al., 2020a, 2020b), but overall results across the first two levels for cohorts in Sri Lanka and Trinidad indicate the course is well-received by extension officers and that the course offers useful skills and techniques that can be applied in practice.

Participants in our cohorts submit IAPs when they complete the workshop. The IAP asks if they intended to apply their learning with a capping project and what activities from the course workbook they will choose. The completed IAPs provided evidence responsive to Kirkpatrick level 3 (behavior) in assessing participants' intent to act on the training. Most participants indicated an intent to act, but only about 25% of participants across the two cohorts submitted a full or partially completed capping project report based on their IAP. Follow-up interviews were conducted with all participants and revealed several obstacles to acting on the IAP, including resource constraints and administrative hurdles, which will be examined as a future priority for the project.

Assessing Kirkpatrick level 4 (impact) is more complicated. Following the principles of our technology stewardship model, a whole community approach is necessary to account for numerous interacting factors, including the profile and motivation of individual technology stewards, local history, and social dynamics, as well as the political, economic, and environmental context within which rural development is taking place. A full assessment of impact is planned for future cohorts using longitudinal and more field-intensive research.

For the initial phase of this project, the selected submitted capping project reports described below provide insight on the ways in which technology stewardship training is applied by extension officers toward fostering ICs as they work toward enabling more inclusive participation in digitalization efforts with their communities of practice.

*Table 8.1* Summary of capping project reports

| Country | Tech stewards | CE activity | Create a campaign goal statement | Conduct campaign and evaluation |
|---|---|---|---|---|
| Trinidad | Adelle | ✔ | ✔ | |
| | Muriel and Marvin | ✔ | ✔ | |
| | Antoinette | | ✔ | ✔ |
| Sri Lanka | Arunasiri | ✔ | | |
| | Pathirange | ✔ | | |
| | Suranjan | ✔ | | |
| | Darshana | | ✔ | |
| | Weragoda | | ✔ | |
| | Chanushka | ✔ | ✔ | |
| | Wijethilake | ✔ | ✔ | |
| | Pradeep | ✔ | ✔ | ✔ |

**Comparative results from Sri Lanka and Trinidad**

We received 11 capping project reports from 40 participants across the two cohorts. Participants chose to complete a single activity or combination of activities from the course workbook. We instructed participants to keep the reports brief and use whatever format they felt most comfortable conveying their efforts. Submitted reports varied in their content and detail, but all included a description of the capping project activity or activities and a discussion of results. The documents included written text, graphs, or charts, and sometimes photographs.

Based on the evidence provided in the submitted reports, we organized the submissions into comparator groups, reflecting various activities and the scope of effort for each participant. For example, reports from comparator group 1 provide evidence showing participants applying the course material in a CE activity but did not include a campaign goal statement in the submitted report. Those in comparator group 2 presented a campaign goal statement in the submitted report, but some did not carry out the preceding CE activity. Finally, reports submitted by those in comparator group 3 described plans for, or results from, a small-scale campaign with their CoP. Of the 11 reports, only one participant submitted a report that included activities from all three categories (Table 8.1).

The chapter presents descriptions from four participants, and they are included here to illustrate the range of activities submitted in the capping project reports.

*Sri Lanka: fostering awareness of choice with rice producers*

Suranjan is an instructor with the Department of Agriculture. He was a member of the 2018 technology stewardship cohort at the University of Peradeniya. His capping project identified "Seed paddy producers in the Galle District" as the CoP. He consulted with smallholders to discuss barriers to communication using a problem / opportunity tree activity guided by the course workbook. The activity invited these smallholders to discuss and articulate difficulties they have communicating with extension services. Suranjan then helped them to identify specific ICs that could

*Figure 8.3* Images included in Suranjan's report.

address some of the challenges. Among these was improving confidence in using mobile phones for text messaging as a potentially less intrusive and lower-cost alternative to voice calls (Figure 8.3).

Suranjan's project report was based on this consultation and did not develop a specific campaign goal. However, it demonstrates how the technology steward can foster *awareness* and *sense* of choice among community members concerning digital technology. The results of the consultation provide further direction for a steward like Suranjan to introduce targeted digital literacy activities to address gaps in ICs capabilities prior to introducing a micro-level innovation campaign with text messaging.

### Trinidad: enhancing a sense of choice for fisheries training

Muriel and Marvin work with the Caribbean Fisheries Training and Development Institute. They were in the 2019 cohort at the University of the West Indies. Their capping project identified seafood technology training assistants as the CoP. Using the training material as a guide, they conducted a community consultation and goal-setting activity that resulted in the campaign objective: "to improve access and availability [of training video] anywhere/anytime by using ICT to curate content" for the "STFTP trainers group" (Figure 8.4).

While their capping project activity did not conduct an entire campaign, creating a campaign objective is an essential step in identifying essential ICs required for inclusive participation in producing and sharing content with digital technology. In addition, the activity is intended to create *awareness* and an enhanced *sense* of choice among community members by engaging them in discussions about how digital technology can support a training program with participatory video. The next step for these technology stewards will be to invite members of this CoP to choose a suitable digital tool for a micro-level innovation campaign.

### Trinidad: exploring use and achievement of choice with cocoa researchers

Antoinette is a researcher and outreach coordinator with Cocoa Research Centre in Trinidad, who participated in the 2019 cohort. The Centre provides extension services for local farmers and is part of a wider CoP that includes members involved

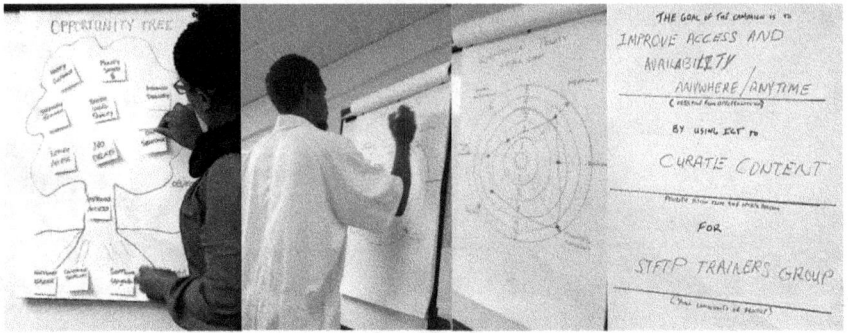

*Figure 8.4* Images included in the capping project report.

in cocoa research, production, and marketing. Her capping project report did not include a CE activity but focused instead on a pre-established campaign goal statement to "improve management of [Cocoa Research Centre] through [an] increase in effective communication with members of staff." This goal reflects a perceived need to foster ICs within the organization that will contribute to its role within the CoP. Her report mentioned the need to address specific communications capabilities such as "sharing files, discussions, synchronous and asynchronous communication, communicating updates" using digital technology.

The significance of this capping project is that Antoinette was able to apply the stewardship training to enable the *use* of choice with this CoP. In this case, she chose the commercial platforms Trello and WhatsApp for a campaign carried out with World Cocoa and Chocolate Day Expo. In a follow-up email with the research team, she describes some valuable insights gained from the capping project experience:

> I think [communications for] the event could have been better managed with ICT, but our team is somewhat in need of convincing (let's say) with regard to the efficacy of it … so I used it and mostly observed others reactions to whenever it was mentioned—so it wasn't a proper campaign … I think maybe a less complicated undertaking would be a better candidate for testing out an ICT with my colleagues.

These remarks illustrate how digital-resistant practices may be revealed through a campaign, as was the case here perhaps in part because the technology steward initiated the campaign without consulting the members of the CoP in the choice of digital tools. The training program advises technology stewards to respect these sites of resistance and use them as opportunities to deliberate on choice rather than continue to push a specific technology solution.

Antoinette's capping project report results were based on an evaluation planning guide included in the course materials. Tech stewards are trained to choose metrics, identify data sources and participatory methods for assessing *achievement* of choice concerning a campaign goal.

Technology stewards collect and analyze results with community members to provide insights on points of failure or resistance and to make a case for acquiring resources necessary to continue or scale up a campaign based on its successful aspects. In this case, Antoinette's evaluation concluded that a "less complicated undertaking" was needed for a future campaign effort, which would serve to guide future consultations with the community members in the choice of technology and its implementation in a subsequent campaign.

*Sri Lanka: negotiating use of choice among extension officers at the Talawakelle Techno Park*

Pradeep participated in the 2018 cohort at the University of Peradeniya. He submitted a nine-page capping project report about a month after completing the course and was the only participant among our cohorts that carried out an entire campaign. Pradeep described his CoP as the advisory and extension division officials from the Talawakelle Techno Park, which is part of the Tea Research Institute of Sri Lanka.

His community assessment concluded that assigning maintenance activities on a shared basis among the extension officers might address some of the current workload concerns. He presented the following campaign goal in his report:

> The goal of the campaign is to improve the maintenance of the TRI Tea Techno park area by using ICT to [support] social networking for sharing the responsibilities ... between the officials of TRI Talawakelle advisory and extension division officials.

Pradeep's report is remarkable because it shows him working through all the activities in the course workbook, from CE to campaign evaluation planning. Moreover, he carried out a two-week campaign with the CoP, implementing a shared digital calendar for his CoP, while gathering data for the evaluation. Pradeep reported mixed results for the campaign, with some members choosing to continue with the shared calendar while others resorted to previous methods. A more detailed presentation of his capping project report is available elsewhere (Gow et al., 2020b).

Pradeep's experience shows the importance of the technology steward's role in enabling use and achievement of choice with a CoP. In this case, the campaign also revealed that community members could differ in their choice of digital technology for scheduling purposes. As such, it brings to light in-group differences, as identified by Kleine (2013, p. 30), that can occur when negotiating individual versus collective choice concerning digital practices. Advising technology stewards on techniques to negotiate and foster a diversity of digital practices within a CoP will be an important consideration as we continue working with future cohorts.

## Conclusion and future directions

This chapter reports on theory and practice that informs a training program for agricultural extension officers in the Global South designed to support a plurality of inclusive DTPs. It began with a conceptual discussion to examine how the framing of DA versus DIA helps articulate a role for public extension and advisory services in supporting inclusive innovation practices. Next, a theoretical framework was presented that integrates the communities of practice literature with the Capabilities Approach to inform the design of a TSTP that has been introduced and assessed with extension practitioners in Sri Lanka and the Caribbean. The reception and results of the TSTP have been encouraging, and we included four descriptive cases to illustrate some of their immediate impact on practice within extension officers. It is hoped these insights will contribute to participatory research involving agricultural extension and advisory services. In particular, the project demonstrates the possibilities for introducing micro-level innovation through participatory action research with extension officers. Extension officers play a key role in the agricultural innovation system and can give voice to smallholders and other marginalized groups to promote greater inclusivity in digitalization processes.

As the training materials will be continually revised, the next significant step for the project will be to introduce more support for the capping project activity that will enable us to carry out a longitudinal and in-depth study of technology stewardship in practice. Technology stewards will need to negotiate a complex terrain of power relations and inclusivity with smallholders that include gender and age considerations. As such, we will be introducing a new training module to address inclusivity in the context of micro-level innovation efforts and evaluation. In addition, the project will consider the possibility of encouraging specific campaigns for marginalized groups within agricultural CoP to understand better how the Choice Framework may, or may not, contribute to a plurality of digitalization efforts that align with local needs and aspirations.

Going forward, scaling up successful micro-level innovations within a plurality of digitalization pathways will be challenging. It is best viewed as a multi-faceted process "that catalyzes three interconnected, and complementary pathways—technology development, capacity development; and policy influence—overseen by inter-disciplinary research and trans-disciplinary networks" (Florey et al., 2020, p. 238). In addition to basic digital literacy, resource constraints and limited institutional support in the Global South are significant barriers and project cohorts have so far relied heavily on a "use what we have" technology acquisition strategy. In effect, this often means looking to incorporate social media or "freemium"-based platforms and digital solutions because they are available at little to no cost. However, the trade-offs in control over data ownership and use, privacy and surveillance, and limited interoperability across platforms also place limits on inclusivity. As the project continues in the next phase, steps will be taken to introduce open-source platforms and tools in line with the Digital Impact Alliance's principles for digital development (Wilson, 2021). While any choice of technology poses its own technical barriers and sustainability issues, open-source solutions will offer smallholders and other marginal groups additional choice and present opportunities to control and manage their digital data (Gow, 2020). Finally, it will be important to pay close attention to digital-resistant practices, both within public extension services and with smallholders, particularly considering that various hybrid digital-analogue configurations will continue to be important for agricultural communities of practice in the Global South.

# References

Achiume, E. T. (2020). *Racial discrimination and emerging digital technologies: A human rights analysis: Report of the Special Rapporteur on contemporary forms of racism, racial discrimination, xenophobia and related intolerance\**. Report no. A/HRC/44/57. United Nations Office of the High Commissioner on Human Rights. https://documents-dds-ny.un.org/doc/UNDOC/GEN/G20/151/06/PDF/G2015106.pdf?OpenElement

Anderson, J. R. (2008). *Agricultural advisory services*. The World Bank. https://openknowledge.worldbank.org/handle/10986/9041

Ayre, M., Mc Collum, V., Waters, W., Samson, P., Curro, A., Nettle, R., Paschen, J.-A., & Reichelt, N. (2019). Supporting and practising digital innovation with advisers in smart farming. *NJAS – Wageningen Journal of Life Sciences, 90–91*(1), 1–12.https://doi.org/10.1016/j.njas.2019.05.001

Benson, A., & Jafry, T. (2013). The state of agricultural extension: An overview and new caveats for the future. *The Journal of Agricultural Education and Extension, 19*(4), 381–93. https://doi.org/10.1080/1389224X.2013.808502

Blum, M. L., Cofini, F., & Sulaiman, V. R. (Authors), Lawrence, T. (Ed.) (2020). *Agricultural extension in transition worldwide: Policies and strategies for reform*. FAO. https://doi.org/10.4060/ca8199en

Bronson, K. (2019). Looking through a responsible innovation lens at uneven engagements with digital farming. *NJAS – Wageningen Journal of Life Sciences, 90–91*(1), 1–6. https://doi.org/10.1016/j.njas.2019.03.001

Bronson, K., & Knezevic, I. (2019). The digital divide and how it matters for Canadian food system equity. *Canadian Journal of Communication, 44*(2), 63–68. https://doi.org/10.22230/cjc.2019v44n2a3489

Brunori, G., Proost, J., & Rand, S. (2020). Small farms and innovation. In B. Gianluca & G. Stefano (Eds.), *Innovation for sustainability* (Vol. 25, pp. 171–91). Emerald Publishing Limited.

Ceballos, F., Kannan, S., & Kramer, B. (2020). Impacts of a national lockdown on smallholder farmers' income and food security: Empirical evidence from two states in India. *World Development, 136*, 105069. https://doi.org/10.1016/j.worlddev.2020.105069

Chevalier, J. M., & Buckles, D. J. (2019). *Participatory action research: Theory and methods for engaged inquiry*. Taylor & Francis.

Cline, T. (2019, February, 21). Gender equality: Ag-tech's potential to boost women's empowerment. SAP Africa News Center. https://news.sap.com/africa/2019/02/gender-equality-ag-techs-potential-to-boost-womens-empowerment/

Daum, T. (2019). ICT applications in agriculture. In P. Ferranti, E. M. Berry, & J. R. Anderson (Eds.), *Encyclopedia of food security and sustainability* (pp. 255–60). Elsevier.

Davis, K. (2015). The new extensionist: Core competencies for individuals. GFRAS Brief no. 3. Global forum for rural advisory services. http://www.g-fras.org/en/knowledge/gfras-publications.html?download=358:the-new-extensionist-core-competencies-for-individuals

Dekker, R. (2017). Frame ambiguity in policy controversies: Critical frame analysis of migrant integration policies in Antwerp and Rotterdam. *Critical Policy Studies, 11*(2), 127–45. https://doi.org/10.1080/19460171.2016.1147365

Dlamini, M. M., & Worth, S. (2019). The potential and challenges of using ICT as a vehicle for rural communication as characterised by smallholder farmers. *Asian Journal of Agricultural Extension, Economics & Sociology, 34*(3), 1–10. https://doi.org/10.9734/ajaees/2019/v34i330202

Doherty, N. F., Coombs, C. R., & Loan-Clarke, J. (2006). A re-conceptualization of the interpretive flexibility of information technologies: Redressing the balance between the social and the technical. *European Journal of Information Systems, 15*(6), 569–82. https://doi.org/10.1057/palgrave.ejis.3000653

Fanzo, J. (2017). From big to small: The significance of smallholder farms in the global food system. *The Lancet Planetary Health, 1*(1), e15–16. https://doi.org/10.1016/S2542-5196(17)30011-6

Feenberg, A. (2005). Critical theory of technology: An overview. *Tailoring Biotechnologies, 1*(1), 47–64. https://www.sfu.ca/~andrewf/books/critbio.pdf

Ferrero y De Loma-Osorio, G., & Zepeda, C. (2019). Planning and managing for human development: What contribution can the capability approach make? In D. A. Clark, M. Biggeri, & A. A. Frediani (Eds.), *The capability approach, empowerment and participation: Concepts, methods and applications* (pp. 311–38). Palgrave Macmillan UK.

Fielke, S., Taylor, B., & Jakku, E. (2020). Digitalisation of agricultural knowledge and advice networks: A state-of-the-art review. *Agricultural Systems, 180*, 102763. https://doi.org/10.1016/j.agsy.2019.102763

Florey, C., Hellin, J., & Balié, J. (2020). Digital agriculture and pathways out of poverty: The need for appropriate design, targeting, and scaling. In J. Donovan, D. Stoian, & J. Hellon (Eds.), *Value chain development and the poor: Promise, delivery, and opportunities for impact at scale* (pp. 229–44). Practical Action Publishing.

Food and Agriculture Organization (FAO) (2013). Food and Agriculture Organization of the United Nations e-learning centre: Social media for development. https://www.fao.org/3/i5960e/i5960e.pdf

Food and Agriculture Organization (FAO) (2020). Farm family knowledge platform: Small-scale fisheries and aquaculture and family farming. [Website.] https://www.fao.org/family-farming/themes/small-scale-fisheries/en/

Fraser, A. (2020). The digital revolution, data curation, and the new dynamics of food sovereignty construction. *The Journal of Peasant Studies, 47*(1), 208–26. https://doi.org/10.1080/03066150.2019.1602522

Ganpat, W. (2013). *The history of agricultural extension in Trinidad and Tobago.* Randle.

Ganpat, W. G., Ramjattan, J., & Strong, R. (2016). Factors influencing self-efficacy and adoption of ICT dissemination tools by new extension officers. *Journal of International Agricultural and Extension Education, 23*(1), 72–85. https://doi.org/10.5191/jiaee.2016.23106

Gigler, B. S. (2015). *Development as freedom in a digital age: Experiences from the rural poor in Bolivia.* https://elibrary.worldbank.org/doi/abs/10.1596/978-1-4648-0420-5

Gow, G. A. (2020). Alternative social media for outreach and engagement: Considering technology stewardship as a pathway to adoption. In M. Adna (Ed.), *Using new media for citizen engagement and participation* (pp. 160–80). IGI Global.

Gow, G., Chowdhury, A., Ramjattan, J., & Ganpat, W. (2020a). Fostering effective use of ICT in agricultural extension: Participant responses to an inaugural Technology Stewardship Training Program in Trinidad. *The Journal of Agricultural Education and Extension, 26*(4), 335–50. https://doi.org/10.1080/1389224X.2020.1718720

Gow, G., Dissanayake, U., Jayathilake, C. K., Kumarasinghe, I., Ariyawanshe, K., & Rathnayake, S. (2020b). *Technology stewardship training for agricultural communities of practice: establishing a participatory action research program in Sri Lanka.* Paper presented at the 11th International Development Informatics Association Conference, United Nations University Institute on Computing and Society, Macau. Abstract. 86–87. March 25–27. https://collections.unu.edu/eserv/UNU:7689/FINAL_-_IDIA2020_Proceedings_-_UNU.pdf

Gow, G., Hambly Odame, H., Dissanayake, U., Jayasinghe-Mudalige, U. K., McMahon, R., Waidyanatha, N., & Jayathilake, C. (2020c). *Introduction to technology stewardship for agricultural extension and advisory services.* https://doi.org/10.29173/oer4

Haenssgen, M. J., & Ariana, P. (2018). The place of technology in the capability approach. *Oxford Development Studies, 46*(1), 98–112. https://doi.org/10.1080/13600818.2017. 1325456

Hall, A., & Dijkman, J. (2019). *Public agricultural research in an era of transformation: The challenge of agri-food system innovation.* CGIAR Independent Science and Partnership Council (ISPC) Secretariat and Commonwealth Scientific and Industrial Research Organisation (CSIRO).

Hayes, N., & Westrup, C. (2012). Context and the processes of ICT for development. *Information and Organization, 22*(1), 23–36. https://doi.org/10.1016/j.infoandorg.2011.10.001

Hearn, G. N., & Foth, M. (2007). Communicative ecologies: Editorial preface. *Electronic Journal of Communication, 17*(1–2), 2–6. https://eprints.qut.edu.au/8171/1/8171.pdf

Heeks, R., Foster, C., & Nugroho, Y. (2014). New models of inclusive innovation for development. *Innovation and Development, 4*(2), 175–85. https://doi.org/10.1080/2157930X. 2014.928982

Hellin, J., & Fisher, E. (2019). The Achilles heel of climate-smart agriculture. *Nature Climate Change, 9*(7), 493–94. https://doi.org/10.1038/s41558-019-0515-8

Ibrahim, S. (2017). How to build collective capabilities: The 3C-model for grassroots-led development. *Journal of Human Development and Capabilities, 18*(2), 197–22. https:// doi.org/10.1080/19452829.2016.1270918

Kalkanci, B., Rahmani, M., & Toktay, L. B. (2019). The role of inclusive innovation in promoting social sustainability. *Production and Operations Management, 28*(12), 2960–82. https://doi.org/10.1111/poms.13112

Karanasios, S. (2014). Framing ICT4D research using activity theory: A match between the ICT4D field and theory? *Information Technologies & International Development, 10*(2), 1–17.

Kirkpatrick, D. L. (1994). *Evaluating training programs: The four levels.* Berrett-Koehler Publishers.

Kleine, D. (2010). ICT4WHAT? – Using the choice framework to operationalise the capability approach to development. *Journal of International Development, 22*(5), 674–92. https://doi.org/10.1002/jid.1719

Kleine, D. (2013). *Technologies of choice? ICTs, development, and the capabilities approach.* MIT Press.

Klerkx, L. (2020). Advisory services and transformation, plurality and disruption of agriculture and food systems: Towards a new research agenda for agricultural education and extension studies. *The Journal of Agricultural Education and Extension, 26*(2), 131–40. https://doi.org/10.1080/1389224X.2020.1738046

Klerkx, L., Jakku, E., & Labarthe, P. (2019). A review of social science on digital agriculture, smart farming and agriculture 4.0: New contributions and a future research agenda. *NJAS – Wageningen Journal of Life Sciences, 90–91*(1), 1–16. https://doi.org/10.1016/ j.njas.2019.100315

Krishnadas, R., & Renganathan, R. (2021). Strategic use of agriculture information system by agripreneurs in rural economies: Current role and future prospects. In L. Sachithra & S. Darshana (Eds.), *Rural entrepreneurship and innovation in the digital era* (pp. 179–95). IGI Global. https://doi.org/10.4018/978-1-7998-4942-1.ch010

Lajoie-O'Malley, A., Bronson, K., van der Burg, S., & Klerkx, L. (2020). The future(s) of digital agriculture and sustainable food systems: An analysis of high-level policy documents. *Ecosystem Services, 45*, 101183. https://doi.org/10.1016/j.ecoser.2020.101183

Lie, R., & Servaes, J. (2015). Disciplines in the field of communication for development and social change. *Communication Theory, 25*(2), 244–58. https://doi.org/10.1111/comt.12065

Lowder, S. K., Skoet, J., & Raney, T. (2016). The number, size, and distribution of farms, smallholder farms, and family farms worldwide. *World Development, 87*, 16–29. https://doi.org/10.1016/j.worlddev.2015.10.041

Mansell, R., & Steinmueller, W. E. (2020). *Advanced introduction to platform economics.* Edward-Elgar.

Matos, F., Vairinhos, V., Salavisa, I., Edvinsson, L., & Massaro, M. (2020). Introduction. In F. Matos, V. Vairinhos, I. Salavisa, L. Edvinsson, & M. Massaro (Eds.), *Knowledge, people, and digital transformation: Approaches for a sustainable future* (pp. 1–6). Springer International Publishing.

Mohapatra, S. (2020). Gender differentiated economic responses to crises in developing countries: Insights for COVID-19 recovery policies. *Review of Economics of the Household.* https://doi.org/10.1007/s11150-020-09512-z

Narine, L., & Harder, A. (2019). Extension officer's sdoption of modern information communication technologies to interact with farmers of Trinidad. *Journal of International Agricultural and Extension Education, 26*(1), 17–34. https://doi.org/10.5191/jiaee.2019.26103

Nature (2020, October 12). Ending hunger: Science must stop neglecting smallholder farmers. *Nature, 586*, 336. https://doi.org/10.1038/d41586-020-02849-6

Norton, G. W., & Alwang, J. (2020). Changes in agricultural extension and implications for farmer adoption of new practices. *Applied Economic Perspectives and Policy, 42*(1), 8–20. https://doi.org/10.1002/aepp.13008

O'Donnell, S., & Beaton, B. (2018). A "Whole-community" approach for sustainable digital infrastructure in remote and northern First Nations. *Northern Public Affairs* (October), 34–37. http://susanodo.ca/wp-content/uploads/2018/10/2018-ODonnell-Beaton.pdf

Oreglia, E., & Srinivasan, J. (2016). ICT, Intermediaries, and the transformation of gendered power structures. *MIS Quarterly, 40*(2), 501–10.

Patrón, P. (2019). Power and deliberative participation in Sen's capability approach. In D. A. Clark, M. Biggeri, & A. A. Frediani (Eds.), *The capability approach, empowerment and participation: Concepts, methods and applications* (pp. 55–74). Palgrave Macmillan UK.

Pereira, L., Wynberg, R., & Reis, Y. (2018). Agroecology: The future of sustainable farming? *Environment: Science and Policy for Sustainable Development, 60*(4), 4–17. https://doi.org/10.1080/00139157.2018.1472507

Phillips, P. W. B., Relf-Eckstein, J.-A., Jobe, G., & Wixted, B. (2019). Configuring the new digital landscape in western Canadian agriculture. *NJAS – Wageningen Journal of Life Sciences, 90–91*(1), 1–11. https://doi.org/10.1016/j.njas.2019.04.001

Pigford, A.-A. E., Hickey, G. M., & Klerkx, L. (2018). Beyond agricultural innovation systems? Exploring an agricultural innovation ecosystems approach for niche design and development in sustainability transitions. *Agricultural Systems, 164*, 116–21. https://doi.org/10.1016/j.agsy.2018.04.007

Pinch, T. J., & Bijker, W. E. (1984). The social construction of facts and artefacts: Or how the sociology of science and the sociology of technology might benefit each other. *Social Studies of Science, 14*(3), 399–41. http://www.jstor.org/stable/285355

Raj, S., & Bhattacharjee, S. (2017). Social media for agricultural extension. *National Institute of Agricultural Extension Management: Extension Next Bulletin no. 1.* http://www.manage.gov.in/publications/extnnext/March2017.pdf

Renken, J. (2019). What motivates ICT4D champions? In P. Nielsen & H. C. Kimaro (Eds.), *Information and communication technologies for development: Strengthening southern-driven cooperation as a catalyst for ICT4D* (pp. 307–18). Springer.

Renken, J., & Heeks, R. (2013). *Conceptualising ICT4D project champions.* Paper presented at the Proceedings of the Sixth International Conference on Information and Communications Technologies and Development. Cape Town, South Africa. Notes. 2, 128–31. https://doi.org/10.1145/2517899.2517928

Robeyns, I. (2017). *Wellbeing, freedom and social justice: The capability approach re-examined.* Open Book Publishers.

Rotz, S., Duncan, E., Small, M., Botschner, J., Dara, R., Mosby, I., Reed, M., Fraser, E. D. G. (2019a, February). The politics of digital agricultural technologies: A preliminary review. *Sociologia Ruralis, 59*(2), 203–29. https://doi.org/10.1111/soru.12233

Rotz, S., Gravely, E., Mosby, I., Duncan, E., Finnis, E., Horgan, M., LeBlanc, J., Martin, R., Neufeld, H. T., Nixon, A., Pant, L., Shalla, V., & Fraser, E. (2019b, May). Automated pastures and the digital divide: How agricultural technologies are shaping labour and rural communities. *Journal of Rural Studies, 68*, 112–22. https://doi.org/10.1016/j.jrurstud.2019.01.023

Saint Ville, A. S., Hickey, G. M., Locher, U., & Phillip, L. E. (2016). Exploring the role of social capital in influencing knowledge flows and innovation in smallholder farming communities in the Caribbean. *Food Security, 8*(3), 535–49. https://doi.org/10.1007/s12571-016-0581-y

Shepherd, M., Turner, J. A., Small, B., & Wheeler, D. (2020). Priorities for science to overcome hurdles thwarting the full promise of the 'digital agriculture' revolution. *Journal of the Science of Food and Agriculture, 100*(14), 5083–92. https://doi.org/10.1002/jsfa.9346

Shilomboleni, H., Pelletier, B., & Gebru, B. (2020). ICT4 Scale in smallholder agriculture: Contributions and challenges. *Information Technologies & International Development, 16*, 47–65. https://itidjournal.org/index.php/itid/article/download/1743/1743-5705-2-PB.pdf

Srnicek, N. (2016). *Platform capitalism.* Polity.

Steinke, J., van Etten, J., Müller, A., Ortiz-Crespo, B., van de Gevel, J., Silvestri, S., & Priebe, J. (2020). Tapping the full potential of the digital revolution for agricultural extension: An emerging innovation agenda. *International Journal of Agricultural Sustainability, 19*(5–6), 549–65. https://doi.org/10.1080/14735903.2020.1738754

Stewart, J., & Hyysalo, S. (2008). Intermediaries, users and social learning in technological innovation. *International Journal of Innovation Management, 12*(03), 295–325.

Sulaiman, V, R., & Davis, K. (2012). The "New extensionist": Roles, strategies, and capacities to strengthen extension and advisory services. *Journal of International Agricultural and Extension Education. 21*(3), 6–18 DOI: 10.5191/jiaee.2014.21301.

Sulaiman, V. R., Hall, A., Kalaivani, N. J., Dorai, K., & Reddy, T. S. V. (2012). Necessary, but not sufficient: Critiquing the role of information and communication technology in putting knowledge into use. *The Journal of Agricultural Education and Extension, 18*(4), 331–46.

Tacchi, J. (2015). Ethnographic action research: Media, information and communicative ecologies for development initiatives. In H. Bradbury (Ed.), *The SAGE handbook of action research* (3rd ed., pp. 220–29). https://doi.org/10.4135/9781473921290

Tacchi, J., Slater, D., & Hearn, G. N. (2003). *Ethnographic action research: A user's handbook developed to innovate and research ICT applications for poverty eradication.* UNESCO.

United Nations Trade and Development Conference (UNCTAD) (2018). *Leapfrogging: Look before you leap. Policy Brief no. 71.* United Nations Conference on Trade and Development. https://unctad.org/system/files/official-document/presspb2018d8_en.pdf

Wahyunengseh, R. D., Hastjarjo, S., Mulyaningsih, T., & Suharto, D. G. (2020). Digital governance and digital divide: A matrix of the poor's vulnerabilities. *Policy & Governance Review, 4*(2), 152–66. https://doi.org/10.30589/pgr.v4i2.296

Walsham, G. (2020). South-south and triangular cooperation in ICT4D. *The Electronic Journal of Information Systems in Developing Countries, n/a*(n/a), e12130. https://doi.org/10.1002/isd2.12130

Wanigasundera, W. A. D. P., & Atapattu, N. (2019). Extension reforms in Sri Lanka: Lessons and policy options. In S. C. Babu & P. K. Joshi (Eds.), *Agricultural extension reforms in South Asia* (pp. 79–98). Academic Press/Elsevier.

Wenger, E., McDermott, R., & Snyder, W. (2002). *Cultivating communities of practice.* Harvard Business School Press.

Wenger, E., White, N., & Smith, J. D. (2009). *Digital habitats: Stewarding technology for communities.* CPSquare.

Wilson, K. (2021, January 12). *"Digital Beacons": The digital impact alliance strategic plan.* Digital Impact Alliance. https://dial.global/wp-content/uploads/2021/01/DIAL_Strategy_FINAL.pdf

Winslow, L. (2017). Frame analysis. In M. Allen. (Ed.), *The SAGE encyclopedia of communication research methods* (pp. 584–86). SAGE. https://doi.org/10.4135/9781483381411

Wiseman, L., Sanderson, J., Zhang, A., & Jakku, E. (2019). Farmers and their data: An examination of farmers' reluctance to share their data through the lens of the laws impacting smart farming. *NJAS – Wageningen Journal of Life Sciences, 90–91*(1), 1–10. https://doi.org/10.1016/j.njas.2019.04.007

Yin, R. K. (2003). *Case study research: Design and methods.* SAGE.

Zuboff, S. (2018). *The age of surveillance capitalism: The fight for the future at the new frontier of power.* Profile Books.

# 9 School closures on rural education

## Examining remote learning experiences during the COVID-19 pandemic in Chile

*Isabel Pavez Andonaegui and Catalina Farías*

## Introduction

The COVID-19 pandemic and subsequent school closures positioned digital inclusion and inequalities at the center of the discussion of education around the world (Dube, 2020; Khilnani et al., 2020; Nguyen et al., 2020). Remote learning revealed persisting gaps in access to and use of technologies for children in rural areas, as vulnerable populations have limited access to technology, connectivity, and digital skills (Dube, 2020; United Nations, 2020). This situation, as the Economic Commission for Latin America and the Caribbean (ECLAC) found, is expected to increase inequities in the educational sphere and have consequences in areas such as social inclusion and economic development (Economic Commission for Latin America and the Caribbean [ECLAC] et al., 2020). Latin America's education system, as was found by the United Nations International Children's Emergency Fund (UNICEF), was severely impacted by the pandemic, facing a record of 158 days of school closures on average (United Nations International Children's Emergency Fund [UNICEF], 2021). The area in which this research was conducted had recently started the school year when pandemic-related restrictions were implemented. While online classes seemed like the natural choice for continuing to educate children in urban areas, rural communities faced more challenges (Martínez et al., 2021). The literature has shown that rural communities presented lower Internet adoption rates and higher barriers to access to and use of technologies than urban areas (Agostini & Willington, 2010; Correa et al., 2017; LaRose et al., 2007; Townsend et al., 2013). This is true for areas in which the main sources of access to information and communication technologies (ICTs) are schools, as these entities serve as equalizers that increase opportunities for full participation in society through access to and training in technologies (Formichella et al., 2020; González-Betancora et al., 2021; Salinas & Sánchez, 2009). Rural school closures were problematic for teachers, students, and families. Schools were seen as providing a path toward better opportunities in life and were viewed as key institutions in rural communities, particularly for mothers (Castro, 2012; Ortega Arias & Cárcamo, 2018).

This study presents an analysis of the experiences and reflections of 12 dyads of seventh-grade students and their mothers ($N = 24$). The goal was to explore how

DOI: 10.4324/9781003282075-12

school closure and remote learning impacted rural communities' access to and use of ICTs. The individuals who participated in this study were also beneficiaries of the government program "Yo elijo mi PC" (I Choose My Computer), which provided a laptop and one year of free Internet access to seventh graders from vulnerable backgrounds. However, by the time of the study and due to the pandemic, the distribution of the computers was severely delayed. Therefore, participants did not have them during their school year. Then, the interviews were conducted before the students received their computer and one year after schools closed and remote learning began.

## Conceptual framework

### Rurality and Internet access

Rurality is usually tied to a geographic gap limiting quality Internet access. In Latin America, an urban inhabitant is 15–41% more likely to have an Internet connection than someone living in a rural area (Galperin, 2017). Although Chile stands out as one of the few Latin American countries that has a solid history of connectivity initiatives and achievements, rurality is still its Achilles' heel, as urban-rural inequities persist (Subsecretaría de Telecomunicaciones de Chile, 2021). For example, only three out of every ten rural households had access to a computer compared to six in ten urban households (Martínez et al., 2021). The presence of children and adolescents in the rural home is one of the most important factors impacting Internet access, which means that their presence contributes a distinctive element (Pavez & Correa, 2018). Rural populations are more likely to have low socioeconomic status and to engage in manual labor (Correa et al., 2017). They have less exposure to and less experience using more complex technologies and acquiring digital skills (Skerratt, 2008).

Distance and isolation characterize rural communities (LaRose et al., 2011), and these elements became even more prominent during the pandemic. Both adults and children were forced to change their routines and find new ways to interact with and connect to the world. The Internet made connectivity levels and access to devices even more important. Although the digital divide has decreased considerably in Chile (Subsecretaría de Telecomunicaciones de Chile, 2021), children living in rural areas face the greatest disadvantages due to long-standing social and structural inequities (Rojas & Poveda, 2018). The most popular device for providing Internet and technology access to children in Latin America remains the smartphone (Trucco & Palma, 2020). This is important because the evidence showed that computers and laptops offered a wider range of activities and functionalities. As such, mobile-only users are digitally included but as members of a less proficient class of users who mainly engaged in entertainment and communication activities (Napoli & Obar, 2014). It is thus important to raise questions about what technology children in rural environments are accessing, how they do so, and why.

Since 2016 the data have shown steady growth in Internet usage in the Global South. Chile was one of the leaders of this trend, with 87.5% penetration. Mobile phones were the most popular device for accessing the Internet in Chile and across Latin America (Trucco & Palma, 2020). This was consistent with evidence that showed that mobile connections were on the rise, particularly among low socioeconomic groups (Donner et al., 2011). This was also a trend among the younger population. Data from a comparative study in Brazil, Chile, Costa Rica, and Uruguay indicated that mobile phones were the most widespread Internet access mode from home among children ages 9–17 (Cabello et al., 2020). However, smartphones provided somewhat limited access, particularly regarding entertainment and communication use (Green & Haddon, 2009; Mascheroni, 2013). Researchers found that this type of use increased the gaps among Internet users, particularly among the most vulnerable cohorts, impacting their digital skill levels and leaving most women behind (Martínez-Cantos, 2017; Napoli & Obar, 2014). Similar results were found in children. Cabello and colleagues (2020) argued, "The most widespread form of inclusion … is at the same time the one that shows the least positive results in the opportunities for the use of technologies, as well as in the skills for use" (p. 49). This was an example of a complex scenario that existed in this field because, as other scholars have found, Internet access was not the answer to the digital divide, and the mobile-only approach was perpetuating gaps (Katz, 2017).

*Digital inclusion and user agency*

The concept of digital inclusion delved into the importance of providing tools to disadvantaged people who could use technologies to develop skills that allowed them to reach their potential at a social, economic, and educational levels (Helsper, 2012; Van Dijk, 2005). The idea of meaningfulness goes beyond the access gap, assuming that the connectivity was not the only element that contributed to or could ameliorate the digital divide and that it was merely the first step in exploring the use of technology from the user perspective. Appropriation delves into the notion of owning technology in a way that makes sense to the user, and it was developed through the Social Construction of Technology research field (Hutchby, 2001; Sørensen, 1994). Technological devices can be adjusted to meet user needs based on their interactions. For instance, while some view smartphones as the world's largest library, others see them as a problem and source of stress due to their pervasiveness (Boczkowski, 2021; Pavez & Correa, 2020). These perceptions can determine how users approach and incorporate technologies into their lives. This also touches on the idea of affordances, as the advantages and opportunities offered by technology are not standardized and depend on the user's needs, expectations, and capacity (Humphreys et al., 2018). This concept allowed us to understand the "functional and relational aspects which frame, while not determining, the possibilities for agentic action in relation to an object" (Hutchby, 2001, p. 444). It elucidated how devices could be shaped through interactions with humans and how they are used (Davis & Chouinard, 2016; Evans et al., 2016; Hutchby, 2001).

The context in which participants engaged in their activities was a key factor for understanding their use of the Internet and technological devices. Evidence showed that geographic characteristics (Cullen, 2001), economic factors (Galperin, 2017;

Hampton et al., 2020; LaRose et al., 2007; Townsend et al., 2013), education level (Gonzales, 2015; Hampton et al., 2020), and the presence of children in the household (Gonzalez & Katz, 2016; Katz, 2010) directly influenced the degree to which individuals used technology and the Internet. As smartphones can be considered metamedia devices that contained features like cameras, multi-media players, and audio and video recorders (Humphreys et al., 2018), it was important to understand how these features affected their affordances. Research has shown that age was one of the most persistent factors related to how people adopt and use technologies (Boczkowski, 2021; Morris & Venkatesh, 2006). Therefore, we can expect to find differences between the two generations of participants in this study.

### The problems faced by rural communities in a connected country

Chile had an ambitious telecommunications agenda. In the late 1990s, the Ministry of Education developed the "Enlaces" (Links) policy, which focused on incorporating ICTs into the educational system. In 2009, it created "Yo elijo mi PC" (I Choose My Computer), a program that gave a laptop, a one-year mobile Internet plan, and digital educational resources to vulnerable seventh-grade students who attended public schools (JUNAEB Ministerio de Educación, n.d.a.). The Telecommunication Development Fund was created during the same period in order to increase telecommunication services coverage in low-income rural or urban areas. The fund focused on residents who were isolated due to geographic or technological conditions. The "Agenda Digital" (Digital Agenda) included over 50 policies focused on expanding Internet access, connectivity, digital government, and education. It remained in place for 16 years.

Despite these efforts, Chile's rural areas were still at a disadvantage. While 89.1% of the urban households had Internet access, this was true for only 76.7% of rural households. Rural communities' access was mainly limited to mobile connections, a trend that has steadily increased over time (Subsecretaria de Telecomunicaciones de Chile, 2017). This contradictory outlook of successful digital policies, high Internet penetration mainly through mobiles, and persistent rural inequalities presented challenges offered an opportunity to explore the technological experiences of the vulnerable population when remote learning and online classes became mandatory. When the COVID-19 pandemic began in Chile in early 2020 in March, most schools had only been open for three days (UNICEF, 2021). The government declared a national emergency and ordered the closure of all schools. Data from Chile's Ministry of Education (MINEDUC, 2020) showed that 3,317 rural schools closed, affecting more than 300,000 students.

Although the Ministry provided online content that allowed educational communities to continue to deliver essential contents remotely (Mineduc, 2020), the digital gap undermined the quality of education received by unprivileged communities across the country (Eyzaguirre et al., 2020). Nationally, the evidence showed that among rural and urban students, only 66% of them were able to attend online classes (Educación 2020, 2020). School closures had a significant social impact on rural families. Studies showed that schools were viewed as key institutions in rural communities (e.g., Salinas & Sánchez, 2009). They were important spaces for establishing relationships with other community members, especially for mothers (Núñez et al., 2014), and all public school students in Chile received free meals at school (Eyzaguirre et al., 2020). The aim of this chapter is to explore the experiences with technology and reflections on the importance of smartphone access for rural students and their mothers during school closures as well as the importance of Internet access for participating in online learning in an adverse digital environment.

## Methods

This research was based on a qualitative approach. Its focus was on indepth descriptions of the participants' experiences and contexts. As it gives the participants a voice, it can be loosely included under the umbrella of participatory research (PR) given that "ontologically and axiologically, PR is characterized by efforts to decenter the power of the 'expert' researcher and the powerful more broadly, addressing those who have traditionally been marginalized from knowledge production" (Rosen, 2021, p. 4). This methodological design allowed researchers to develop a relationship of trust with the participants, which can be helpful with the disclosure of sensitive information (Ummel & Achille, 2016). For a holistic approach, dyad interviews were conducted because they provided a better understanding of the relationship and dynamics between the mothers and their children (Kendall et al., 2010). This instrument provided a more complete understanding of the phenomenon under study thanks to the gathering of information from different but complementary sources (Ummel & Achille, 2016). In this study, 24 indepth semi-structured interviews were conducted with dyads of 12 mothers and their seventh-grade children. The participants are described in Table 9.1.

We first recruited study participants by sending an open invitation to 15 rural schools in central Chile. Three schools from rural towns in three different districts participated. The towns had fewer than 6,000 habitants and the main economic

*Table 9.1* Description of participants in the study.

| | Age | Number of children in household | Children interviewed | Age | Computer in household |
|---|---|---|---|---|---|
| Mother 1 | 40 | 2 | Child 1 | 11 | No |
| Mother 2 | 30 | 3 | Child 2 | 12 | No |
| Mother 3 | 33 | 3 | Child 3 | 11 | Yes |
| Mother 4 | 40 | 2 | Child 4 | 12 | No |
| Mother 5 | 49 | 1 | Child 5 | 12 | No |
| Mother 6 | 32 | 4 | Child 6 | 12 | No |
| Mother 7 | 39 | 1 | Child 7 | 12 | No |
| Mother 8 | 39 | 4 | Child 8 | 11 | Yes |
| Mother 9 | 33 | 2 | Child 9 | 11 | No |
| Mother 10 | 32 | 3 | Child 10 | 11 | No |
| Mother 11 | 37 | 4 | Child 11 | 11 | No |
| Mother 12 | 31 | 2 | Child 12 | 12 | No |

activities were related to agriculture and local trade. Also, 85% of students attending these schools were from low-income families and were considered highly vulnerable (JUNAEB, n.d.b.). Initial contact was mediated by the school principals, who provided families' contact information for those in the seventh grade and coordinated the first encounter. A snowball sampling approach was then used among students in the class to enroll the rest of the participants. The first families interviewed provided new contact information for mothers of children at the same school who might be interested in participating. Informed consent was approved in advance by the host institution's Ethical Committee, and the respective forms were signed by the children and their parents.

The interviews were conducted between December 2020 and May 2021. Due to the pandemic, which led to quarantines in Chile and school closures, the interviews were conducted face-to-face and through WhatsApp calls, phone calls, or video calls depending on each participant's technological skills. The interviews were transcribed and analyzed using N'Vivo software and a process of codification, condensation, and interpretation (Kvale & Brinkman, 2009). The main topics of interest were identified through our review of the literature on rural communities, education, and digital inclusion, but there was also a process of emergent coding that addressed the specificities of the context and participants. For example, the participants had very different levels of access to technology. In most cases, children had their own devices, although two used their parents' smartphones to access the Internet. Only three of the 12 households owned a computer.

## Findings

### *Geography and gender as barriers*

It was well-known that one of the main factors that explained the limited use of the Internet and digital technologies in rural communities was the areas' geographic characteristics (Cullen, 2001). This was certainly the case for the respondents and areas in this study. Although participants reported that they valued the calm, safety, and better quality of rural life for their children and did not see any disadvantages, they consistently reported connectivity issues. Both the students and their mothers stated that their main challenge during the pandemic was using their town's 3G Internet connections, which proved unstable. This was particularly problematic for those living near a hill, as it affected the quality of the connection. For example, Child 11, an 11-year-old girl (Child 11, 11) who was often unable to connect to her online classes, explained:

I have a bad signal and I live in front of a giant hill ... . The signal was sometimes cut off during school hours, so I had to leave the meeting and re-enter. Sometimes they [teachers] told me to just skip class because of my challenges accessing the Internet.

In other cases, signal instability was related to external factors such as power outages (Mother 7, 39):

> The power suddenly goes out and we are left without Internet .... The antenna's [signal] falls off. The Internet doesn't come back. The antenna uses electricity .... Sometimes it takes all day to fix it. It was hard for my oldest daughter to attend remote classes ... in the end she had to tell the teacher that she couldn't participate because the power went out.

Gender roles also were part of the exchanges. Mothers and children reported that patriarchal dynamics hampered girls' access to devices, something that did not affect their male counterparts. The United Nations Educational, Scientific and Cultural Organization found that it was common for parents to expect girls to take on domestic roles in rural communities even though this could prevent them from getting an education (United Nations Educational, Scientific and Cultural Organization [UNESCO], 2020, 2021). Women and girls were often responsible for more household chores than men and boys. Female student participants reported that they were responsible for childcare, cooking, and cleaning when they were not attending classes or doing homework. Some participants claimed that their household responsibilities did not leave them any free time to use their smartphone (Child 8, 11):

> I almost never have free time [to use the cell phone] because I have to take care of my brothers. I also clean at home, or I cook for my brothers, and I take care of our pets.

In some cases, rural parents treated their children differently in terms of access to technological tools. Boys tended to be introduced to technology first, and girls faced more restrictions when it came to accessing devices (EQUALS Skills Coalition & UNESCO, 2019). In this study, mothers recognized that their daughters helped them more with household chores, but saw this as natural. Girls saw these tasks as related to their gender because they involved family matters (Bianchi et al., 2000; Goldscheider et al., 2015):

> She [Child 6] helps me set the table, tidy up the house, sweep, clean the furniture, those kind of things .... The boys are only made to tidy up their toys when they finish using them.
>
> (Mother 6, 32)

> She helped me around the house, did the cleaning, and did her homework in the afternoon. She has to do her part and, when I need to do things, I tell her to take care of the younger girl, hang up her clothes, and sweep.
>
> (Mother 3, 33)

*Development of digital skills: "I can learn by myself" and the role
of older siblings*

Due to their familiarity with smartphones, students argued that they did not see connecting to online classes or interacting with teachers and students as challenges. They showed high levels of confidence and empowerment related to their digital skills and ability to use their smartphone. Most of the young participants learned to use their devices by watching others. In this context, older siblings or cousins helped boost technology adoption and digital skills development. Mothers played a less prominent role in these areas. However, during the pandemic, both groups learned to use new applications such as Zoom or Google Meet. They also had to perform new tasks such as filling out an online form or printing a document. When asked about the challenges they could face when the computer arrived, participants stated that they were confident that they would learn to use it by themselves, even though the majority of our respondents had never used one before outside of school. Their responses included: "The Internet is easy," and "I learn right away just like other people."

Mothers and children did not feel that school had played a fundamental role in the development of students' digital abilities. They argued that schools only teach the basics (even though schools received negative feedback about changing the curriculum). Furthermore, participants reported that they were bored during computer classes because instructors only showed them videos. The students who participated in our survey assured us that they did not learn anything new in computer classes:

> Computer class is more to search for things… . We always look up answers using Google.
>
> (Child 2, 12)

One interesting finding of our research was that participants spent more time on their smartphones during school closures. This allowed them to access and engage in other activities, such as selling things on Instagram, filling out online forms, and making video calls through WhatsApp. The mothers did report becoming aware of technology dependency. In some families, this led to the establishment of new rules. Parental control was focused on smartphone use and checking the websites and social media platforms that their children visited. In one case, a mother removed the sim card from her daughter's smartphone to keep her from accessing WhatsApp:

> We bought her a cell phone without a chip so that she could only watch videos. This year she started using social media, but just Instagram …. She doesn't have WhatsApp because she doesn't have a phone number.
>
> (Mother 3, 33)

> I check the children's cell phones. If they have a password, they have to share it with me…. I do this for fear of who might be on the other side of a call or chat more than anything, I don't know, I could start exchanging

messages with a child and say I'm 12 years old. That's what worries me more than anything else.

(Mother 1, 40)

The exercise of parental control reported focuses on technical aspects, such as checking websites, setting time limits for phone use, and even confiscating devices at night. Students reported that they do not turn to their mothers if they wanted to talk about how to use the Internet safely or when they do not know how to do something. They preferred to ask an older sibling or cousin when they want to learn new things on the Internet. Given that it is more common to have relatives as neighbors in rural settings, family members were often listed as a first choice. In this study, and as was also the case in the literature, participants tended to rely on classmates only in the absence of an older sibling (Salinas & Sánchez, 2009).

### The perceived affordances of smartphones and computers

Although most of the mothers in the study reported that their children were "too young to have a mobile," they allowed them to have one. In some cases, it was a device that had been discarded due to an upgrade or a gift from another family member. While the mothers saw smartphones as a source of entertainment for their children rather than a tool that they could use to communicate, they did state that being able to contact them through calls or WhatsApp was necessary due to distances in rural environments. The mothers also gained independence from their children when they were given smartphones, and it ended the constant nagging to use the parents' devices for social media or online gaming.

Smartphones were also key for allowing students to continue their education during school closures. In the case of the three rural schools covered by this study, online classes were rare due to significant Internet problems, so schools distributed printed materials on a weekly basis. The students had to complete the homework and submit it at the end of each month, at which point they would receive a new packet. Students and their mothers used smartphones to search for information in order to complete assignments or contact the teacher through WhatsApp. This method was combined with online classes in cases in which the students and teacher managed to access a stable connection.

This shift in the use of smartphones changed the mothers' perception of the benefits of having the device. They eventually came to see smartphones as fundamental tools for their children's education. This was also true for WhatsApp, as groups were created between parents and teachers, while the students chatted directly with teachers to address any questions. In some cases, teachers video chatted with their students individually using WhatsApp (Child 8, 11):

We completed the homework before the classes and then reviewed the assignments in class. We had to send photos of our work. If I didn't understand something, I would ask my monitor [teacher assistant] through WhatsApp, and she would ask the teacher.

Despite their mothers' perceptions, the children stated that their smartphone use did not change during the pandemic. They did report using it more for educational purposes, but the main function of the device was still entertainment and communication. Young people played video games, accessed social media, chatted or video chatted with their friends, and used applications to edit and make videos. Phrases such as "I use it a lot. I have games on my phone and use it to talk with my classmates" (Child 12, 12) and "I play [on my cell phone] and watch music videos and TV series. I have TikTok, Snapchat, Google, Instagram, and YouTube" (Child 3, 11) confirmed that the affordances of smartphones were different for mothers and their daughters even when they share situational, physical, and technological contexts (Humphreys et al., 2018).

Only three families had previous access to personal computers, but the participants' testimonies suggested that they have high expectations for the device children were supposed to receive during the school year. They claimed that they would use it for educational purposes only because of its functionality. The activities that they anticipated engaging in were online classes, searching for information, completing homework assignments, and improving school performance. The participants noted that computers would allow the girls to do things that they could not do on smartphones and complete those tasks more quickly, such as printing a document or working with Word or Excel:

> The computer is important to be able to do homework, for classes, well, for everything, because when she grows up, she will want to go to university, and she will use a computer.... It will benefit her a lot.
>
> (Mother 2, 30)

> I think I will do better at school because I will be able to store the school files there, use Zoom, and listen well, and all that.
>
> (Child 12, 12)

The mothers and children who participated in this study have different perceptions of the main purpose of the smartphone. As such, they associate it with different uses. Both groups had similar expectations about the educational affordances of personal computers.

### Conclusion

The aim of this study was to explore how rural school closures influenced access to and uses and perceptions of the Internet, particularly through smartphones, among seventh graders and their mothers. Twenty-four participants were interviewed: 12 mothers with school-aged children and their 12 children who were selected to receive a computer through government program "Yo elijo mi PC" (I Choose My Computer). The analysis confirmed that after one year of school closure, the main challenges faced by the families were connectivity problems, such as network instability, inability to connect to online classes, and power outages. This is consistent with previous research that showed that geographic factors are one of the main reasons for lower Internet and technology adoption in rural communities (Boase, 2010; Correa & Pavez, 2016; Cullen, 2001). Students reported that there was a gender barrier for girls that hampered equal access to the Internet and devices (EQUALS Skills Coalition & UNESCO, 2019). Gender roles have a significant impact on the participants' family dynamics, as women were associated with household tasks (e.g., childcare), while men were thought to perform activities outside of the home (Goldscheider et al., 2015). The mothers interviewed for this study stated that they are not aware of the gender differences in the family dynamic. The girls tended to be aware of them, though they did not question it. Girls were expected to help mothers to complete domestic tasks while boys' responsibilities were limited to tidying up their personal things. In some cases, the number of household chores that they were required to perform left girls no time to access their smartphones.

The testimonies confirmed that views of smartphone affordances are different for the mothers and students and were influenced by context. For instance, prior to the pandemic, mothers believed that smartphones were mainly useful for entertaining their children. They now see them as useful for educational purposes. Students feel empowered by using mobile phones but view such devices as useful mainly for entertainment and social media. Although few participants had access to a computer, the affordances of that kind of device are mainly perceived as educational. The participants claimed that when the new computer arrives, they will use it only for school and not entertainment or communication because they use smartphones for the latter. It would be interesting, however, to study in both groups to see if the perceived affordances of these devices remain unchanged over time or change once they add the computer to their media ecology.

The respondents' testimonies also revealed the role older siblings or cousins played in the development of students' digital skills. Children tended to ask older cousins or siblings for help when they wanted to engage in a new activity on the Internet. Mothers do not play a prominent role in this context. This increases students' empowerment and self-confidence, as they argue that they can solve most problems on their own.

The sampling and results highlight the importance of participatory research in rural environments. Rural communities present contextual factors that condition the set of challenges that they face. Research that gives participants a voice, especially those from vulnerable backgrounds, helps to move the discussion beyond rural versus urban, and focuses on their unique contexts. This study does present limitations. The results of this research do not allow us to generalize about Chile's entire rural population. Nevertheless, the dyad interviews allowed us to take a holistic approach and arrive at a more complete understanding of the technological family dynamics that surround mothers and children. The presence of gender roles and the way they affected access to and the adoption of technologies for young girls must be analyzed by future researchers. Conducting interviews with male family members could help researchers to further understand this phenomenon. From the affordances side, scholars should visit participants again after they receive a computer and explore how this device influenced their practices and perceptions, both educational and otherwise.

## Acknowledgments

The authors would like to thank Professor Teresa Correa and the reviewers for their valuable comments and insights. This work was funded by National Commission for Scientific and Technological Research (Fondecyt) Grant N. 11200039 and the Millennium Nucleus Grant to Improve the Mental Health of Adolescents and Youths, Imhay, Millennium Science Initiative Program—NCS2021_081.

## References

Agostini, C. A., & Willington, M. (2010). Radiografía de la brecha digital en Chile: ¿Se justifica la intervención del Estado? *Revista de Estudios Públicos, 119*, 5–32. https://www.estudiospublicos.cl/index.php/cep/article/view/390/588

Bianchi, S., Milkie, M., Sayer, L., & Robinson, J. (2000). Is anyone doing the housework? Trends in the gender division of household labor. *Social Forces, 79*(1), 191–28. https://doi.org/10.1093/sf/79.1.191

Boase, J. (2010). The consequences of personal networks for Internet use in rural areas. *American Behavioral Scientist, 53*(9), 1257–67. https://doi.org/10.1177/0002764210361681

Boczkowski, P. J. (2021). *Abundance: On the experience of living a world of information plenty.* Oxford University Press.

Cabello, P., Claro, M., & Dodel, M. (2020). Modalidades de acceso material a Internet y su relación con las habilidades y prácticas digitales. In D. Trucco & A. Palma (Eds.), *Infancia y adolescencia en la era digital. Un informe comparativo de los estudios de Kids Online del Brasil, Chile, Costa Rica y el Uruguay* (pp. 41–53). CEPAL.

Castro, A. (2012). Familias rurales y sus procesos de transformación: estudio de casos en un escenario de ruralidad en tensión. *Psicoperspectivas. Individuo y Sociedad, 11*(1), 180–203. https://doi.org/10.5027/psicoperspectivas-Vol11-Issue1-fulltext-172

Correa, T., & Pavez, I. (2016). Digital inclusion in rural areas: A qualitative exploration of challenges faced by people from isolated communities. *Journal of Computer-Mediated Communication, 21*(3), 247–63. https://doi.org/10.1111/jcc4.12154

Correa, T., Pavez, I., & Contreras, J. (2017). Beyond access: A relational and resource-based model of household Internet adoption in isolated communities. *Telecommunications Policy, 41*(9), 757–68. https://doi.org/10.1016/j.telpol.2017.03.008

Cullen, R. (2001). Addressing the digital divide. *Online Information Review, 25*(5), 311–20. https://doi.org/10.1108/14684520110410517

Davis, J. L., & Chouinard, J. B. (2016). Theorizing affordances: From request to refuse. *Bulletin of Science, Technology & Society, 36*(4), 241–48. https://doi.org/10.1177/0270467617714944

Donner, J., Gitau, S., & Marsden, G. (2011). Exploring mobile-only Internet use: Results of a training study in urban South Africa. *International Journal of Communication, 5*, 574–97. https://pubs.cs.uct.ac.za/id/eprint/706/1/750-4634-1-PB.pdf

Dube, B. (2020). Rural online learning in the context of COVID 19 in South Africa: Evoking an inclusive education approach. *Multidisciplinary Journal of Educational Research, 10*(2), 135–57. https://doi.org/10.17583/remie.2020.5607

Economic Commission for Latin America and the Caribbean (ECLAC) and the Regional Bureau for Education in Latin America and the Caribbean of the United Nations Educational, Scientific and Cultural Organization (OREALC/UNESCO Santiago) (2020). *Education in the time of COVID-19.* COVID-19 Report ECLAC-UNESCO. United Nations. http://repositorio.cepal.org/bitstream/handle/11362/45905/S2000509_en.pdf?sequence=1&isAllowed=y

Educación 2020 (2020). *#EstamosConectados. Testimonios y experiencias de las comunidades educativas.* https://educacion2020.cl/wp-content/uploads/2020/04/Informe-Final-Encuesta-EstamosConectados-E2020.pdf

EQUALS Skills Coalition & UNESCO (2019). *I'd blush if I could. Closing gender divides in digital skills through education.* EQUALS & UNESCO. https://unesdoc.unesco.org/ark:/48223/pf0000367416/PDF/367416eng.pdf.multi

Evans, S. K., Pearce, K. E., Vitak, J., & Treem, J. W. (2016). Explicating affordances: A conceptual framework for understanding affordances in communication research. *Journal of Computer-Mediated Communication, 22*(1), 35–52. https://doi.org/10.1111/jcc4.12180

Eyzaguirre, S., Le Foulon, C., & Salvatierra, V. (2020). Educación en tiempos de pandemia: antecedentes y recomendaciones para la discusión en Chile. *Estudios Públicos, 159*, 111–80. https://doi.org/10.38178/07183089/1430200722

Formichella, M. M., Alderete, M. V., & Di Meglio, G. (2020). Nuevas tecnologías en los hogares: ¿Hay una recompensa educativa? Evidencias para Argentina. *Education in the Knowledge Society, 21.* https://doi.org/10.14201/eks.23553

Galperin, H. (2017). Why are half of Latin Americans not online? A four-country study of reasons for Internet non-adoption. *International Journal of Communication, 11*, 3332–54.

Goldscheider, F., Bernhardt, E., & Lappegård, T. (2015). The Gender Revolution: A framework for understanding changing family and demographic behavior. *Population and Development Review, 41*(2), 207–39. https://doi.org/10.1111/j.1728-4457.2015.00045.x

Gonzales, A. L. (2015). Disadvantaged minorities' use of the Internet to expand their social networks. *Communication Research, 44*(4), 467–86. https://doi.org/10.1177/0093650214565925

Gonzalez, C., & Katz, V. (2016). Transnational family communication as a driver of technology adoption. *International Journal of Communication, 10*, 2683–703. link.gale.com/apps/doc/A534020704/LitRC?u=anon~6ac39a98&sid=googleScholar&xid=e8df13fb

González-Betancora, S., López-Puig, A., & Cardenal, M. E. (2021). Digital inequality at home. The school as compensatory agent. *Computers & Education, 168*, 104195. https://doi.org/10.1016/j.compedu.2021.104195

Green, N., & Haddon, L. (2009). *Mobile communications. An introduction to new media.* Bloomsbury Publishing.

Hampton, K., Fernandez, L., Robertson, C., & Bauer, J. M. (2020). Repercussions of poor broadband connectivity for students in rural and small town Michigan. *TPRC48: The 48th Research Conference on Communication, Information and Internet Policy.* http://dx.doi.org/10.2139/ssrn.3749644

Helsper, E. J. (2012). A corresponding fields model for the links between social and digital exclusion. *Communication Theory, 22*(4), 403–26. https://doi.org/10.1111/j.1468-2885.2012.01416.x

Humphreys, L., Karnowski, V., & von Pape, T. (2018). Smartphones as metamedia: A framework for identifying the niches structuring smartphone use. *International Journal of Communication, 12*, 2793–809. http://ijoc.org/index.php/ijoc/article/view/7922

Hutchby, I. (2001). Technologies, texts and affordances. *Sociology, 35*(2), 441–56. https://doi.org/10.1177/s0038038501000219

JUNAEB Ministerio de Educación (n.d.a.). *Becas TIC 2022*. https://www.junaeb.cl/becas-tic

JUNAEB Ministerio de Educación (n.d.b.). *IVE*. https://junaeb.cl/live

Katz, V. (2010). How children of immigrants use media to connect their families to the community. *Journal of Children and Media, 4*(3), 298–315. https://doi.org/10.1080/17482798.2010.486136

Katz, V. (2017). What it means to be "under-connected" in lower-income families. *Journal of Children and Media, 11*(2), 241–44. https://doi.org/10.1080/17482798.2017.1305602

Kendall, M., Murray, S. A., Carduff, E., Worth, A., Harris, F., Lloyd, A., Cavers, D., Grant, L., Boyd, K., & Sheikh, A. (2010). Use of multiperspective qualitative interviews to understand patients' and carers' beliefs, experiences, and needs. *British Medical Journal, 340*, 196–99. https://doi.org/10.1136/bmj.b4122

Khilnani, A., Schulz, J., & Robinson, L. (2020). The COVID-19 pandemic: New concerns and connections between eHealth and digital inequalities. *Journal of Information, Communication and Ethics in Society, 18*(3), 393–403. https://doi.org/10.1108/JICES-04-2020-0052

Kvale, S., & Brinkman, S. (2009). *InterViews: Learning the craft of qualitative research interviewing* (2nd ed.). Sage.

LaRose, R., Gregg, J. L., Strover, S., Straubhaar, J., & Carpenter, S. (2007). Closing the rural broadband gap: Promoting adoption of the Internet in rural America. *Telecommunications Policy, 31*(6–7), 359–73. https://doi.org/10.1016/j.telpol.2007.04.0

LaRose, R., Strover, S., Gregg, J. L., & Straubhaar, J. (2011). The impact of rural broadband development: Lessons from a natural field experiment. *Government Information Quarterly, 28*(1), 91–100. https://doi.org/10.1016/j.giq.2009.12.013

Martínez, Y., Mata, S., & Vega, M. (2021). *Diagnóstico sobre las brechas de inclusión digital en Chile*. Banco Interamericano de Desarrollo. http://dx.doi.org/10.18235/0003032

Martínez-Cantos, J. L. (2017). Digital skills gaps: A pending subject for gender digital inclusion in the European Union. *European Journal of Communication, 32*(5), 419–38. https://doi.org/10.1177/0267323117718464

Mascheroni, G. (2013). Parenting the mobile Internet in Italian households: Parents' and children's discourses. *Journal of Children and Media, 8*, 440–56. https://doi.org/10.1080/17482798.2013.830978

MINEDUC. (2020). *Educación en pandemia. Principales medidas del Ministerio de Educación en 2020*. Ministerio de Educación. https://www.mineduc.cl/wp-content/uploads/sites/19/2021/01/BalanceMineduc2020.pdf

Morris, M. G., & Venkatesh, V. (2006). Age differences in technology adoption decisions: Implications for a changing work force. *Personnel Psychology, 53*(2), 375–403. https://doi.org/10.1111/j.1744-6570.2000.tb00206.x

Napoli, P., & Obar, J. (2014). The emerging mobile Internet underclass: A critique of mobile Internet access. *The Information Society, 30*(5), 323–34. https://doi.org/10.1080/01972243.2014.944726

Nguyen, M. H., Gruber, J., Fuchs, J., Marler, W., Hunsaker, A., & Hargittai, E. (2020). Changes in digital communication during the COVID-19 global pandemic: Implications for digital inequality and future research. *Social Media + Society, 6*(3). https://doi.org/10.1177/2056305120948255

Núñez, C. G., Solis, C., & Soto, R. (2014). ¿Qué sucede en las comunidades cuando se cierra la escuela rural? Un análisis psicosocial de la política de cierre de las escuelas rurales en Chile. *Universitas Psychologica, 13*(2), 615–25. https://doi.tg/10.11144/Javeriana.UPSY13-2.qscc

Ortega Arias, M. D., & Cárcamo Vásquez, H. (2018). Relación familia-escuela en el contexto rural. *Miradas desde las familias. Educación, 27*(52), 98–18. https://revistas.pucp.edu.pe/index.php/educacion/article/view/19920

Pavez, I., & Correa, T. (2018). Resistance, opportunities and tensions. The role of children and young people in Internet adoption of isolated rural communities. In G. Mascheroni, C. Ponte, & A. Jorge (Eds.), *Digital parenting. The challenges for families in the Digital Age* (pp. 41–49). The International Clearinghouse on Children, Youth and Media. https://norden.diva-portal.org/smash/get/diva2:1265024/FULLTEXT02.pdf

Pavez, I., & Correa, T. (2020). "I don't use the internet": Exploring perceptions and practices among mobile-only and hybrid internet users. *International Journal of Communication, 14*, 2208–26.

Rojas, E., & Poveda, L. (2018). *Estado de la banda ancha en América Latina y el Caribe.* Comisión Económica para América Latina y el Caribe. CEPAL.

Rosen, R. (2021). Participatory research in and against time. *Qualitative Research.* https://doi.org/10.1177/14687941211041940

Salinas, A., & Sánchez, J. (2009). Digital inclusion in Chile: Internet in rural schools. *International Journal of Educational Development, 29*(6), 573–82. https://doi.org/10.1016/j.ijedudev.2009.04.003

Skerratt, S. (2008). The persistence of place: The importance of shared participation environments when deploying ICTs in rural areas. In G. Rusten & S. Skerrat (Eds.), *Information and communication technologies in rural society. Being rural in a digital age* (pp. 83–106). Routledge.

Sørensen, K. H. (1994). Adieu Adorno: The moral emancipation of consumers. 1–11. https://www.researchgate.net/publication/325628170_Technology_in_use_Two_essays_on_the_domestication_of_artefacts

Subsecretaria de Telecomunicaciones de Chile. (2017, Diciembre). Novena encuesta Accesos y Usos de Internet. https://www.subtel.gob.cl/wp-content/uploads/2018/07/Informe_Final_IX_Encuesta_Acceso_y_Usos_Internet_2017.pdf

Subsecretaria de Telecomunicaciones de Chile. (2021). Especial Análisis Tráfico Internet Marzo 2020–Junio 2021. https://www.subtel.gob.cl/wp-content/uploads/2021/09/PPT_Series_JUNIO_2021_V0.pdf

Townsend, L., Sathiaseelan, A., Fairhurst, G., & Wallace, C. (2013). Enhanced broadband access as a solution to the social and economic problems of the rural digital divide. *Local Economy, 28*(6), 580–95. https://doi.org/10.1177/0269094213496974

Trucco, D., & Palma, A. (2020). Políticas y estadísticas regionales: el contexto para la infancia y adolescencia en la era digital. In D. Trucco & A. Palma (Eds.), *Infancia y adolescencia en la era digital. Un informe comparativo de los estudios de Kids Online del Brasil, Chile, Costa Rica y el Uruguay* (pp. 23–40). CEPAL.

Ummel, D., & Achille, M. (2016). How not to let secrets out when conducting qualitative research with dyads. *Qualitative Health Research, 26*(6), 807–15. https://doi.org/10.1177/1049732315627427

United Nations (UN) (2020, August). *Policy brief: Education during COVID-19 and beyond.* United Nations. https://www.un.org/sites/un2.un.org/files/sg_policy_brief_covid-19_and_education_august_2020.pdf

United Nations Educational, Scientific and Cultural Organization (UNESCO) (2020, Version 2, July 2020). *COVID-19 response-remediation. Helping students catch up on lost learning, with a focus on closing equity gaps*. https://unesdoc.unesco.org/ark:/48223/pf0000373766/PDF/373766eng.pdf.multi

United Nations Educational, Scientific and Cultural Organization (UNESCO) (2021). *#HerEducationOurFuture. Keeping girls in the picture during and after the COVID-19 crisis*. https://unesdoc.unesco.org/ark:/48223/pf0000375707

United Nations International Children's Emergency Fund (UNICEF) (2021). *COVID-19 and school closures one year of education disruption*. https://data.unicef.org/resources/one-year-of-covid-19-and-school-closures/

Van Dijk, J. A. G. M. (2005). *The deepening divide: Inequality in the information society*. Sage.

# 10 Are rural smallholders ready for agricultural digitalization? Farmer (In)competencies and the political economy of access in digital agricultural extension and advisories in Northern Ghana

*Abdul-Rahim Abdulai*

## Introduction

The unfolding digitalization—application of digital innovations—of agricultural processes in rural smallholder systems in Africa is still hampered in many ways (African Union Commission & Organisation for Economic Co-operation and Development [OECD], 2021; Duncan et al., 2021; Kim et al., 2020; Tsan et al., 2019a). Notably, digital agricultural extension and advisories services (AEAS), where farmers are provided with information and knowledge through digital mediums, are fraught with wide-ranging barriers to adoption, use, and scalability (Barnett et al., 2019; Duncombe, 2016). According to Emeana et al. (2020), digital services have not been scalable due to low adoption and use by farmers. Other researchers have attributed the limited usage to skills problems, weak business practices of nascent start-ups, and poor enabling environments (Duncan et al., 2021; Mwangi & Kariuki, 2015; Shilomboleni, 2020; Trendov et al., 2019). While these studies highlight existing barriers to farmers' participation (Alabi, 2016; Ali, 2012; Salemink et al., 2017), there are inadequate discussions concerning access issues. Issues remain as to how they influence the ability of farmers to benefit from digitalization (Ribot & Peluso, 2003), and their underlying political-economic configurations. This discussion extends the literature by assessing the competencies of farmers and the resulting effects on digital access and power dynamics.

The experiences of smallholders in rural Northern Ghana are drawn on to extend ongoing discussions on this topic. Farmers in the area have been at the center of digital advisories, including radio-based climate and agronomic programs, SMS-based weather and price alerts, call-based agronomic and nutritional services, and Interactive Voice Response (IVR) systems for extension (Etwire et al., 2017; Hidrobo et al., 2021; Mohammed, 2019). This research draws on surveys and focus group discussions to assess farmers' perceptions and experiences with these services, which are helpful in understanding issues of innovations (Barrett & Rose, 2020; Bojang et al., 2020; Makate et al., 2017). Farmers' self-assessments of competencies in various digital tasks are employed to show limitations in literacies and how that may influence their ability to benefit from digital AEAS.

DOI: 10.4324/9781003282075-13

In the discussion it is demonstrated that beneficial access for rural farmers in Africa was still a far-reaching goal in digital advisories due mainly to digital incompetencies emanating from political and economic dynamics. It is argued there is a need for a re-focus of access in digitalization from one that concentrates on coverage statistics to one that focuses on farmers' ability and power to benefit from services.

## Digitalization and digital advisories

Digitalization of agriculture is common in Africa's smallholder systems (Food and Agriculture Organization of the United Nations & International Telecommunication Union, 2022). Digitalization describes the application of various digitally enabled tools and services within the agri-food space. It involves tools, software, and data and the services created from them to support any process in food and agriculture (Tsan et al., 2019a). In the literature, the issue of digitalization of agriculture and how it looks within smallholder systems is examined (see Abdulai, 2022a, 2022b, 2022c; Duncan et al., 2021). This chapter's essential elements of interest are digital advisories. Specifically, the Centre for Technical and Rural Development has classified digital services into five main areas: advisory and information services, market linkages, supply chain management services, financial access services, and macro agricultural intelligence services (see Tsan et al., 2019b).

One of the main areas of digitalization is digital advisories that refers to the application of any form of information communication technologies (ICT) in disseminating information and knowledge to farmers (Naika et al., 2021; Tsan et al., 2019b). Advisories may include deploying ICT tools such as radio and mobile phones to newer devices such as satellite imagery and mobile applications in agricultural advisories. The use of social media and other new media are central. Also of interest is the application of any form of digital tools or systems to facilitate delivery of agricultural knowledge and information to farmers. Digital advisories provide information on topics that include agronomic practices, weather, and market knowledge. Examples include market information systems and services, early warning tools for weather and climate advisory or pest and disease control, precision advisories, and livestock and farm management software. In the range of digitalized services, digital advisories may require the least technical ability from farmers as they are mainly at the receiving end of information and knowledge.

While many digital advisories exist, smallholders' primary use forms are radio programs, SMS alerts, call centers, IVRs, and social media. For advisories, farmers are at the receiving end of information and only require the skills to utilize such services (Emeana et al., 2020; McCampbell et al., 2021; Prabha, 2021). Accessing digitalization and digital advisories may require primarily language and digital skills, including reading and understanding English and using, at a minimum, a mobile phone to undertake varied tasks such as calling, SMS, IVR, and browsing.

## Political economy and access of digitalization for smallholders

Engagement by governments, non-government organizations, international donors, and private cooperation in agricultural digitalization raises political-economic questions. According to Birner et al. (2021), private and corporate actors, as well as agribusinesses, dominate the digital agriculture ecosystem. In a 2020 report by the World Bank on the state of disruptive innovations in Africa (Kim et al., 2020), it was noted that the International Finance Corporation (IFC), the Meltwater Foundation, Ahl Venture Partners, the Global System for Mobile Communications Association (GSMA), and the United States Agency for International Development (USAID) were the most prominent financiers (Kim et al., 2020, p. 23). These entities are a mix of venture capital and development partners with varying motives. In recent years, influential technology corporations have become interested in digitalization of which Google, IBM, Alibaba, and Microsoft are examples. Microsoft's FarmBeats typifies how the entities' activities may hold relevance for information and advisory services. FarmBeats aims to leverage artificial intelligence (AI) and data technologies to increase farmers' knowledge of their farms and environment to ensure precision in resource use. Besides the big corporations, digitalization is driven by the many private, mostly youth-led, technology and business start-ups spread across Africa (Tsan et al., 2019b). All of these entities have varying motives and interests.

Agricultural digitalization has political and economic interests and incursions. Elsewhere, I have argued that these entities promote and control the narratives around agricultural digitalization as they position it as transformative for farmers, including the ability to enhance access to agricultural information and knowledge (Abdulai, 2022a). Meanwhile, such propositions may mask the smallholders' realities, thereby hindering true transformation (Abdulai, 2022a; McCampbell et al., 2021). Importantly, these interests overlook the many challenges that still hinder farmers' participation and engagement in innovations, including the cost of services, poor Internet access, cultural barriers, and low capacities (Akullo & Mulumba, 2016; Beza et al., 2018; Francis & Addom, 2014; Salemink et al., 2017), as well as socio-economic issues such as gender, age, and education (Mwangi & Kariuki, 2015). These barriers may have consequences for inclusive access, which political economists and critical scholars have long pondered regarding emerging innovations (Basu & Chakraborty, 2011; Bronson & Knezevic, 2016; Carolan, 2018; Rotz et al., 2019).

To explore the concept of access, this discussion draws on definition as "the *ability* to derive benefits from things" (Ribot and Peluso, 2003, p. 150). It entails inherent or derived ability to exercise control and influence over resources. This definition includes control and "powers—embodied in and exercised through various mechanisms, processes, and social relations—that affect people's "ability to benefit from resources" (Peluso & Ribot, 2020, p. 300). These powers can result from the material, cultural, and political-economic activities configuring resource access (Peluso & Ribot, 2020, p. 300). The focus on *"ability 'to '"* shifts attention from entitlements to inherent or derived powers people may have to benefit from resources.

This power-based approach to access offers a nuanced view of people exerting control over how they benefit from resources (Myers & Hansen, 2020). Access is helpful in understanding exclusivity in digitalization, as defined by how structures may inhibit people from benefiting from a phenomenon (Hall, 2015). I draw on this concept to discuss farmers' inherent challenges hindering their ability to benefit from digitalization and advisories while exploring deep political-economic structures that underpin such issues. The issue of competencies, the skills, and abilities to undertake a task, are also important as prior studies have raised concerns about smallholder (digital) literacy (Birner et al., 2021; McCampbell et al., 2021). Also noted are farmers' self-perceptions of competencies in using a phone and undertaking specific digital tasks needed for direct engagement with digital advisories and extensions.

## Methodology

### Study context

This research is situated in the Savelugu Municipality in Northern Ghana (see Figure 10.1), which has Savelugu as the capital. The municipality is in the northern part of the region and has a total land area of about 2022.6 km². As of 2020, the municipal population was about 38,074, with 70% male and 30% female, and average household size of 5.8 persons (Ministry of Local Government & Rural Development, 2014; Savelugu Municipal Assembly, 2020).

Map prepared by Marie Puddister, Geography, Environment, and Geomatics Department, University of Guelph, Canada.

The municipality is mainly rural, and the majority of its population identifies agriculture as their primary occupation. Due to the agricultural potential in the area, the municipality has benefited from many digital agricultural activities. It was one of the first areas to pilot digital agricultural service provision through the Vodafone Farmers' Club projects, which provided smallholders with agronomic and nutritional information for an improved quality of life (Hidrobo et al., 2021). It is also a long-term beneficiary of Farm Radio services, which are currently running in some of the study's communities. The district also has the operational presence of Esoko Ghana, one of Ghana's largest digital service providers, offering farmers phone-based weather and market information to improve their decision-making. The presence of diverse digital AEAS makes this area an ideal case for understanding rural smallholders' challenges to digital adoption and utilization of digital advisories.

### Data collection and analysis

The research covered smallholders from five selected rural communities in the municipality: Libiga, Zaazi, Kodohizegu, Kpalung Yapala, Kulnaadaayili, Ying, and Duko (see Table 10.1; Figure 10.1). The communities were randomly selected from past and present Farm Radio International and Vodafone Farmers' Club projects.

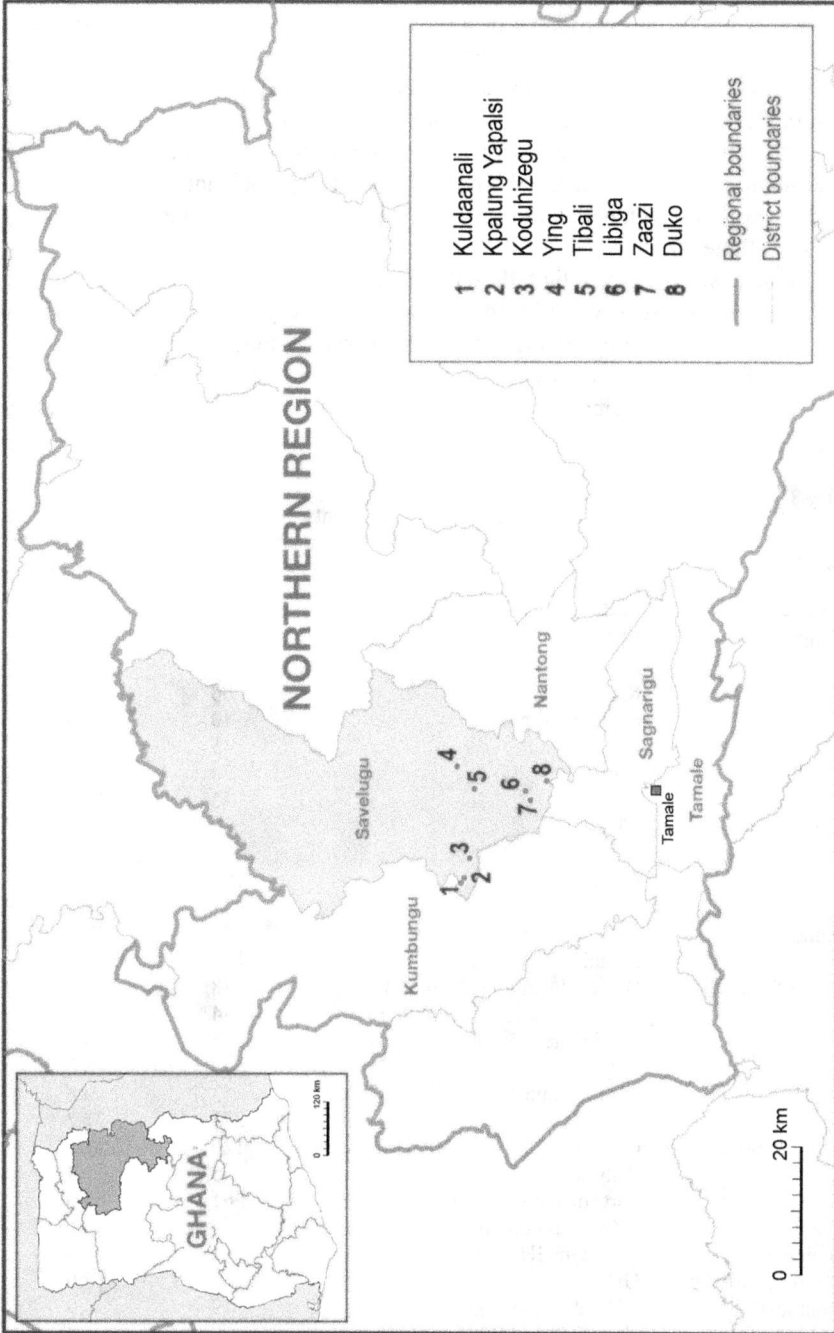

*Figure 10.1* Map of study areas.

The study's goal was to assess the perceptions and experiences of rural people regarding emerging digital agriculture. These communities presented an opportunity to view people who have or are currently experiencing digitalization and non-beneficiaries in such communities. The sample size determination assumed that much of the population in the district are farmers. At an estimated population of 150,000, a minimum number of 383 was assumed at a 95% confidence interval based on Krejcie and Morgan's (1970) sample size determination. A total of 386 rural smallholders were surveyed. The distribution of participants among the communities was based on population size and people's availability during data collection (see Table 10.1).

The data was collected digitally with the help of trained enumerators from the research area and supervised by the author. The surveys were conducted with a structured questionnaire, capturing farmers' characteristics, experiences, and perceptions. Enumerators were assigned specific communities and distributed to sections of villages. Participant selection was systematically made with a set pattern

*Table 10.1* Farmer and farm characteristics of respondents

| Variable | Options | Frequency N = 386 | Percentage |
|---|---|---|---|
| Communities | Libiga | 71 | 18.4 |
| | Zaazi | 55 | 14.2 |
| | Koduhizegu | 60 | 15.5 |
| | Kpalung Yapalsi | 40 | 10.4 |
| | Kulnaadaali | 45 | 11.7 |
| | Ying | 70 | 18.1 |
| | Duko | 45 | 11.7 |
| Age | 15–24 | 38 | 9.5 |
| | 25–40 | 206 | 53.5 |
| | 41–60 | 109 | 28.4 |
| | 61+ | 33 | 8.6 |
| Gender | Male | 224 | 42 |
| | Female | 162 | 58 |
| Level of education | Basic education (complete) | 28 | 7.3 |
| | Basic education (incomplete) | 44 | 11.4 |
| | Certificates / vocational | 3 | 0.8 |
| | High school | 33 | 8.5 |
| | Higher education | 3 | 0.8 |
| | No education | 275 | 71.2 |
| Farming model | Only feeding the family (subsistence) | 120 | 31.1 |
| | Only for sale (commercial) | 5 | 1.3 |
| | Part for family and part for sales (semi-commercial) | 261 | 67.6 |
| Income (self-reported per annum) | less than GHC 1,000 | 291 | 75.4 |
| | GHC 1,001–2,000 | 64 | 16.6 |
| | GHC 2,001–3,000 | 19 | 4.9 |
| | GHC 3,001–4,000 | 7 | 1.8 |
| | GHC 4,001–5,000 | 2 | 0.5 |
| | GHC 5,001+ | 3 | 0.8 |

of one household in every other third house. The survey participants included different genders, ages, farming models, income from farming, etc. (see Table 10.1).

The survey data was supplemented with three focus group discussions at Libiga, Kuldaanaali, and Kuduhizegu. One focus group had 12 participants, and the other two had six participants each. One was a mixed focus group, another was female only, and another was male only. Participants were selected based on their active participation in Farm Radio International and Esoko Ghana digital advisory programs and their availability during the discussion. The survey's data was analyzed through the Statistical Package for Social Services (SPSS). The background information of farmers was analyzed with descriptive statistics, while chi square was used for more inferential insights. For qualitative data, analysis was done using N'Vivo 12 involving a process recording, audio coding and drawing out of extracts to support the quantitative data findings (Mortelmans, 2019; Siccama & Penna, 2008). The outcomes of the recordings and the surveys form the basis for this discussion.

## Results

### *The digital profiles of rural smallholders*

The research examined selected digital characteristics of rural smallholders across four variables established through a preliminary literature review: mobile phone ownership, access to cellular Internet, awareness of digital services, and current or historical usage of digital services (see Table 10.2). The purpose was to understand rural and smallholder preparedness for digitalization.

Mobile phone ownership and usage was high among rural smallholders, with about 96% owning mobile devices. This high usage was unsurprising, as previous research revealed a growing mobile economy in rural Africa due partly to the availability of cheap phones (GSMA, 2017, 2020b). Common mobile usage offers opportunities for innovative digital services to reach smallholders. However, access to the Internet through cellular, the most feasible option available, was limited in these communities due to the high percentage of feature phone usage (85%), poor

*Table 10.2* Digital profile of respondents

| Variable | Options | Frequency | Percentage |
|---|---|---|---|
| Mobile phone ownership | No phone | 16 | 4.1 |
| | Feature phone | 330 | 85.5 |
| | Smart phone | 23 | 6 |
| | Both | 17 | 4.4 |
| Access to cellular Internet | Yes | 34 | 8.8 |
| | No | 352 | 91.2 |
| Awareness of digital services in the area | Yes | 282 | 73.1 |
| | No | 104 | 26.9 |
| Experience with digital services | Yes | 221 | 53.3 |
| | No | 165 | 42.7 |
| | **Total** | **386** | **100** |

networks, and cost of purchasing data. This finding confirms that Internet access is still uneven across Africa with rural areas highly disadvantaged (Evans, 2018; Tsan et al., 2019b). These, among other factors, like cost and (digital) illiteracy (Tables 10.2 and 10.3), also explain the low ownership of smartphones among the rural population. Smallholders' ability to participate in digitalization may be hampered without smartphones and cellular Internet access. However, most farmers were aware of existing services in their communities. About 53.3% had experience participating in digital AEAS, including climate advisory radio programs, SMS-based agronomic, weather, nutritional alerts, and call centers. Experiences with digital advisories were influenced by innovative solutions outside the scope of the Internet, including digital integrated radio programs, call centers, SMS, and IVR systems.

## What do rural smallholders perceive as the barriers to accessing digital AEAS?

The study also examined challenges currently undermining smallholders' ability to utilize and benefit from digital solutions. Figure 10.2 shows the issues rural people perceive as barriers to successful agricultural digitalization and utilization of digital advisory services.

The three commonly perceived challenges to using digital services were limited knowledge and capabilities to use digital services (83.16% of respondents), the expensive nature of certain services (77.98% of respondents), and poor network and Internet systems in rural communities (47.9% of respondents).

The primary limitation respondents noted as hindering their utilization was limited knowledge and capabilities in digitalization. Capabilities in this context was

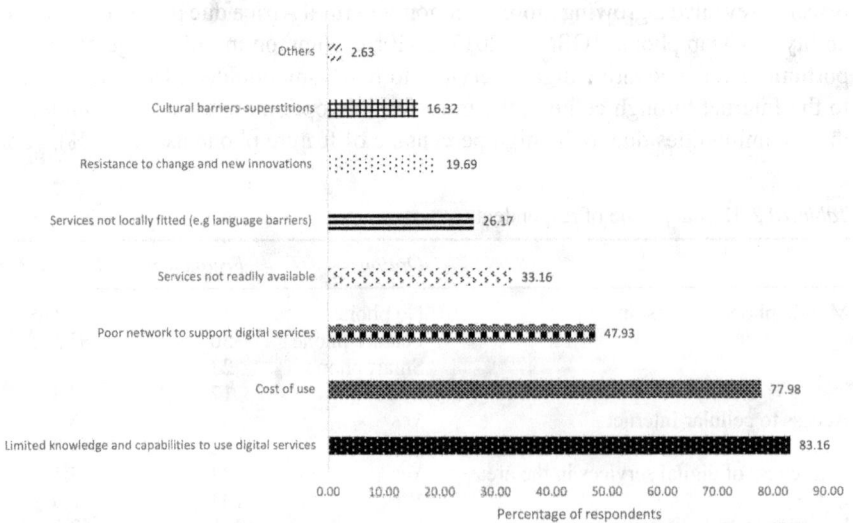

*Figure 10.2* Perceived challenges to rural digitalization.

narrowly defined in relation to competence to engage with services. Farmers were particular about their limitations in using digital services due to English and digital illiteracies. As evident in Table 10.1, over 70% of respondents had never been to school. This undermines any potential to use digital devices or understand digital messages delivered personally. As one respondent noted, "we cannot read and write. Even if I want to call, I have to tell someone to do it for me" (Damba, Kulnaadayili). Another respondent stated, "nobody in my household knows the book. It's hard for us to use these things when we can't understand what is sent" (Nyab Jahanfo, Libiga). The inability to fully utilize digital services was common among many respondents.

## How do smallholders assess their digital competencies?

Having established that literacy and competencies are the main challenges small-holders perceived as hindering their participation in digitalization, I used farmers' self-assessments to examine their abilities to undertake some basic digital tasks. In the study, 386 farmers were asked to assess their ability to undertake basic digital tasks needed for digital participation (Food and Agriculture Organization of the United Nations & International Telecommunication Union, 2022; GSMA, 2020a; Tsan et al., 2019a). The competencies were assessed using a five-point Likert scale ranging from very low level of (no) competence (1), where a farmer had no ability to independently undertake any form of the digital task to a high level of competence, to a very high level wherein the farmer could independently and confidently perform any form of the specific digital activity (5). Table 10.3 shows farmers' self-assessed competencies in nine basic tasks.

*able 10.3* Farmers' self-assessed digital competencies

| Digital skill area: Farmers' ability to: | Very low level of (no) competence (%) | Low level of competence (%) | The average level of competence (%) | Moderately high level of competence (%) | High level of competence (%) |
|---|---|---|---|---|---|
| Place calls | 23.3 | 4.1 | 2.1 | 30.8 | 39.8 |
| Receive and read SMS | 74.9 | 4.4 | 4.4 | 12.4 | 11.7 |
| Send SMS messages | 67.1 | 2.8 | 1.0 | 11.1 | 10.1 |
| Send audio messages | 75.9 | 4.9 | 3.1 | 7.0 | 10.6 |
| Follow IVR | 70.7 | 8.0 | 6.5 | 6.0 | 8.8 |
| Browse the Internet for information | 81.6 | 1.8 | 4.7 | 4.7 | 7.3 |
| Use social media | 82.4 | 2.3 | 4.1 | 3.9 | 7.3 |
| Use an independent phone apps for my activities | 82.4 | 3.4 | 2.3 | 4.9 | 7.0 |
| Use a computer | 87.6 | 2.1 | 3.6 | 2.3 | 4.4 |

*Jote.* 1 = Very low level of competence—no experience in the skill area; 2 = Low level of competence—limited xperience in the skill area; 3 = Average level of competence—some experience in the skill area; 4 = Moderately igh level of competence—good experience in the skill area; 5 = High level of competence—extensive experience 1 the skill area.

Most farmers assessed themselves to have very low (no) competencies across many digital skill areas, mostly due to illiteracy. Besides the ability to place and receive phone calls, farmers had low competence in all other digital skills. The high competence in a phone call is explainable by the ubiquity of the tool in Africa. Unlike other tasks, such as sending SMS, using social media, or browsing, which require formal education and English literacy, operating a phone can be done with little education. This situation is even made easier by the availability of feature phones (85.5% of farmers owned only feature phones), which can be operated with low skill levels and English literacy. This finding confirms that phone usage remains the most common and accessible for rural engagement in the digital space in Africa (Aker & Mbiti, 2010; Evans, 2018; GSMA, 2020b). However, smallholders generally assessed themselves as limited in higher skills in digital activities, including social media, browsing, app use, and computer operation. These digital activities currently require higher skills and literacy in English. The inability to undertake higher digital tasks means farmers are unlikely to participate in advisories provided through those mediums.

Having established that the most accessible, in terms of availability and competencies, tool for farmers was the mobile phone, farmers were assessed as to how they perceived their general abilities to use a mobile phone for farming activities and the variations across genders and ages. Farmers were asked to self-assess their ability to use a phone on a scale of 1–5. Low competence is 1, the only ability is answering phone calls, whereas 5 is an excellent usage of all forms without external help (see Table 10.4).

Most farmers (60.8%) perceived themselves as having a very low level of competence or experience using the phone for farming activities. The chi-square analysis showed a significant relationship between age and gender and farmers' perceived ability. Male farmers and farmers aged 15–24 were more likely to have a high self-assessment of competencies in phone usage. Female farmers and those

*Table 10.4* Farmers' self-assessment on the ability to use mobile phone

| Variable | Options | Very low level of (no) competence (%) | Low level of competence (%) | Average level of competence (%) | Moderately high level of competence (%) | High level of competence (%) | Chi-square test |
|---|---|---|---|---|---|---|---|
| Age | 15–24 | 37.8 | 10.8 | 16.2 | 24.3 | 10.8 | $X^2 = 25.66$ |
|  | 25–40 | 58.7 | 9.2 | 14.6 | 10.7 | 6.8 | $P = 0.012$ |
|  | 41–60 | 67.9 | 3.7 | 18.3 | 7.3 | 3.7 | $V = 0.149$ |
|  | 61+ | 78.8 | 3.0 | 15.2 | 0.0 | 3.0 |  |
| Gender | Male | 48.7 | 7.6 | 19.6 | 14.3 | 9.8 | $X^2 = 40.18$ |
|  | Female | 77.2 | 7.4 | 10.5 | 4.3 | 0.6 | $P < 0.001$ |
|  |  |  |  |  |  |  | $V = 0.323$ |
| Total |  | 60.8 | 7.3 | 15.8 | 10.1 | 6.0 |  |

*Note.* 1 = Very low level of competence—no experience in the skill area; 2 = Low level of competence—little experience in the skill area; 3 = Average level of competence—some experience in the skill area; 4 = Moderately high level of competence—good experience in the skill area; 5 = High level of competence—extensive experience in the skill area above 60 years of age were more likely to have low competencies in phone usage. These findings are unsurprising since males have more opportunities for education than females in rural Africa. Likewise, access to education has been growing steadily in these areas thereby presenting younger populations with educational opportunities to gain competencies that older people did not have growing up. These findings show that there still are unequal barriers in access to digitalization as male farmers have an advantage regarding innovations over females and the younger farmers over older ones. Ultimately, the finding that most farmers do not perceive themselves as competent in phone usage reveals that ownership of digital tools does not translate into usage or even the ability to use them. Many farmers had phones they could not sufficiently use to participate in digitalization without the help of third parties. Hence, farmers' phone ownership and competencies may have implications for access.

### Discussion: do smallholders have access and power in digital advisories?

The results highlight that the typical smallholders' perceived limited knowledge and capabilities as the main barrier to utilizing digitalization and advisories. Many smallholders could not use the phone for calling at best. Other digital skills such as browsing, using social media, operating a computer, and following IVR were outside their capabilities. Female and older farmers were more likely to have this limitation. The low digital competencies restrict engagement and participation in higher-level digital AEAS delivered through SMS, IVR, social media, and online communities (Fabregas et al., 2019; Osei et al., 2018; Tamene et al., 2020). These findings further emphasize the low digital literacy and digital divides that exist within African and smallholder systems (Food and Agriculture Organization of the United Nations & International Telecommunication Union, 2022; McCampbell et al., 2021).

Within the ongoing digital divide are underlying concerns of access, the ability to benefit from things (Ribot & Peluso, 2003, p. 153), which political economy scholars have long explored in resource use (Carolan, 2018). For example, the emergence of digitalization and the newer "things" such as data and code is noted to raise access questions, including who can use, benefit, and exercise sovereignty over them (Carolan, 2018). Farmers' experiences unravel novel ways of understanding their ability to benefit from digital advisories. The limited ownership of digital tools hampers access as people may not be able to utilize and benefit from what they do not have. Likewise, as only about 85.5% of the farmers had only feature phones, many are excluded from higher forms of advisories carried through apps, social media, and other smart information platforms. The inability to undertake simple digital tasks, including sending SMS, following IVR, or browsing the Internet also severely hinders digital benefits (McCampbell et al., 2021). This lack of literacy (in English) also means that farmers may not benefit from digitalization even if they have the tools and skills because they cannot understand the information delivered.

These limitations of farmers open critical conversations about access in digitalization. So, uncritically pointing to the ubiquity of mobile phones or the growing registration of farmers in services as indications of readiness for digitalization (Fabregas et al., 2019; Osei et al., 2018) masks critical access concerns. While such optimism may be justifiable, the mere ownership of a mobile phone does not always translate into the ability to benefit from digital advisories. Likewise, the presence of services or even 'participation' by farmers doesn't guarantee benefits.

Farmers may still lack power as their conditions undermine control over how to benefit from services. Without the prerequisite skills and competencies, a vital access element, the power (control) over benefiting, is missing. So, linked to access concerns is how existing and emerging structures reinforce or create power imbalances and unfair distribution of resource benefits in food systems (Anderson & Leach, 2019; Anderson et al., 2019). This is an avenue of the political economy of digital access. I have argued that digitalization may entrench old structures while creating newer classes within smallholder systems (Abdulai, 2022a). This argument is built on to emphasize that the (digital) literacy challenges of smallholders are outcomes of the neo-liberal-created educational divides (Bomah, 2014).

As developing economies are growingly absorbed by capitalist ideals, governments have, not without protest, largely abdicated educational interventions that could enhance rural literacies (Bomah, 2014; Geo-Jaja, 2004; Patnaik, 1997). Restricted access to teachers, long distances to schools, and poor education infrastructures, all of which are influenced by political-economic interactions, are a few of the challenges that short change rural literacy opportunities (Anlimachie & Avoada, 2020; Senadza, 2012). For example, Anlimachie and Avoada (2020) found in Ghana rural areas trailed urban centers in all measured indicators of access, including school enrolment, access to improved schools and sanitation, percentage of trained teachers, and gender parity. Partly these challenges arise because rural communities do not have the economic ability to invest in literacy, thereby leaving smallholders neglected by the public and private political and economic interests that shape education. This situation has left rural folks with high (digital) illiteracies further disempowering and subjecting them to weak positions in digitalization. By extension, farmers are treated to tokenism of digitalization as participation (or lack thereof) is effectively determined by the external donors and private actors that dominate the digitalization ecosystem (Birner et al., 2021).

Ultimately, this cycle will confine marginalized smallholders to a position of powerlessness to effectively contribute to the direction of digital AEAS and digitalization futures. This pattern will create uneven benefits, where, say, digitally illiterate smallholders are left to the mercy of the few elite (educated) farmers, who may capture any potential benefits due to their digital advantages. Smallholders are disempowered by their conditions, which, in turn, undermine meaningful participation and benefits of digital innovation as emerging inequalities (Abdulai, 2022a; Carolan, 2020; Duncan et al., 2021; Klerkx et al., 2019; Rose & Chilvers, 2018), therefore, leave illiterate farmers, women, and older farmers, among others, at the margins. This critical lack of access opens spaces for interrogating digital divides within and between the wide diversity of rural farmers.

## Conclusion and the way forward for inclusive digital advisories

This discussion confirms that many smallholders do not have formal literacies and lack competencies in basic digital tasks, which ultimately leave them with low levels of (digital) literacies (Food and Agriculture Organization of the United Nations & International Telecommunication Union, 2022; McCampbell et al., 2021). The findings raise critical political-economic questions for smallholder digitalization as the issues of access shape smallholders' power in the digital agriculture space. This underscores a need to reshape the conversations around access to digitalization and digital AEAS. We must ask questions such as: Do farmers have beneficial access to digital advisories?; Who has the power to benefit from digital services?; and What are the sources of such powers? Understanding farmers' conditions, preparedness, and, more importantly, their power to benefit from digital innovation is essential for inclusive digital advisories and digitalization futures.

The insights on access and power reflected in this chapter have implications for farmers' engagement with AEAS in the post-COVID-19 age. The onset of the COVID-19 pandemic triggered attention to and broadened the use of digital tools for agricultural information sharing, delivering AEAS to farmers, and keeping two-way communication between farmers and extension services (Baffoe-Bonnie et al., 2021; Davis, 2020; Food and Agriculture Organization of the United Nations, 2021). However, this discussion has shown that embedded in the growing attention are structural limitations and inequities that can hinder meaningful participation by farmers in the post-COVID-19 era. The analysis provides a reality check to digital advisories to not lose sight of lingering issues of access and power imbalances that can derail any potential progress in digitalization futures. Thoughtful interventions are needed if farmers would become active participants in digital AEAS in the post-COVID-19 extension spaces.

Measures that enhance smallholders' digital experiences, their access, and power within digitalization is crucial. Such measures must reduce challenges and put power in the hands of farmers. It is proposed that there are two strategies toward these goals. *In the short-term*, stakeholders must continue to work on designing a local language digital AEAS based on the tools that require no direct formal or English literacy from farmers, such as radio and phone calls. Not all digital devices are suitable for smallholders due to their limited literacies. Hence, it is essential to create services that primarily utilize tools smallholders own and can use. Farmers would be better placed to utilize and benefit from advisories by making this attempt.

*In the medium- to long term*, governments and emerging private extension entities should work toward developing innovative digital literacy programs for smallholder farmers. Non-formal education schemes and farmers' training programs could provide the essential platforms for achieving this goal. In particular, as the

Food and Agriculture Organization of the United Nations (FAO-UN) has outlined (Food and Agriculture Organization of the United Nations, 2021, p. 3), smallholders can be empowered toward digital literacies by creating farmer participatory education and training courses on digital literacy; developing the necessary infrastructures and facilities, such as farmer digital vocational and digital training school; and developing guidelines and training curricula that enhance the assessment of digital literacies based on their current needs. With such encompassing strategies, farmers would be bought to speed on the essential digital competencies needed to thrive in digitalization.

## Funding

This research received funding and support from the International Development Research Centre of Canada.

## References

Abdulai, A.-R. (2022a). A new Green Revolution (GR) or Neoliberal entrenchment in agri-food systems? Exploring narratives around digital agriculture (DA), food systems, and development in Sub-Sahara Africa. *Journal of Development Studies, 0*(0), 1–17. https://doi.org/10.1080/00220388.2022.2032673

Abdulai, A.-R. (2022b). *The digitalization of agriculture and the (un) changing dynamics of rural smallholder farming systems in Ghana, Sub-Sahara Africa*. [Doctoral dissertation. University of Guelph.]

Abdulai A.-R. (2022c). Toward digitalization futures in smallholder farming systems in Sub-Sahara Africa: A social practice proposal. *Frontiers in Sustainable Food Systems, 6*, 866331. https://doi.org/10.3389/fsufs.2022.866331

African Union Commission & Organisation for Economic Co-operation and Development (2021). *Africa's development dynamics 2020: Digital transformation for quality jobs*. Organisation for Economic Co-operation and Development. OECD Publishing. https://doi.org/10.1787/0a5c9314-en

Aker, J. C., & Mbiti, I. M. (2010). Mobile phones and economic development in Africa. *Journal of Economic Perspectives, 24*(3), 207–32. https://doi.org/10/dr58kc

Akullo, W. N., & Mulumba, O. (2016). Making ICTs relevant to rural farmers in Uganda: A case of Kamuli district. *Revista Brasileira de Planejamento e Desenvolvimento, 5*(2), 291–304. https://doi.org/10/ggt3sv

Alabi, O. (2016). Adoption of information and communication technologies (ICTs) by agricultural science and extension teachers in Abuja, Nigeria. *Journal of Agricultural Education, 57*(1), 137–49. https://doi.org/10/ggt3sp

Ali, J. (2012). Factors affecting the adoption of information and communication technologies (ICTs) for farming decisions. *Journal of Agricultural & Food Information, 13*(1), 78–96. https://doi.org/10/gg93c7

Anderson, M., & Leach, M. (2019). Transforming food systems: The potential of engaged political economy. *IDS Bulletin, 50*(2), Article no. 2. https://doi.org/10.19088/1968-2019.123

Anderson, M., Nisbett, N., Clément, C., & Harris, J. (2019). Introduction: Valuing different perspectives on power in the food system. *IDS Bulletin, 50*(2), Article no. 2. https://doi.org/10.19088/1968-2019.114

Anlimachie, M. A., & Avoada, C. (2020). Socio-economic impact of closing the rural-urban gap in pre-tertiary education in Ghana: Context and strategies. *International Journal of Educational Development, 77,* 102236. https://doi.org/10.1016/j.ijedudev.2020.102236

Baffoe-Bonnie, A., Martin, D. T., & Mrema, F. (2021). Agricultural extension and advisory services strategies during COVID-19 lockdown. *Agricultural & Environmental Letters, 6*(4), e20056. https://doi.org/10.1002/ael2.20056

Barnett, I., Faith, B., Mitchell, B., & Sefa-Nyarko, C. (2019). *External evaluation of mobile phone technology-based nutrition and agriculture advisory services in Africa: Mobile phones, agriculture, and nutrition in Ghana: Qualitative follow-up study report.* Institute of Development Studies, UK.

Barrett, H., & Rose, D. C. (2020). Perceptions of the fourth agricultural revolution: 'What's in', 'what's out', and what consequences are anticipated? *Sociologia Ruralis, 62*(2), 162–89. https://doi.org/10.1111/soru.12324

Basu, P., & Chakraborty, J. (2011). New technologies, old divides: Linking internet access to social and locational characteristics of US farms. *GeoJournal, 76*(5), 469–81. https://doi.org/10/dgpd2x

Beza, E., Reidsma, P., Poortvliet, P. M., Belay, M. M., Bijen, B. S., & Kooistra, L. (2018). Exploring 'farmers' intentions to adopt mobile Short Message Service (SMS) for citizen science in agriculture. *Computers and Electronics in Agriculture, 151,* 295–310. https://doi.org/10/gd2jwh

Birner, R., Daum, T., & Pray, C. (2021). Who drives the digital revolution in agriculture? A review of supply-side trends, players and challenges. *Applied Economic Perspectives and Policy, n/a.* https://doi.org/10.1002/aepp.13145

Bojang, F., Traore, S., Togola, A., & Diallo, Y. (2020). Farmers perceptions about climate change, management practice and their on-farm adoption strategies at rice fields in Sapu and Kuntaur of the Gambia, West Africa. *American Journal of Climate Change, 9*(1), 1–10. https://doi.org/10.4236/ajcc.2020.91001

Bomah, K. B. (2014). Digital divide: Effects on education development in Africa. *Conference in LYIT Dept. of Computing: Technical Writing Presentation.* Letterkenny Institute of Technology, Ireland.

Bronson, K., & Knezevic, I. (2016). Food studies scholars can no longer ignore the rise of big data. *Canadian Food Studies/La Revue Canadienne Des Études Sur l'alimentation, 3*(1), 9. https://doi.org/10/ggt3s3

Carolan, M. (2018). 'Smart' farming techniques as political ontology: Access, sovereignty and the performance of Neoliberal and not-so-Neoliberal worlds. *Sociologia Ruralis, 58*(4), 745–64. https://doi.org/10.1111/soru.12202

Carolan, M. S. (2020). Automated agri-food futures: Robotics, labor and the distributive politics of digital agriculture. *Journal of Peasant Studies, 47*(1), 184–207. https://doi.org/10.1080/03066150.2019.1584189

Davis, K. (2020). Agricultural education and extension in a time of COVID. *Journal of Agricultural Education and Extension, 26*(3), 237–38. https://doi.org/10.1080/1389224X.2020.1764224

Duncan, E., Abdulai, A.-R., & Fraser, E. D. G. (2021). Modernizing agriculture through digital technologies: Prospects and challenges. In *Handbook on the human impact of agriculture.* Elgar online. https://www.elgaronline.com/view/edcoll/9781839101731/9781839101731.00018.xml

Duncombe, R. (2016). Mobile phones for agricultural and rural development: A literature review and suggestions for future research. *European Journal of Development Research, 28*(2), 213–35. https://doi.org/10/f8gc3m

Emeana, E. M., Trenchard, L., & Dehnen-Schmutz, K. (2020). The revolution of mobile phone-enabled services for agricultural development (m-Agri Services) in Africa: The challenges for sustainability. *Sustainability, 12*(2), 485. https://doi.org/10.3390/su12020485

Etwire, P. M., Buah, S., Ouédraogo, M., Zougmoré, R., Partey, S. T., Martey, E., Dayamba, S. D., & Bayala, J. (2017). An assessment of mobile phone-based dissemination of weather and market information in the Upper West Region of Ghana. *Agriculture & Food Security, 6*(1), 1–9. https://doi.org/10.1186/s40066-016-0088-y

Evans, O. (2018). Digital agriculture: Mobile phones, Internet and agricultural development in Africa. *Actual Problems of Economics, 7–8*(205–206), 76–90. https://EconPapers.repec.org/RePEc:pra:mprapa:90359

Fabregas, R., Kremer, M., & Schilbach, F. (2019). Realizing the potential of digital development: The case of agricultural advice. *Science, 366*(6471). https://doi.org/10.1126/science.aay3038

Food and Agriculture Organization of the United Nations. (2021). *Empowering smallholder farmers to access digital agricultural extension and advisory services* (p. 8). Food and Agriculture Oganization of the United Nations.

Food and Agriculture Organization of the United Nations & International Telecommunication Union. (2022). *Status of digital agriculture in 47 sub-Saharan African countries*. Food and Agriculture Organization of the United Nations, Rome, Italy.

Francis, J., & Addom, B. J. (2014). Modern ICTs and rural extension: Have we reached the tipping point. *Rural21 Publication*. https://www.rural21.com/fileadmin/downloads/2014/en-01/rural2014_01-S22-24.pdf

Geo-Jaja, M. A. (2004). Decentralisation and privatisation of education in Africa: Which option for Nigeria? *International Review of Education, 50*(3), 307–23. https://doi.org/10.1007/s11159-004-2625-3

GSMA. (2017). *The mobile economy sub-Saharan Africa 2017*. GSMA.

GSMA. (2020a). *Digital agriculture maps 2020 state of the sector in low and middle-income countries*. GSMA. https://www.gsma.com/r/wp-content/uploads/2020/09/GSMA-Agritech-Digital-Agriculture-Maps.pdf

GSMA. (2020b). *The mobile economy Sub-Saharan Africa 2020*. GSMA. https://www.gsma.com/mobileeconomy/wp-content/uploads/2020/09/GSMA_MobileEconomy2020_SSA_Eng.pdf

Hall, D. (2015). Land's essentiality and land governance. In K. Pistor & O. De Schutter (Eds.), *Governing access to essential resources* (pp. 49–66). Columbia University Press. https://www.jstor.org/stable/10.7312/pist17278.4

Hidrobo, M., Palloni, G., Aker, J., Gilligan, D., & Ledlie, N. (2021). Paying for digital information: Assessing 'farmers' willingness to pay for a digital agriculture and nutrition service in Ghana. *Economic Development and Cultural Change*, 713974. https://doi.org/10/gnc9h4

Kim, J., Shah, P., Gaskell, J. C., Prasann, A., & Luthra, A. (2020). *Scaling up disruptive agricultural technologies in Africa*. World Bank. https://doi.org/10.1596/978-1-4648-1522-5

Klerkx, L., Jakku, E., & Labarthe, P. (2019). A review of social science on digital agriculture, smart farming and agriculture 4.0: New contributions and a future research agenda. *NJAS-Wageningen Journal of Life Sciences, 90–91*(1), 1–16. https://doi.org/10/gg7vs6

Krejcie, R. V., & Morgan, D. W. (1970). Determining sample size for research activities. *Educational and Psychological Measurement, 30*(3), 607–10. https://doi.org/10/ggb4r2

Makate, C., Makate, M., & Mango, N. (2017). Smallholder farmers' perceptions on climate change and the use of sustainable agricultural practices in the Chinyanja Triangle, Southern Africa. *Social Sciences, 6*(1), 30. https://doi.org/10/ggt349

McCampbell, M., Adewopo, J., Klerkx, L., & Leeuwis, C. (2021). Are farmers ready to use phone-based digital tools for agronomic advice? Ex-ante user readiness assessment using the case of Rwandan banana farmers. *Journal of Agricultural Education and Extension, 0*(0), 1–23. https://doi.org/10.1080/1389224X.2021.1984955

Ministry of Local Government & Rural Development. (2014). *Sector medium term development plan (2014–2017). Building vibrant district assemblies for economic growth and development.* https://new-ndpc-static1.s3.amazonaws.com/CACHES/PUBLICATIONS/2016/07/15/Ministry+of+Local+Govrnment&Rural+Development.pdf

Mohammed, H. N. (2019). Digital tools and smallholder agriculture: The role of ICT-enabled extension services in rural smallholder farming in northern Ghana. [Master's thesis. International Institute of Social Studies.]

Mortelmans, D. (2019). Analyzing qualitative data using NVivo. In H. Van den Bulk, M. Puppis, K. Donders, & L. Van Audenhove (Eds.), *The Palgrave handbook of methods for media policy research* (pp. 435–50). Springer.

Mwangi, M., & Kariuki, S. (2015). Factors determining adoption of new agricultural technology by smallholder farmers in developing countries. *Journal of Economics and Sustainable Development, 6*(5), 208–26. https://core.ac.uk/download/pdf/234646919.pdf

Myers, R., & Hansen, C. P. (2020). Revisiting a theory of access: A review. *Society & Natural Resources, 33*(2), 146–66. https://doi.org/10.1080/08941920.2018.1560522

Naika, M. B., Kudari, M., Devi, M. S., Sadhu, D. S., & Sunagar, S. (2021). Digital extension service: Quick way to deliver agricultural information to the farmers. In C. Galanakis (Ed.), *Food technology disruptions* (pp. 285–23). Academic Press Ltd.

Osei, C., Yeboah, A., Arthur, F., Agbedanu, E., & Chidiac, S. (2018). *Digital platforms for agro-advisory and business service delivery: Lessons from scaling-up of AgroTech in Ghana.* Agrotech Policy Paper, 1–15. https://idl-bnc-idrc.dspacedirect.org/bitstream/handle/10625/57113/IDL-57113.pdf?sequence=2&isAllowed=y

Patnaik, U. (1997). Political economy of state intervention in food economy. *Economic and Political Weekly, 32*(20/21), 1105–12. https://www.jstor.org/stable/4405416

Peluso, N. L., & Ribot, J. (2020). Postscript: A theory of access revisited. *Society & Natural Resources, 33*(2), 300–06. https://doi.org/10.1080/08941920.2019.1709929

Prabha, D. (2021). An analytical study of mobile agro advisories among the farmers. *Medicon Agriculture & Environmental Sciences, 1*(1), 26–31. https://themedicon.com/MCAES-21-003.pdf

Ribot, J. C., & Peluso, N. L. (2003). A theory of access. *Rural Sociology, 68*(2), 153–81. https://doi.org/10.1111/j.1549-0831.2003.tb00133.x

Rose, D. C., & Chilvers, J. (2018). Agriculture 4.0: Broadening responsible innovation in an era of smart farming. *Frontiers in Sustainable Food Systems, 2.* https://doi.org/10.3389/fsufs.2018.00087

Rotz, S., Gravely, E., Mosby, I., Duncan, E., Finnis, E., Horgan, M., LeBlanc, J., Martin, R., Neufeld, H. T., Nixon, A., Pant, L., Shalla, V., & Fraser, E. (2019). Automated pastures and the digital divide: How agricultural technologies are shaping labour and rural communities. *Journal of Rural Studies, 68,* 112–22. https://doi.org/10/ggkjzk

Salemink, K., Strijker, D., & Bosworth, G. (2017). Rural development in the digital age: A systematic literature review on unequal ICT availability, adoption, and use in rural areas. *Journal of Rural Studies, 54,* 360–71. https://doi.org/10/gbz54b

Savelugu Municipal Assembly. (2020). Composite budget for 2020–2023 programme based budget estimates for 2020 Savelugu Municipal. Ministry of Finance, Ghana. https://www.mofep.gov.gh/sites/default/files/composite-budget/2020/NR/Savelugu.pdf

Senadza, B. (2012). Education inequality in Ghana: Gender and spatial dimensions. *Journal of Economic Studies, 39*(6), 724–39. https://doi.org/10.1108/01443581211274647

Shilomboleni, H., Pelletier, B., & Gebru, B. (2020). ICT4 Scale in smallholder agriculture: Contributions and challenges. *Information Technologies & International Development, 16,* 47–65. https://itidjournal.org/index.php/itid/article/download/1743/1743-5705-2-PB.pdf

Siccama, C. J., & Penna, S. (2008). Enhancing validity of a qualitative dissertation research study by using N'Vivo. *Qualitative Research Journal, 8*(2), 91–103. https://doi.org/10.3316/QRJ0802091

Tamene, L., Abera, W., & Erkossa, T. (2020). *Digital solutions to transform agriculture: Lessons and experiences in Ethiopia.* https://hdl.handle.net/10568/111778

Trendov, N. M., Varas, S., Zeng, M. (2019). *Digital technologies in agriculture and rural areas—Status report.* Food and Agriculture Organization of the United Nations. Licence: cc by-nc-sa3.0 igo. http://www.fao.org/3/ca4985en/ca4985en.pdf

Tsan, M., Totapally S., Hailu M., & Addom, B. K. (2019a). *The digitalisation of African agriculture report 2018–2019: Executive summary.* CTA/Dalberg Advisers. https://cgspace.cgiar.org/bitstream/handle/10568/103198/Executive%20Summary%20V4.5%20ONLINE.pdf?sequence=1&isAllowed=y

Tsan, M., Totapally, S., Hailu, M., & Addom, B. (2019b). *The digitalisation of African agriculture report, 2018–2019.* CTA/Dalberg Advisers. https://hdl.handle.net/10568/101498

# 11 Redefining the use of ICTs as tools for empowering farmers and rural communities

## A Malawian case study

*Fally Masambuka-Kanchewa*
*and Mary T. Rodriguez*

## Introduction

Improving access to information among rural communities is critical for promoting rural development (Aker et al., 2016). Experts have identified different practices that can positively impact agricultural output if implemented (e.g., climate responsive practices). Persuasive communication has been widely used as a tool for promoting behavior change when implementing rural development programs and projects (Stratigea, 2011). The presence of Information Communication Technologies (ICTs) has provided essential tools for bridging the information gap (Oyeyinka et al., 2014). In rural development efforts, ICTs are used for provision of extension services, such as commodity prices, new agricultural techniques, and weather forecasting (Hopestone, 2014). There is an increased focus on promoting the use of ICTs to ensure that rural communities have access to information on services such as commodity prices, and employment opportunities (Asunka, 2016). Despite increased deployment of ICT infrastructure in rural communities, there is limited evidence of the contribution of ICTs to their economic development (Asunka, 2016; Pigg & Crank, 2005). Asunka (2016) contends that if the introduction and use of ICTs in rural communities is to be successful, its development and deployment must be following a needs-based approach as opposed to the use of a one-size-fits-all approach. Yet, agricultural extension agents and rural development practitioners using the one-size-fits-all model, as well as ICTs, often fail to engage farmers and communities as creators of innovative practices and owners of their own development. Their primary focus is often one-way communication, delivering information to farmers and rural communities (Masambuka-Kanchewa et al., 2020a). The use of ICTs, in this way, fails to capture their potential as tools to promote innovativeness and to contribute toward empowerment of farmers and rural communities. Implementation of such approaches require that rural development organizations and experts are aware of the limitations of ICTs and the factors that may contribute toward their effective use.

Implementation of the stay-at-home measures and lockdowns worldwide due to the COVID-19 pandemic revealed the extent to which ICTs are underutilized in

DOI: 10.4324/9781003282075-14

agriculture and rural development (Dharmawan et al., 2021). The pandemic added to the already complex and challenging issues facing rural communities. ICTs during that time failed to highlight farmers' innovative approaches in mitigating these challenges and instead were used to merely disseminate information without their input. Farmers should be seen as innovative problem-solvers capable of addressing complex challenges. Communication for development (C4D) should amplify their voices. This chapter's discussion will examine the current use of ICTs in C4D and propose ways to co-create and co-design ICT messaging to strengthen its potential.

### Study context: Malawi

C4D has been implemented in many development projects in many countries on the African continent including Malawi. According to the World Bank Group, Malawi is considered one of the poorest countries in the world. Its economy heavily relies on the agricultural sector, employing nearly 80% of its estimated population of 18.6 million (World Bank, 2018. With most of the country's agricultural production relying on subsistence rain-fed agriculture, Malawi risks increased susceptibility to changes in weather and high food insecurity for its people.

### Theoretical framework

This study was informed by two theories, namely, the diffusion of innovations and media richness theory. The two theories were developed by Westerners with an ethnocentric lens. However, both theories have been explored in various contexts over the past several decades. Our research proposes to strengthen these theories by investigating them together, examining them in a non-Western context, and by pushing the boundaries of the theories by including farmers as active participants in the communication and behavior change processes.

The diffusion of innovations theory describes the processes by which an innovation is communicated through a social system and the factors that affect its adoption or lack thereof (Rogers, 1995). According to Rogers (1976), communication plays a crucial role in ensuring that innovations are diffused into the social system. Rogers indicated that people go through various decision-making stages as they decide to adopt or not adopt a given innovation, namely, "knowledge, persuasion, decision, implementation, and confirmation" (Rogers, 2004, p. 20). During the various stages, specific communication channels are used to obtain information about the innovation and its use to potential adopters. This occurs due to mass communication or interpersonal communication, depending on the decision-making stage and what type of information is needed (Rogers, 2003). The importance of communication in this process cannot be understated. Communication is an essential component for rural development (Stratigea, 2011).

Provision of information on improved technologies is considered as the solution to rural development (Mtega, 2012). In most cases, rural development experts choose communication channels that reach a wide audience within a short period of time, such as mass media without considering the decision stage of the potential adopters (Masambuka-Kanchewa et al., 2020b). This method of communication does not allow for the nuanced information necessary for potential adopters especially during the persuasion decision-making stage, nor does it allow for bidirectional exchange of information.

Rural development service providers have used ICTs as a tool of mass communication to create awareness of new and improved technologies (Food and Agricultural Organization [FAO], 2003). These service providers most often provide scientifically proven information from experts, reducing farmers and rural communities to being passive recipients of information (Kumar & Singh, 2012; Masambuka-Kanchewa et al., 2020b). Farmers and rural community members have valuable, first-hand knowledge and experience that are useful and crucial for promoting sustainable rural development. However, most of this knowledge and experience is left untapped and unshared when information is disseminated via ICTs (Masambuka-Kanchewa et al., 2020a).

Rural development organizations and experts must make choices as to how they communicate their messaging to their stakeholders. Several factors influence their choice of media channels that are used and the content that is shared. Media Richness Theory helps us to understand these decisions. First, the theory guides decisions based on the 'richness' of the information defined as "the ability of information to change understanding within an allotted amount of time" (Daft & Lengel, 1986, p. 560). Therefore, communication messages that can reduce / clarify ambiguous issues in a timely manner are considered 'rich.' On the other hand, communication that requires a long time for the recipient to understand the message or to overcome different perspectives is considered lower in richness. The choice of media is driven, therefore, by the richness of the information.

According to this theory, media channels differ in their capacity to deliver information (Song & Wang, 2011). The type of message and expectations of its impact influences the choice of channels used when delivering the information (Daft & Lengel, 1986). The choice of communication channels or media is influenced by the individual's perception of ambiguity associated with a particular task (Fulk & Boyd, 1991). Most organizations' choice of media channels is influenced by their ability to manage and control the flow of information and ability to complete a particular task (Short et al., 1976). According to Daft and Lengel (1984, 1986), as cited by Fulk and Boyd (1991, p. 409), media richness is characterized by "speed of feedback, variety of communication channels employed, personalness of source, and richness of language." Based on these characteristics, face-to-face communication is considered as one of the richest communication channels (Daft & Lengel, 1986).

Communication channels that provide immediate feedback and information that is relatable to the user are favored by users as they reduce uncertainty accompanied by the message (Liao & Teng, 2018). These factors often go unconsidered when choosing a media channel. In many cases, an organization's task ambiguity and the ability of the media to reduce uncertainties when delivering the message drive the choice of communication channels to be used (Fulk & Boyd, 1991). Therefore, it is not surprising to see that in most rural development efforts, face-to-face communication has been the most preferred and most used communication channel. However, the use of face-to-face communication channel does not allow extension professionals to reach larger audiences in a timely manner (Ndimbwa et al., 2021). The limitations associated with face-to-face communication channels have led to the increased use of ICTs, such as radio and mobile phones to reach large audiences more efficiently (Gakuru et al., 2009; Lwoga, 2009). These ICTs are often used to disseminate information on improved technologies with the aim of persuading rural community members to change their behaviors (Dhaka & Chayal, 2016; Gakuru et al., 2009; Mukherjee, 2011). However, there are few efforts aimed at using communication to understand factors influencing communities and their behaviors.

Aside from the organization's choice of media dissemination channels, farmers also have their preferred methods to receive and engage with information. Syiem and Raj (2015) reported that farmers prefer using mobile phones to communicate because they enable them to seek and get advice from extension agents in real time. They can send photos, ask questions, and get follow-up answers to issues they are facing in their production practices. The demand for real-time information and feedback necessitates the need to understand how ICTs were used during the COVID-19 pandemic when there was limited face-to-face interactions among and between communities and development agents.

**Purpose and specific objectives**

This discussion aims to examine the current use of ICTs in communication for development (C4D) in Malawi and propose ways to co-create and co-design ICT messaging to strengthen its potential to reach and address the needs of the most vulnerable people. Specifically, this chapter seeks to:

- Identify types of ICTs used by agricultural extension and advisory service providers in Malawi during COVID-19 lockdowns.
- Describe factors influencing choice of ICTs during COVID-19 lockdowns.
- Describe factors influencing choice of content disseminated through ICTs during COVID-19 lockdowns.
- Examine how agricultural extension and advisory service providers engaged with their clients using ICTs during the COVID-19 lockdowns.

**Methodology**

In this exploratory case study, we conducted key informant interviews and analyzed various data sources to examine how rural community development organizations in Malawi used ICTs during the COVID-19 pandemic when public gathering in large numbers was restricted. Before commencing the study, a proposal for the study was submitted to and reviewed by the Iowa State University Institutional Review Board (ID-22-068). The study was determined to be exempt.

The study was originally designed to incorporate two phases: a quantitative questionnaire and qualitative content analysis. We began with the survey which was sent to 400 rural development and agricultural extension experts in various organizations in Malawi. These experts were part of a WhatsApp group for the Malawi Forum for Rural Advisory Services (MaFAAS). The forum brings together people working for various agricultural extension and rural advisory services providers in Malawi to "strengthen agricultural extension and advisory service through information sharing and action to achieve professionalization, standardization, and quality assurance" (Ibenu, 2023).

A total of 24 respondents from different rural development organizations responded to the questionnaire. Only seven respondents completed all the questions in the survey. The low response rate may be attributed to two reasons: survey fatigue and information overload. During the same period of survey administration, there were several other surveys requesting responses. Despite absence of research on the impact of online survey fatigue in agricultural extension and advisory services, De Koning et al. (2021) reported that survey fatigue contributed to a low response rate and data quality during COVID-19 amongst their population of potential respondents. The WhatsApp group was open for every member to share information at any time. As such, there was an influx of messaging sent daily. The researcher observed that within an hour, more than ten messages in the form of links, audio messages, videos, or documents were shared on the platform. This information overload may have overwhelmed individuals and they may have missed the survey link. Efforts were made to share the link at least once every week for a month to increase the response rate but proved futile.

The few responses that were obtained were used to identify questions that were included in the key informant interviews. Three online, one-on-one interviews were held on Zoom and audio recorded for transcription purposes. We took a purposive, convenience sample to interview representatives from three major agricultural extension organizations in Malawi. The organizations included one public organization known as the Department of Agricultural Extension Services (DAES) and two non-governmental organizations: Farm Radio Trust (FRT) and Self-Help Africa. The DAES is responsible for coordinating all agricultural extension activities in the country. Self-Help Africa is an international, non-governmental organization involved in several agricultural extension and rural development projects. FRT is a non-governmental organization that focuses on dissemination of agricultural information using ICTs. To ensure that the phenomenon was thoroughly examined, data source triangulation was conducted (Denzin, 1970). Thus, a content analysis of six radio programs and three video clips from the three organizations were examined. Triangulation was conducted between the interviews from the representative of each organization and the content that was shared through radio programs of video clips to identify synergies. During each interview, the participants were asked to share samples of materials that they disseminated during the pandemic. The participants chose the materials that they shared depending on what was available. It was observed that only one organization, DAES, shared video clips alongside the radio programs while the rest only shared radio programs.

MaxQDA, a computer-assisted qualitative data analysis software, was used when analyzing data from the key informant interviews and for the content analysis of the communication artifacts. Researchers allowed for open, emergent codes to arise during data analysis (Saldaña, 2021). Using thematic analysis, themes and sub-themes were generated based on the objectives of the study.

To ensure the rigor and trustworthiness of our qualitative study, we ensured credibility by conducting member checking with research participants and peer debriefing amongst the two researchers. Theory triangulation was also conducted to ensure that relevant interpretations were made from different perspectives by using two theories when interpreting the results (Denzin, 1970). We have also provided a thick, rich description for transferability of the study. Finally, we used various data sources for triangulation and provide our reflexivity statement for confirmability (Lincoln & Guba, 1985).

As the instruments for data collection and analysis, we provided our reflexivity statement to allow readers to see the lens through which we conducted this work. We are academics with a passion for agricultural development. Combined, we have 25 years of experience working with local famers and communities in various contexts around the world and have expertise in agricultural development, extension education, communication, and behavior change.

## Results

### Types of ICTs used by agricultural extension and advisory service providers in Malawi during COVID-19 lockdowns

Participants were asked to identify the ICTs used by their organizations to reach out to their clients during the COVID-19 pandemic lockdowns in order of frequency of use. Three themes emerged, namely, *types of ICT used, innovative approaches used*, and *challenges*.

### Types of ICT used

The interviews and content analysis revealed that radio was the major ICT used during the COVID-19 pandemic, followed by mobile phones through a mobile text app called 321 and the "Mlimi hotline." Two organizations reported using the "Mlimi hotline," which is an agricultural call center used by farmers and extension agents to have access to agricultural information. When using the call center, farmers and extension agents have access to e-extension agents who can provide them with information from subject matter specialists. On the other hand, 321 is a mobile app which allows farmers to receive pre-recorded agricultural extension advice through text or audio. Other interactive mobile phone platforms were also used as evidenced by the following comment by Participant P:

> We also used the interactive voice response. The interactive voice response was one of the channels that we used, or one of the methods that we used as well to deal with the COVID-19 disruption to information.

*Innovative approaches used*

All of the organizations reported using mobile phones and radio to provide agricultural advisory services and information about COVID-19. The organizations created innovative approaches to enhance the use of mobile phones and radio sets. For example, Participant Z said that "We developed some animation films. These were shared in several WhatsApp groups so that farmers could see various procedures on how to carry out a particular technology."

Another organization reported using an innovative approach where farmers could record their questions and radio programs on the radio set then share with each other. Each farmer was then able to listen to the programs and questions or comments from other farmers without meeting with the rest of the farmers. This was only applicable to farmers who were beneficiaries of a project where they were organized into groups known as "Community ICT hubs" as evidenced in the following comment of Participant P:

> The groups that we work with most of them have radio sets and these radio sets have recording capabilities, so they can record audios. If they have questions, they were able to record the question and share the radio set with the extension officer who would also record the response and share it with the group.

Only farmers within the ICT hubs had access to the radio sets. These ICT hubs were established before COVID-19 and were formally called radio listening clubs where farmers once came together to listen to radio programs and discuss them. Participant P mentioned that the community ICT hubs were comprised of a group of farmers who work on improving access to information among farmers working on a specific enterprise and take access to information seriously.

*Challenges faced*

Despite the increased use of ICTs, when disseminating information during COVID-19, it was observed by Participant P that a lack of ICT tools among the farmers was a major challenge:

> We also discovered that most of the farmers that we have, almost 40% of the farmers that are not in the groups. They do not have radio sets, they have mobile phones ... they're able to listen to radio using the mobile phones.

Apart from not having radio sets, it was also reported that ICT infrastructure in the country was not fully developed as some areas were still not fully connected. Other areas have poor infrastructure and cell service that cannot connect properly as was noted by Participant Z:

> There are some other places where I could say the signal is weak..., I think there has to be a way of maybe subsidizing or assisting these farmers with handsets so that... they can access the messages easily.

## Describe factors influencing choice of ICTs during COVID-19 lockdowns

### Accessibility to the organization

The interviews revealed that all the organizations used ICTs accessible to them. Participant P mentioned:

> production of the radio programs was being done virtual… radio has been pivotal … reaching so many people … does not require or did not require physical interaction within our groups.

This view was echoed by Participant Z who indicated their organization used existing radio programs during the pandemic. While they did search for the best ways to communicate with farmers, they leaned into a previously developed communication strategy as Participant E commented:

> We looked at ways in which we can share information to farmers. …we reached back to our strategy … developed in 2018… radio was the most preferred in relation to print materials.

### Number of people reached

The number of people reached using a specific ICT was also indicated as one of the factors influencing choice of ICTs used. Participant P stated:

> Yeah, radio is the one you can reach to so many people … it doesn't require physical interaction … its ability to communicate without physical interaction as well was pivotal in the time when there were COVID prevention restrictions.

When realizing that not all the farmers have access to radio sets, some organizations turned to farmer groups. They saw this as an opportunity to ensure that various people received their messages. Participant P observed that:

> If you target an individual, you know that you are reaching one person and the possibility of that person to share the information to others is unguaranteed but if you are working with farmers in a group, you are guaranteed that the information will be shared with other people.

### Cost associated with using ICTs

Costs associated with production and dissemination of information was reported as one of the factors that influenced the choice of ICTs used by various organizations. Participant Z stated:

> Availability of funds influenced choice of ICTs used… the pandemic came as a surprise, there was no time to source a lot of funds so that we would carry

out all these activities. And there are some other messages which could best fit maybe on TV as opposed to publications but due to lack of funds we were not able to use the right platform.

To address this impact on the cost of production, the organization entered into collaborations with other organizations to take advantage of the available ICTs. Participant Z commented:

> And some of the initiatives that we were put in place in collaboration with the partners. I'm talking about mainly FRT. We also embarked on a radio campaign in collaboration with FRT, which was titled patsogolo ndi Ulimi mu nyengo ya [moving forward with farming in the midst of COVID] Covid.

The costs did not only affect the types of ICTs used as price but also influenced the type of radio stations used to air the radio programs. Participant E said:

> So we sampled out some farmers, and most of them preferred radio as compared with some materials like print, TV, so that's why when we had COVID-19 we justified the donor that we need to use the radio stations. So, firstly we engaged the National Radio Station, we engaged ZBS, Zodiak Broadcasting Station, but then because the cost was so huge, … so that's why we engaged the community radio station.

### Factors influencing choice of content disseminated through ICTs during COVID-19 lockdowns

*Predetermined content*

The content developed during COVID-19 was obtained from the Guide to Agricultural Production (GAP) and emerging issues such as outbreaks after approval by the National Content Development Committee (NACDC). The NACDC is a committee comprised of various stakeholders involved in agricultural research and advisory services. It is led by the DAES and its role is to ensure harmonization in the development and dissemination of agricultural information in Malawi. Thus, all the stakeholders are required to produce and disseminate content that has been approved by NACDC. Participant Z commented:

> We followed through the agricultural calendar and then on each message of agriculture activities … had to have a message that had to talk about COVID-19, in terms of if farmers are harvesting or they are marketing their crops…. Each message was coming with an advisory on how best to protect themselves as they undertake those farming activities.

Approved content also provided guidance based on ecological zones as indicated by Participant P:

> the key and relevant messages, would be messages, A, B, C, D for that particular area… they make sure that there's whatever materials or whatever

tropical areas that have to be covered, they have to make sure that they even the gap, unless there are issues that are emerging, then a new solution would have to be devised.

Participant E noted that they did not use all of the content recommended by the NACDC because it did not address some of the emerging issues such COVID-19 and Fall Armyworms as evidenced in the following quote:

> So most of the content was coming from NACDC, but there were some is- sues which were emerging... like the Fall Armyworm management, it's not part of the guide, but we had to develop them. So, messages which were emerging, ... we also had to develop and disseminate them.

### Promotion of improved technologies

Content analysis of radio programs from two organizations revealed that most of the information provided was aimed at promoting adoption of improved technolo- gies. As such, it focused on ensuring that farmers adopted new and improved tech- nologies as noted by Participant P:

> We're not going there to get their knowledge, to say, what do they think is the solution, but what is it that they're doing? So that's where we're able to get the information about the technologies that they're using, the locally invented technologies of dealing with the Fall Armyworm.

During the content analysis of communication artifacts, it was observed that two out of the three organizations aired programs that were in line with the policy direc- tion that dictated that only scientifically proven technologies should be promoted in Malawi. The two participants indicated that their organizations only disseminated information about scientifically proven technologies but did not "promote" farm- ers' indigenous or local knowledge, citing policy restrictions as the reason. When asked about this practice, Participant P said:

> We serve as information brokers, so we take what farmers are doing and push it to the Ministry, we tell them that this is how farmers are dealing with the problem. So, we will ask the ministry what to tell us the scientific way of dealing with it [is].

Participant Z also emphasized promotion of scientifically proven technologies:

> Some farmer groups shared with our personnel collecting radio interviews, but some of this content was not broadcast on air pending approval from re- search... the Department of Research hasn't yet come up with a position on some of these innovations.... We need to balance because we could not start distributing the messages, which are not yet research proven.

It was also noted that despite these organizations only airing programs that were scientifically proven, they did realize that in other cases there were no scientifically proven technologies to address farmers' needs. Participant P said the government was in "a fix" when it came to getting research out on an issue it was not prepared to address. Ultimately, they were only able to give recommendations or "precautions that they need to observe to make sure that the problem does not escalate, or it does not threaten the lives of people."

Participant E indicated airing programs that promoted farmers' local and indigenous knowledge was possible because it was a requirement of their organization's funding to promote sharing of indigenous knowledge. This position was not without conflict as Participant E observes:

> this is farmer research … some of the Department of Agriculture extension staff, felt that we are too quick to disseminate what is working at a farmer level. We had conflicts to say this shouldn't be shared on TV, this shouldn't be shared on radio, because we are waiting for the Department of Agricultural Research service to approve, to clear the technologies, especially on Fall Armyworm management.

Content analysis of the two radio programs from this organization indicated farmers were only provided an opportunity to share their experiences when addressing the challenge of Fall Armyworms. In cases where farmers shared experiences on scientifically proven technologies, extension agents or other experts emphasized the importance of the technologies and provided more guidance and direction.

### Sharing success stories

Content analysis of the radio programs and videos indicated that in all of the programs where farmers were featured, there was an emphasis on sharing success stories. In most cases, the farmers shared their success stories from either participating in extension programs or benefits from adopting improved technologies.

## Engagement with the clients using ICTs during COVID-19 lockdowns

The content analysis of radio programs, video clips, and interviews demonstrated that most of the channels used did not provide opportunities for real-time engagement. The programs were not interactive as they did not provide opportunities for the listeners to ask questions. However, only in one of the radio programs from one organization did the presenter urge listeners at the end of the program to call the "Mlimi hotline" if they had any questions. "Our E-extension agents were operating from home which allowed farmers to call the officers any day of the week except Sundays if they have queries, or alerts on what they are facing in the field," Participant P said. The Mlimi hotline was introduced before COVID-19 and during COVID-19. The call center not only assisted farmers to get access to information, but was also used by the farmers to report outbreaks and other emerging issues.

The information obtained from the call centers was used by the organizations to develop content that was disseminated through various ICTs. Participant P said: "Some of the initiatives we used to identify the messages to go to farmers ... they were able to get the messages required for areas, ecological zones, also known as ADDs." It was also reported that during COVID-19, the farmers used the call center to get access to information that they would normally get through in-person meetings.

All the organizations indicated having limited interaction with farmers. However, for FRT, the pandemic provided an opportunity for them to test the effectiveness of the ICTs when engaging with the farmers. Participant P said:

> The COVID-19 pandemic made it difficult to interact with farmers that meet as groups...But the challenge provided us with an opportunity because we are an ICT-based institution ... it was a springboard for us to see how far and how effectively the ICTs could fill the gap that was there in terms of provision of extension services when the traditional government way was challenged. So, it provided us with an opportunity to use the digital platforms to disseminate information and to see how effective it was going to be which proved to be pivotal in as far as filing the gap that was caused by COVID-19.

## Conclusion

In the study, the current use of ICTs in communication for development (C4D) was examined. We found that farmer preferences and voices were mostly ignored and underrepresented. While engagement differed amongst the different organizations, for the most part, farmers were still seen as recipients of information. Whether it was due to government or organizational policy, farmer experiences, knowledge, and practices were not shared with the greater public. Lack of acknowledgment of farmer expertise and know-how reduces farmer self-efficacy and power to assess best practices for urgent and emergent issues. The results indicated presence of a top-down approach when choosing content delivered using the ICTs. Farmers' knowledge and experiences seem not to be valued as the main premise for using ICTs is to disseminate information to farmers to improve adoption of scientifically proven technologies (Dhaka & Chayal, 2016; Gakuru et al., 2009; Mukherjee, 2011). The presence of policies that limit dissemination of indigenous knowledge and practices was identified as one of the factors influencing choice of content disseminated using ICTs. Increased use of ICTs, such as radio as an improved technology transfer tool, limits the ability of the ICTs to be used for amplifying voices of the rural communities and for promoting innovation sharing. Therefore, their potential in promoting innovation and knowledge sharing among rural communities is still untapped. It was observed that presence of funding is critical to promote farmers' engagement, capturing, and documentation of indigenous knowledge.

The results indicated radio was the most used ICT during the lockdowns. One of the factors for increased used of radios was its ability to reach a wide audience and convenience to the organization. To ensure increased coverage, several innovative approaches were applied which led to increased access to information among the community members. In all of the approaches farmers were perceived as information seekers and not sources of information. These findings are in line with the previous studies (Masambuka-Kanchewa et al., 2020b) where it was reported that farmers are perceived as passive recipients and not sources of information or innovators.

The COVID-19 pandemic created stress and fear amongst the members of the community (Najjuka et al., 2021; Semo et al., 2020). Despite the absence of research on the impact of lockdowns on farmers in Malawi, the findings of this study indicated that they were not only dealing with COVID-19-related stress, they were also fighting outbreaks like the Fall Armyworms. While the scientific community had no recommendations or treatment for COVID-19, they were also struggling how to address this detrimental, agricultural outbreak. Farmers indicated that they devised coping mechanisms and had to apply their indigenous knowledge to survive. The ability of the farmers to deal with all these emerging issues at once speaks of their resilience, which required documentation and sharing. The continued focus on using ICTs as a tool for transferring information denied the public an opportunity to learn more about the various communities' survival mechanisms. Liao and Teng (2018) reported that one of the factors influencing choice of communication channels among users is the channel's ability to provide a personal focus and opportunities for feedback. Realizing that during the lockdowns face-to-face meetings were restricted and that members of society had unanswered questions surrounding the pandemic, ICTs could have provided a better platform for promoting this type of engagement.

Rogers' diffusion of innovation (DOI) theory states that the choice of media channel greatly depends on the decision-making stage with mass media being critical when creating awareness of issues or technologies. However, with issues such as the COVID-19 pandemic, using one-way communication via mass media and ICTs such as radio and mobile phones, organizations miss the opportunity to understand the audience's perceptions and knowledge. This is crucial for development and implementation of projects and programs to effectively support and persuade them to adopt any innovation. Using ICTs in ways that provide opportunities for farmers and other members of the community to ask questions and engage with experts or the innovation itself can help dissuade doubt or receive social reinforcement. Dialogue and feedback are also useful to technology developers as it provides insights that may inform improvements on the innovations and improve uptake.

The continued use of communication in development as a tool for disseminating information during the COVID-19 lockdowns speaks about the need for change in approach from using a one-size-fits-all approach to using a needs-based approach. The change in approach will ensure that rural community voices are not just used to confirm or support the impact of improved technologies but also the diversity, resilience, and creativity of the rural communities. Such an approach will even be useful toward strengthening scientific discoveries since these experiences will serve as learning points to guide research and innovation development.

During those unprecedented times, organizations sought to continue to support their stakeholders as best as they could, choosing to disseminate information previously approved and research-based through channels that would reach the most people. We must find ways to empower farmers in the quest for C4D by going beyond the one-way communication and information deficit model. Farmers' knowledge and experiences should be at the center of C4D. Agricultural organizations should balance the dissemination of proven practices with the solutions found through trial and error of citizen scientists, i.e., the farmers testing and retesting ways to keep their livelihoods afloat.

### Modeling ICT use for empowerment

We recommend that practitioners seeking to use ICTs for C4D consider the various stages of Rogers' diffusion of innovation (DOI) decision-making stages and the need for reciprocal communication as called for in Media Richness Theory. To incorporate the various stages of Rogers' DOI and ensure that ICTs used in C4D are rich, we propose the following model (see Figure 11.1).

**Figure 1**

Model for developing and implementing ICTs

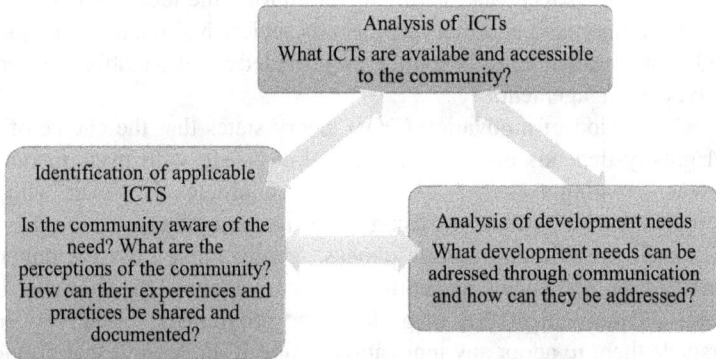

*Figure 11.1* Analysis of ICTs.

To move from a one-size-fits-all approach to a needs-based approach, we propose that every C4D organization should first conduct an ICT asset and needs assessment of each community they serve. This assessment should focus on identifying and describing the community's ICT capacity and needs, available innovations, and gaps. During the process, barriers and opportunities, threats, and strengths of various individual and community assets should be identified, described, and documented. As stated in Media Richness Theory, each of the ICTs available and accessible to the community should analyze the richness of the media by describing its ability to provide information in a language and form that is accessible to various members of the community as well as its ability to promote engagement.

Engagement should be assessed based on the ability of the ICT to provide opportunities for feedback and real-time information. The analysis of ICTs through the Media Richness Theory will facilitate a more participative approach for C4D.

*Analysis of development needs*

Once the community's ICT assets and needs have been identified, C4D practitioners, organizations, and communities should conduct an analysis of each development need. The analysis will identify those needs that can be improved through knowledge and information sharing and how it can be shared. Access to information is crucial for ensuring sustainable development. However, not all development needs can be addressed through communication. Therefore, it is critical for C4D experts to work with communities to analyze each development's need and identify ways of addressing those needs. Each need should be assessed to identify the ICTs best suited to address the need, acknowledging some needs may require use of multiple ICTs.

*Identification of applicable ICTs*

As indicated by Rogers (2003, an individual's decision-making stage influences the effectiveness of each communication channel. It is important to examine and establish the decision-making stage of the community members. Unlike in the DOI model, where the decision stage was analyzed in relation to a specific technology that originated from outside, the decision-making stage should be analyzed based on the impact and urgency of the problem for the intended community. To ensure community voice and ownership in this process, the following steps should be applied when identifying the applicable ICTs.

AWARENESS CREATION STAGE

Practitioners and communities should work together to establish the extent to which the community is aware of the development need, ways of addressing it through communication, and the type of information required to address the problem. Practitioners should facilitate conversations that will allow the community members to describe their level of awareness of the need, its impact on their lives, and how ICTs can be used to address it.

IDENTIFICATION OF PERCEPTIONS TOWARD THE NEED

Practitioners should create opportunities to allow the communities to express their perceptions about the need. Various ICTs should be used to document and share these perceptions. Efforts should also be made to enable different opinions from the community. Most importantly, ICTs that promote dialogue and feedback such as phone-in radio or television programs should be used. Expert voices can be incorporated at this stage. However, more opportunities should be provided to the community members to share their perceptions.

CAPTURING, DOCUMENTATION, AND SHARING OF INNOVATIVE APPROACHES

Practitioners and communities should work together to identify relevant ICTs for capturing, documenting, and sharing innovative approaches to address the need. Choice of ICTs should be determined based on the richness of each tool and the needs of the community. During this process, it is important to ensure that all voices are incorporated without bias. Both experts' and farmers' voices should be included equally.

*Further recommendations*

Any successful C4D activity or program will depend on changes in policies and availability of funding. It is recommended that governments and C4D organizations and practitioners review their policies to ensure that they do not hinder rural communities from sharing their experiences. Instead, they should embrace indigenous knowledge, highlight farmer identified best practices, and disseminate improved technologies. It is essential for governments and other C4D organizations to consider infrastructure development to ensure improved access to ICTs. This will increase access and availability of ICTs for participatory communication laying the foundation for more sustainable development.

Finally, it is proposed that further research should focus on examining farmers' perceptions of the information accessed during COVID-19. Such studies will provide insights on how farmers evaluated these messages and the impact that the messages had on their lives. Future research should focus on identifying farmers coping mechanism by identifying their experiences, practices, and knowledge systems applied during the lockdowns. Conducting such types of research will provide opportunities for documenting farmers' indigenous knowledge, experiences, and practices which can then inform technology development and dissemination.

## References

Aker, J. C., Ghosh, I., & Burrell, J. (2016). The promise (and pitfalls) of ICT for agriculture initiatives. *Agricultural Economics, 47*(S1), 35–48. https://doi.org/10.1111/agec.12301

Asunka, B. A. (2016). The significance of information and communication technology for SMEs in rural communities. *Journal of Small Business and Entrepreneurship Development, 4*(2), 29–38. https://doi.org/10.15640/jsbed.v4n2a4

Daft, R. L., & Lengel R. H. (1984). Information richness: A new approach to managerial information processing and organizational design. In B. Straw & L. L. Cummings (Eds.), *Research in organizational behavior* (Vol. 6, pp. 191–233). Greenwich: JAI Press.

Daft, R. L., & Lengel, R. H. (1986). Organizational information requirements media richness and structural design. *Management Science, 32*(5), 554–71.

De Koning, R., Egiz, A., Kotecha, J., Ciuculete, A. C., Ooi, S. Z. Y., Bankole, N. D. A., Erhabor, J., Higginbotham, G., Khan, M., Dalle, D. U., Sichimba, D., Bandyopadhyay, S., & Kanmounye, U. S. (2021). Survey fatigue during the COVID-19 pandemic: An analysis of neurosurgery survey response rates. *Frontiers in Surgery, 8*, 690680. https://doi.org/10.3389/fsurg.2021.690680

Denzin, N. K. (1970). Strategies of multiple triangulation. In N. K. Denzin (Ed.), *The research act in sociology: A theoretical introduction to sociological method* (pp. 297–313). McGraw-Hill.

Dhaka, B. L., & Chayal, K. (2016). Farmers' experience with ICTs on transfer of technology in changing agri-rural environment. *Indian Research Journal of Extension Education, 10*(3), 114–18. https://seea.org.in/uploads/pdf/v10321.pdf

Dharmawan, L., Muljono, P., Hapsari, D. R., & Purwanto, B. P. (2021). Digital information development in agriculture extension in facing new normal era during COVID-19 pandemics. *Journal of Hunan University Natural Sciences, 47*(12). https://johuns.net/index.php/journal/article/download/482/482-959-1-SM.pdf

Food and Agricultural Organization (FAO) (2003). *Revisiting the "magic box": Case studies in local appropriation of information and communication technologies (ICTs)*. Food and Agricultural Organization of the United States. https://www.fao.org/3/y5106e/y5106e00.htm

Fulk, J., & Boyd, B. (1991). Emerging theories of communication in organizations. *Journal of Management, 17*(2), 407–46. https://www.jstor.org/stable/2631846

Gakuru, M., Winters, K., & Stepman, F. (2009, April 1–2). *Innovative farmer advisory services using ICT. W3C Workshop Africa perspective on the role of mobile technologies in fostering social development, Maputo, Mozambique*. https://www.w3.org/2008/10/MW4D_WS/papers/fara.pdf

Hopestone, K. C. (2014). The role of ICTs in agricultural production in Africa. *Journal of Development and Agricultural Economics, 6*(7), 279–89. https://doi.org/10.58 97/jdae2013.0517

Ibenu, S. (2023). *Malawi forum for agricultural advisory services sets a direction for pluralistic extension system*. African Forum for Agricultural Advisory Services. https://www.afaas-africa.org/malawi-forum-for-agricultural-advisory-services-sets-a-direction-for-pluralistic-extension-system/

Kumar, A., & Singh, K. M. (2012). Role of ICTs in rural development with reference to changing climatic conditions. In K. M. Singh & M. S. Meena (Eds.), *ICT for agricultural development under changing climate* (pp. 17–28). Narenda Publishing House.

Liao, G.-Y., & Teng, C.-I. (2018). How can information systems strengthen virtual communities? Perspective of media richness theory. *PACIS 2018 Proceedings, 34*. https://aisel.aisnet.org/pacis2018/34

Lincoln, Y. S., & Guba, E. G. (1985). *Naturalistic inquiry*. Sage.

Lwoga, E. (2009). Application of knowledge management approaches and information and communication technologies to manage indigenous knowledge in the agricultural sector in selected districts of Tanzania. [Doctoral Dissertation. University of Kwazulu-Natal.]

Masambuka-Kanchewa, F., Lamm, K., & Lamm, A. (2020a). Beyond diffusion of improved technologies to promoting innovation creation and information sharing for increased agricultural productivity: A case study of Malawi and Kenya. *Journal of International Agricultural and Extension Education, 27*(1), 79–92. https://doi.org/10.5191/jiaee.2020.27106

Masambuka-Kanchewa, F., Rodriguez, M., Buck, E., Niewoehner-Green, J., & Lamm, A. (2020b). Impact of agricultural communication interventions on improving agricultural productivity in Malawi. *Journal of International Agricultural and Extension Education, 27*(3), 116–31. https://doi.org/10.4148/2831-5960.1102

Mtega, W. P. (2012). Access to and usage of information among rural communities: A case study of Kilosa District Morogoro Region in Tanzania. *Partnership: The Canadian Journal of Library and Information Practice and Research, 7*(1). https://doi.org/10.21083/partnership.v7i1.1646

Mukherjee, S. (2011). Application of ICT in rural development: Opportunities and chal-
lenges. *Global Media Journal, 2*(2), 1–8. https://www.caluniv.ac.in/global-mdia-journal/
Winter%20Issue%20December%202011%20Students'%20Research/SR1%20-%20
Mukheree.pdf

Najjuka, S. M., Checkwech, G., Olum, R., Ashaba, S., & Kaggwa, M. M. (2021). Depression,
anxiety, and stress among Ugandan university students during the COVID-19 lockdown:
An online survey. *African Health Sciences, 21*(4), 1533–43. https://doi.org/10.4314/ahs.
v21i4.6

Ndimbwa, T., Mwantimwa, K., & Ndumbaro, F. (2021). Channels used to deliver agricultural information and knowledge to smallholder farmers. *IFLA Journal, 47*(2), 153–67.
https://doi.org/10.1177/0340035220951828

Oyeyinka, R. A., Bello, R. O., & Ayinde, A. F. O. (2014). Farmers utilization of farm-radio
programmed for marketing of agricultural commodities in Oyo State, Nigeria. *European
Journal of Business and Management, 6*(35), 58–68.

Pigg, K. E., & Crank, L. D. (2005). Do information communication technologies promote rural economic development? *Community Development, 36*(1), 65–76. https://doi.
org/10.1080/15575330509489872

Rogers, E. M. (1976). Communication, and development: The passing of the dominant paradigm. *Communication Research, 3*(2), 213–40. https://doi.org/10.1177/
009365027600300207

Rogers, E. M. (1995). *Diffusion of innovations* (4th ed.). The Free Press.

Rogers, E. M. (2003). *Diffusion of innovations* (5th ed.). New York: Free Press.

Rogers, E. M. (2004). A prospective and retrospective look at the diffusion model. *Journal
of Health Communication, 9*(S1), 13–19. https://doi.org/10.1080/10810730490271449

Saldaña, J. (2021). *The coding manual for qualitative researchers* (4th ed.). SaRoge.

Semo, B. W., & Frissa, S. M. (2020). The mental health impact of the COVID-19 pandemic:
Implications for Sub-Saharan Africa. *Psychology Research and Behavior Management,
13*, 713–20. https://doi.org/10.2147/PRBM.S264286

Short, J., Williams, E., & Christie, B. (1976). The social psychology of telecommunications
*(No Title).* https://www.gbv.de/dms/ilmenau/toc/020337264.PDF

Song, B., & Wang, W. (2011). Instant messaging continuance: A media choice theory perspective. *Frontiers of Business Research in China, 5*(4), 537–58. https://doi.org/10.1007/
s11782-011-0144-1

Stratigea, A. (2011). ICTs for rural development: Potential applications and barriers involved. *Netcom. Réseaux, Communication et Territoires, 25*(3/4), 179–204. https://doi.
org/10.4000/netcom.144

Syiem, R., & Raj, S. (2015). Access and usage of ICTs for agriculture and rural development
by the tribal farmers in Meghalaya state of North-East India. *Agrarinformatika/Journal of
Agricultural Informatics, 6*(3), 24–41. http://real.mtak.hu/29903/1/190_1024_1_PB_u.pdf

World Bank Group. (2018) *Malawi Systematic Country Diagnostic: Breaking the Cycle
of Low Growth and Slow Poverty Reduction.* World Bank, https://doi.org/10.1596/31131

# 12 Internet use genres

## A lens for analyzing patterns of Internet adoption in rural communities of Canada, Chile, and Vietnam

*Maria Bakardjieva, Isabel Pavez Andonaegui,
Teresa Correa and Trang Pham*

### Introduction

In this chapter, the findings from three empirical studies of Internet adoption in rural communities are considered through the concept of Internet use genres. These studies were conducted in Canada (2004–2005), in Chile (2015–2017), and in Vietnam (2017–2018) when these countries were implementing policies to provide broadband Internet infrastructure in rural areas. Despite the social and cultural differences between these countries, the findings showed clear common patterns in the ways rural residents appropriated the Internet. This led us to look at the specificity of rural settings for explanations of users' action choices. For that purpose, we introduce the concept of Internet use genres.

The study hypothesized that users do not simply plug new media into the meaningful activities they already perform. Users devise new meaningful things to do when they have new media at hand. The study focused on the origin of such emergent media-based practices. If, as advocated by some scholars (Bräuchler & Postill, 2010; Couldry, 2004), media should be seen as inseparable from the practices that surround them, then where should one look for the gestation of these practices without succumbing to overt technological determinism and with due attention to the role of users?

### *User agency*

The question of media user agency has been approached through the lens of the domestication model first proposed by Silverstone and Hirsch (1992). According to this model, users bring new media technologies they encounter in the public world into their domestic sphere where active interpretation and appropriation take place on the users' own terms. Feenberg (1999) has emphasized users' capacity for creative appropriation that represents a democratic rationalization of technologies' shapes and uses. This view brings into focus critical, reflexive, and potentially empowering applications of media technologies, also recognized by researchers of the Social Construction of Technology school (Oudshoorn & Pinch, 2003). We share with these authors an interest in the sources of user agency and

DOI: 10.4324/9781003282075-15

the repertoires of practices that users evolve around these artefacts and systems in their everyday lives.

Miller and Slater's (2000) ethnographic investigation of the uptake of early Internet-based communication media in Trinidad has suggested that user-initiated technology-related practices played a role in what they referred to as the expansion of human capabilities such as pursuing educational, professional, and cultural goals that were previously unattainable. Their term "expansive potential" referred to users' ability to imagine previously unimaginable ways of being and to expand their reach and possibilities for action (Miller & Slater, 2000, pp. 11–12). The concept of Internet use genre is intended as a lens allowing researchers to explore the origins of such practices with a view to the exigencies of the specific social situations in which they arise.

### *Internet use genre as concept and heuristic device*

According to Miller (2015), the concept of genre serves as a "structurational nexus mediating not only intention and exigence, form and substance, but also action and structure, medium and product, the material and the symbolic" (p. 56). Rhetoricians, such as Miller (1984), have transformed the definition of genre to reflect the growing consensus that it should "be centered not on the substance or the form of discourse but on the action it is used to accomplish" (p. 151).

This shift makes it possible to apply the concept outside the sphere of language and rhetoric to social action more generally and with respect to non-linguistic materials and means (see Bakardjieva, 2005). Miller (1984) argued that defining genre as social action requires an examination of the tripartite connection between genre, recurrent situation, and the actor's motive "because human action, whether symbolic or otherwise, is interpretable only against a context of situation and through the attributing of motives" (p. 152). This definition mediates "between private intentions (purpose) and socially objectified needs (exigence)" (Miller, 2015, p. 57). Genre, Miller (1984) insists, "must be located in the social world, neither in a private perception nor in material circumstance" (p. 157), but in a personally and culturally meaningful social need.

Genre theory and history address the question of the origin and evolution of (rhetorical) genres. Miller (2015) has noted that even though genre by definition is something recurring and recognizable, it comprises both stability and change. Genres are "typified rhetorical actions based in recurrent situations" (Miller, 1984, p. 159), which can "change, evolve, and decay" (p. 163).

Voloshinov (1973) offers a powerful account of the co-evolution of social life, human action, and linguistic forms. He refers to "little behavioral genres," of which speech forms are one element. Voloshinov (1973, p. 97) observes:

> Each situation, fixed and sustained by social custom, commands a particular kind of organization of audience and, hence, a particular repertoire of little behavioral genres [The behavioral genre is]... a fact of the social milieu .... It meshes with that milieu and is delimited and defined by it in all its internal aspects.

Voloshinov was focused on the investigation of the diachronic evolution of language. He theorized about genres of speech, i.e., verbal communication. There is a strong argument to be made for applying his notion of genre to behavior generally (or social action in Miller's formulation), and specifically to communication technology use. As Bakardjieva (2005, p. 26) has argued:

> Despite obvious differences in the nature of their materiality and internal organization, both language and technological systems are culturally established, formal structures of means and rules, or as de Certeau puts it, 'ensembles of possibilities ... and interdictions' (1984, p. 98) that the user actualizes in his or her individual concrete operations.

De Certeau (1984) maintains that the notion of the *speech act* is applicable in a sphere much broader than that of verbal communication because it suggested a general distinction between "the forms used in a system and the ways of using a system" (p. 98). Following the same logic, it is proposed that the act of use is to the technological system what the speech act is to language (see Bakardjieva, 2005, p. 27). Hence speech genres represent a useful model for the understanding of the use genres involving other "mediational means" (Wertsch, 1991, p. 28), which facilitate human action including communication technology.

In this study the concept of genre is applied to types of social action that do not rely exclusively on words and grammar, but manipulate multiple digital communication devices and interfaces. These social actions, like their purely verbal and rhetorical counterparts, are organized in typical configurations of form and content, personal intention, and social exigence. They are anchored in recurring social situations. Our comparative analysis of broadband Internet adoption looks for the social situations that shape the Internet use genres that the rural communities in the three countries have in common.

### The emergence of rural use genres

The starting point in the gestation of a new use genre is the local situation of the user anchored in different spheres of social life. It presents exigence (a socially defined need), implicitly or explicitly. Against the backdrop of the perceived exigence, competent members of the culture orient themselves toward certain meaningful goals (intention). They then draw on other elements of their situation such as available materials, tools, norms and know-how to perform action that would bring them closer to the intended result.

When communication technologies are present in such local situations, they are drawn into actors' responses to exigencies and pursuit of intentions. Because social situations recur and are collectively defined, typical combinations of exigence, intention, formal, and substantive elements, including communication technologies, can blend together to produce typical genres of social action. Seen from the point of communication media, these genres can be defined as *use genres*. In this model, there is adequate recognition of the situation and exigence that are outside of the control of the actor. It also gives due consideration to the actor's intention and their meaningful action choice. The interplay of these different forces leads to the emergence of ways of use that could be conservative or innovative, enforced or freely imagined. When these ways of using enter into the cultural action repertoire of a community and are stabilized by way of social reproduction, they become recognized use genres. Because communication technology is adaptable, the characteristic exigencies and intentions that shape actors' responses to diverse situations lead to the proliferation of multiple use genres.

Thus, Internet use genres are products of social situations. They are shaped by exigencies and intentions related to work and leisure, to social interaction in the school or office, in the friends' group or political organization. The Internet does not possess the infinite variety of elements and flexibility of forms that language does. Yet, it does offer its own building blocks and rules for organizing them that allow a wide spectrum of actions to be imagined and performed. Internet use genres are the members of this multitude that standout as socially recognizable, recurrent, and finely adapted to typical situations in different social milieus.

Rurality is one such milieu. It should not be analyzed as being opposite to or in comparison to urban areas that represent another type of milieu. There is a need to look at how technologies are adopted in rural communities from a situated and holistic viewpoint (Williams & Cutchin, 2002). The challenges that rural communities face are different due to the nature of their location and composition. For instance, the majority of their inhabitants have manual jobs. Therefore, there are fewer opportunities to explore digital technologies and to develop skills in their everyday lives (Park, 2017). The communities involved in our studies tended to have fewer young families with children, which was due to the lack of educational opportunities (Correa & Pavez, 2016).

The fieldwork from all three studies was performed when each respective country's government was pushing a broadband connectivity policy targeting rural areas. The studies had a common objective, which was to understand how the Internet was adopted in rural areas. The secondary analysis of the existing data allowed us to elaborate on the concept of Internet use genres and to identify and label genres associated with the context of rurality observed in all three datasets.

The rural contexts we had examined shared spatial, demographic, and economic features, such as relative remoteness and isolation from urban industrial centers and communication infrastructure, substantive outward migration of residents, especially young people; limited range of productive industries and other sources of employment and livelihood; and limited educational and consumption opportunities. Broadband Internet infrastructure sponsored by the government of each country was being introduced into these contexts during the periods when our studies took place.

The first project was conducted in 2004–2005 in Canada, a country with a population of 35 million. In the province of Alberta, a large-scale provincial government initiative called the Alberta SuperNet was introduced at that time and led to the construction of a broadband optical network that reached into the remote communities of the sparsely populated province (see Bakardjieva, 2008; Bakardjieva & Williams, 2010). Rural residents in four communities, who were early Internet adopters (34 participants), were invited to join focus groups to discuss how they used the new medium in their everyday lives. These accounts provided detailed descriptions of the particular ways of using the Internet these early rural adopters had enacted or initiated and the circumstances and intentions behind their choices.

The second study took place in rural Chile. With a population of 17 million people, Chile had been developing national Information Communication Technology (ICT) policies since 1992 (Kleine, 2013). Between 2010 and 2011, the Telecommunication Development Fund had subsidized a 3G connection to remote areas that had remained out of reach. The study showed that 63% of the people from these recently connected isolated rural communities had never used the Internet before (Correa et al., 2017). The data were collected through multiple methods (interviews, face-to-face survey, focus groups) in 22 isolated communities between 2014 and 2017. Overall, 1,000 people participated in the survey with 48 and 60 participating in interviews and focus groups, respectively.

The study in Vietnam was conducted between 2016 and 2017. In 2011, this country of 95 million people approved a national rural information and communication plan that included building a $400 million broadband network infrastructure that included mobile, fixed, and satellite connections. Its goal was to cover all the administrative units in rural Vietnam where about 70% of the population, roughly 67 million people, lived (Pham, 2021). The data collected and analyzed in this project included 44 individual and 17 group interviews with a total of 79 rural Internet users.

For the purposes of this chapter, a secondary analysis of the individual and group interviews in the three datasets was conducted in order to trace the formation of the Internet use genres that were observed in all three countries. The analysis aimed to capture (1) what rural residents were saying they were doing on the Internet (actions, form, and content); (2) why they had chosen these particular ways of using the medium (exigence and intention); (3) how their use practices had evolved (situated appropriation); and (4) how those ways of use fitted into their everyday lives (situation).

In the following section, these genres are delineated with attention to variations related to salient aspects of users' life situations, such as economic well-being, age, marital status, employment, and geographical and cultural characteristics of the local community (Pavez & Correa, 2020). We discuss the common conditions and parallel motivations that had led to the emergence of what we consider as characteristic rural Internet use genres These are genres that could likely be found in rural contexts across the world because they respond to the specific life situations faced by rural residents.

## Rural Internet use genres across Canada, Chile, and Vietnam

### *Internet-powered business*

The limited local business opportunities characteristic of the rural areas in all three countries created an exigence for rural entrepreneurs to look for ways to expand their business operations and customer base. This led to a combination of elements including form, content, and intent into a use genre that could be labeled *Internet-powered business*. In the Canadian case, this genre was emerging out of external pressure on rural farmers and small business owners to step up the dynamic of their communication with clients and suppliers. Alice (rancher, aged 53) (not her real name) described the situation and the exigence it created:

> We have a pure breed catalogue, we have a website. We try to market to people who wouldn't ordinarily come to this part of the country for a sale .... Before the Internet, if we had sent out our catalogue to all these people, they would telephone us and say 'I like that number [refers to a pure breed bull], what does he look like?' And then you send them a picture, and if they did not contact us at least three weeks before our sale, there was no way that we were going to get that video on time .... About two years ago we had a sale and two days before the sale people were telephoning: 'Could you send us a picture on the Internet from front to front, back to side ....' And [I] took the pictures and people ended up driving up for our sale and there were figures up to about $10,000 .... That was the highest selling that we have ever had.

This exigence was further exacerbated by the expansive digitization of businesses in urban areas and the demand on rural entrepreneurs to keep up. Yet, dial-up connectivity was failing to deliver the necessary speed, so quality connections and broadband adoption were actively pursued. Julia (accountant, aged 52) spoke about that experience:

> I find that because of the Internet my business associates and my customers have an increased expectation of response. It used to be that they would phone me during business hours and if they couldn't contact me, they would not expect to hear from me until the next day. That is no longer true. I can be working at one in the morning, and I have customers that are emailing me knowing that I work late and are actually hopeful that I am going to be responding to them instantly in the morning. So, it makes us all immediately available to all people who we are in association with, be that our families or our business associates.

In rural Chile, villagers with business interests in tourism such as renting vacation cabins started to hear demands about access to WiFi, bank transfers, and online booking—Internet-facilitated operations that they knew nothing about. This was a perceived exigence for surviving economically, as customers were increasingly keen to be able to have information, pictures, and reviews from the places put up for rent, and to have access to the Internet during their sojourn. Therefore, in order to stay in business and be competitive these Internet non-users felt the need to develop the digital side of their entrepreneurship.

Although doing business over the Internet is common for urban dwellers, rural residents had to evolve their versions of this genre, which was not immediately available or intuitive. Even though the exigence was there, they needed to figure out how to put together form and content in ways that suited their local situations. This was clear in the case of a Chilean artisan from a rural community located in a remote region on the border with Argentina. He realized that he could not make a living from what he sold only during the high tourist season. Guided by his teenage son who was using Facebook to socialize with peers, he learned how to navigate the functionalities of the platform so that he could promote his products widely and year-round, connect with other artisans, and look for opportunities and invitations to present his work at craft fairs in other parts of the country. A Vietnamese manager of a garment factory found herself in a similar situation as she faced the challenge of dealing with customers in Korea concerning orders and clothing samples. The situation required unconventional solutions, which led her to adopt videoconference technology to discuss business with these customers in virtual meetings.

These situations related to the challenge of pursuing economic activities in rural contexts immersed in the increasingly digitalized business and consumption environment presented similar social exigences. They prompted rural users to pick particular appropriation paths that resulted in the stabilization of the Internet-powered business genre. The Internet allowed local farmers and entrepreneurs to establish an online presence and to tap into a wider consumer population and business opportunities. This motivated the adoption of the technology as it supported the pursuit of already meaningful goals. It also expanded users' aspirations. In a situation of remoteness from major economic centers, relative isolation and reliance on a small number of traditional trades, this genre was particularly meaningful and enticing for rural dwellers.

Consequently, this genre, characterized by particular forms (online shop-windows; digital communication with suppliers and customers), content (business information and transactions), intention (to reach new markets and grow operations), and exigence (the limited scope of local possibilities in an isolated place), was taken up by more rural users in all of the three countries.

### Fostering new Internet-enabled lifestyles

Enticed by information and activity made available by the Internet, rural residents drew on the medium to craft new lifestyles that differed from local custom. They picked up new ideas regarding nutrition, entertainment, and personal relationships. These lifestyles were not experienced and navigated equally by people from different generations. Young people wanted change, while their parents and grandparents more often preferred to hold on to their respective rural traditions. Members of the older generations sometimes saw Internet technology as a threat to traditional values, practices, and hierarchies of knowledge and authority. In some Chilean communities, such as Carrizal in the south, this idea of technology as a disruptor was mostly promoted by older people who were alarmed that children almost stopped playing football in the plaza and would rather sit in close groups with heads down over a screen. These new practices were received with uneasiness as most of the non-users feared the young would get addicted to Internet content over which the adults had no control. Older people also expressed fears about the togetherness of family life and how screens were driving their children away. These negative attitudes prompted fear and non-use that were the opposite to the enthusiastic engagement with the medium by children and youth who embraced the opportunity for communication with peers and friends that was not always easy in their rural everyday lives.

Rural Vietnam saw a similar polarization of experiences and use genres, as traditions and values, particularly in the passing down of knowledge and advice from the elderly, constituted a part of rural cultural identity. For instance, there were beliefs that pregnant women should follow rigorous naturist diets to maintain their own strength and their newborns' health. Yet, information and availability of new food products and diets made the younger generation in rural Vietnam aware of new nutritional options. One new mother did not follow the established nutrition rules concerning postpartum diet and instead searched for advice online. Her father-in-law looked upon this kind of behavior disapprovingly:

> Progressive young people do not listen to the experience that older adults have. [The young people] consider science as the highest authority and [traditional beliefs] as untrustworthy. For example, [the daughter-in-law] eats something older people believe she should abstain from [because] Google … tells her she can eat it while breastfeeding …. As a result, health problems arise in raising our own children.

Presented with the chance to expand or contrast traditional knowledge with information found on the Internet, young people in rural communities were choosing new lifestyles. However, for that father-in-law traditional experience was superior to modern knowledge. Despite the valid question regarding users' ability to discriminate between sources and assess the reliability of information adequately, this elderly respondent was unhappy with the sense of empowerment experienced by the younger generations. These intergenerational tensions demonstrate that new use genres are not unproblematic or easily accepted and may cause conflict. They create new configurations of social relations as the arrival of technology in more isolated places meets with uneven degrees of competence and openness to change among users in different socio-biographical situations.

In Canada, where the aging of the rural population is also a problem, the older rural residents interviewed, unlike those in Chile and Vietnam, saw in the Internet a promise for the invigoration of their communities. They believed that reliable access could facilitate a movement back to the rural communities by young professionals. It would enable these professionals and their families to settle in the community thereby changing the community's demographic and economic landscape. In the focus groups there were young professionals who were using the Internet to devise a previously difficult-to-achieve lifestyle, one that was more widely appreciated and spread years later during the COVID-19 pandemic. They had developed their careers in urban centers but yearned for the advantages of rural life. For them broadband connectivity was a way to return to the countryside while retaining their employment opportunities and links with worldwide networks of colleagues.

### Internet-augmented education

The situation of rural families with school-aged children generated exigencies related to educational resources and aspirations. In Chile, distances between schools and houses, and in some places harsh weather conditions, made it difficult for children to stay late at school or return in the afternoon to complete their homework when online access to sources was needed. The access to these educational tools was defined as a need that led families to embrace Internet-augmented education at home. This use genre involved education-related websites, applications, and content as well as interpersonal communication between students and overlapped with socializing on social networking sites and online gaming. Children were not the only beneficiaries of home access because laptops and computers were usually placed in common spaces in the home. Consequently, parents and other adults in the household also started to make use of them for practical reasons and to satisfy curiosity. Thus, even if the main motive for bringing the connection home was children's education, it also helped advance the education of the adults, as one in four stated that they acquired digital skills with the help of their children (Correa,Pavez and Contreras, 2017).

The situation of families with school children in Canadian rural communities was similar. Young people growing up in this environment had difficulty accessing high schools and limited choices of educational courses at that level. Receiving formal education in a remote and scarcely populated area as exigence, combined with the resolute intention of parents to give their children the best possible learning opportunities, made the Internet a critical source of educational content and activities. It made it possible for students to pursue their interests and grow their talents through online courses while remaining close to their families and contributing as workers on the family farm. Adults used the Internet connection to upgrade their education and skills through interactive formats leading to trade certificates and professional degrees.

The genre of Internet-augmented education was prominently manifested in Vietnamese villages. The most common jobs in these communities, farming and fishing, were performed by low-paid workers who spent their entire lives toiling on their designated family land. As a result, some parents and grandparents were strongly motivated to invest in opportunities for their children to study online in the hope of breaking the cycle of poverty by opening more profitable employment options other than farming or fishing. For example, a rural family whose livelihood had been based on fishing for generations subscribed to the Internet and purchased a computer, three smartphones, and a television for the sake of their four children—two of them college students. The mother explained: "The job is very tough, and we do not want them to follow our parents' suit." The Internet connection was considered by the parents as an educational investment.

In Vietnam, parents preferred to provide Internet access for their children at home rather than let them use the medium at Internet cafés so that they could monitor what the young people did online and how much time they spent on it. This was because Internet cafés, where young people congregated, had evolved into gaming hubs with high-speed Internet and favorite games pre-installed on computers.

## Conclusion

This chapter examined the factors behind Internet adoption and use patterns in rural communities in Canada, Chile, and Vietnam. It proposed Internet use genres as a concept that helps to understand the social and economic complexities involved. The concept offers an analytical lens that aids the investigation of the cultural appropriation of digital communication technologies at the level of use in specific social contexts. The study's comparative examination traced the emerging Internet use genres in the rural communities in each of the three countries to elements of the participants' relevance systems and socio-biographical situations that were shaped by the geographical, socio-economic, and cultural conditions of rurality. Despite the distinct social and political profiles of these societies, we found that similar use genres evolved. This is an indication that use genres respond to recurring local situations and are highly malleable and subject to user agency. It is the study's argument that being able to identify, classify, and anticipate the evolution of these genres on the part of development agencies could encourage rural end-users' participation in technological development.

Internet use genres are constituted by forms and content of technology use that respond to the perceived exigences and systems of relevance organizing those rural users' lifeworlds. These systems become a cognitive map that helps rural inhabitants to make sense of the new technologies they encounter and to envision new possibilities for building capacity and self-realization. Sometimes local situations and systems of relevance become sources of fear and consternation. Consequently, the meaning-making process and the respective appropriation of the Internet is not linear. Rather, it navigates a complex terrain of overlapping and often contradictory pressures created by personal and local circumstances that are shaped by larger social landscapes. For instance, the networking of the farm and small business, the crafting of new lifestyles, and the pursuit of education via the Internet all emerged as Internet use genres with characteristic forms and content shaped by various degrees of intention and exigence. They were a product of the individual and cultural adaptation of the functionalities of broadband communication technologies to the perceived needs and projected goals of rural inhabitants. By drawing a path among situational elements, exigence, intent, form, and content, the concept of internet use genres offered a solid analytical framework that revealed the driving forces of practice-making and the spread of the resultant ways of using (genres) across communities of similarly positioned users. It helped us recognize and balance systemic imperatives and subjective motives, or in other words the roles of structure and agency in the formation of prevalent social practices around new technologies.

Our goal here was to illustrate the working of this concept within a particular social context and with respect to a selected communication technology, i.e., the Internet. Being able to identify, classify, and anticipate the evolution of Internet use genres into widely shared and normatively sanctioned practices could give researchers and policymakers a nuanced understanding of the dynamics of adoption processes in rural communities. This is important because there is a common assumption that rural areas would follow the path of urban communities. Rural areas, in reality, have their own technological, social, and cultural circumstances that produce specific exigencies and motivations. Taking the development of use genres as an entry point focuses attention on the situated rationality of the choices and decisions guiding Internet appropriation in these social localities. It requires approaching the introduction of new communication technologies in these places in an open dialogue with their inhabitants and careful attention to these users' own practical inventions and innovations as they confront and discover the affordances of new technologies in various life situations. The inclusive development of rural regions could benefit from such organic forms of participation by diverse users that go beyond formal procedures and limited cooperation with local elites. We propose that thinking in terms of use genres could benefit the exploration of emerging social practices involving other new technologies and artefacts in a variety of contexts.

The lens of use genres is particularly suited to exceptional circumstances such as the ones created by the COVID-19 pandemic and ensuing quarantines. The COVID-19 restrictions created new exigencies and transformed the sense-making driving personal choices and actions involving communication technologies. Users undertook massive experimentation and adaptation of communication media with a view to the situated rationalities of work, family life, and survival. These innovative use genres have to be studied as they are examples of meaningful deployment of digital media that users have found empowering in the face of adversity.

Furthermore, these genres can be adapted for the purposes of research. They could offer researchers ideas and solutions to challenges, such as remote access, building trust with respondents, and getting to understand their situated needs. Online communication with familiar and unfamiliar others is now routine for many people in remote and marginalized communities. To make the best of a hard situation, researchers could capitalize on the lessons learned and work with such communities more frequently and examine their issues more deeply and more extensively.

## References

Bakardjieva, M. (2005). *Internet society: The Internet in everyday life*. Sage.

Bakardjieva, M. (2008). Making sense of broadband in rural Alberta, Canada. *Observatorio (OBS\*)*, *2*(1), 33–53. https://obs.obercom.pt/index.php/obs/article/view/81/146

Bakardjieva, M., & Williams, A. (2010). Super network on the prairie: The discursive framing of broadband connectivity by policy planners and rural residents in Alberta, Canada. *Culture Unbound: Journal of Current Cultural Research*, *2*, 153–75. http://www.cultureunbound.ep.liu.se/v2/a10/

Bräuchler, B., & Postill, J. (Eds.) (2010). *Theorising media and practice* (Vol. 4). Berghahn Books.

Correa, T., & Pavez, I. (2016). Digital inclusion in rural areas: A qualitative exploration of experiences and challenges faced by people from isolated communities. *Journal of Computer Mediated Communication, 21*(3), 247–63. https://doi.org/10.1111/jcc4.12154

Correa, T., Pavez, I., & Contreras, J. (2017). Beyond access: A relational and resource-based model of household Internet adoption in isolated communities. *Journal of Telecommunications Policy, 41*, 757–68. https://doi.org/10.1016/j.telpol.2017.03.008

Couldry, N. (2004). Theorising media as practice. *Social Semiotics, 14*(2), 115–32. https://doi.org/10.1080/1035033042000238295

De Certeau, M. (1984). *The practice of everyday life.* University of California Press.

Feenberg, A. (1999). *Questioning technology.* Unpublished manuscript.

Kleine, D. (2013). *Technologies of choice?: ICTs, development, and the capabilities approach.* MIT Press.

Miller, C. (1984). Genre as social action. *Quarterly Journal of Speech, 70*(2), 151–67. https://doi.org/10.1080/00335638409383686

Miller, C. (2015). Genre as social action (1984), revisited 30 years later (2014). *Letras and Letras, 31*(3), 56–72. https://doi.org/10.14393/LL63-v31n3a2015-5

Miller, D., & Slater, D. (2000). *The Internet: An ethnographic approach.* Berg.

Oudshoorn, N., & Pinch, T. (2003). Introduction: How users and non-users matter. In N. Oudshoorn & T. Pinch (Eds.), *How users matter: The co-construction of users and technologies* (pp. 1–25). MIT Press.

Park, S. (2017). Digital inequalities in rural Australia: A double jeopardy of remoteness and social exclusion. *Journal of Rural Studies, 54*, 399–407. https://doi.org/10.1016/j.jrurstud.2015.12.018

Pavez, I., & Correa, T. (2020). "I don't use the Internet": Exploring perceptions, experiences and practices among mobile-only and hybrid internet users. *International Journal of Communication, 14*, 2208–26.

Pham, T. (2021). Broadband Internet rollout in rural Vietnam: From policy to everyday use. [Doctoral dissertation. University of Calgary.] https://prism.ucalgary.ca/handle/1880/113455?show=full

Silverstone, R., & Hirsch, E. (Eds.) (1992). *Consuming technologies: Media and information in domestic spaces.* Routledge.

Skerratt, S. (2008). The persistence of place: The importance of shared participation environments when deploying ICTs in rural areas. In G. Rusten & S. Skerrat (Eds.), *Information and communication technologies in Rural Society. Being rural in a digital age* (pp. 83–106). Routledge.

Voloshinov, V. (1973). *Marxism and the philosophy of language.* Seminar Press.

Wertsch, J. V. (1991). *Voices of the mind: A sociocultural approach to mediated action.* Harvard University Press.

Whitacre, B. (2010). The diffusion of internet technologies to rural communities: A portrait of broadband supply and demand. *American Behavioral Scientist, 53*(9), 1283–303. https://doi.org/10.1177/0002764210361684

Williams, A., & Cutchin, M. (2002). The rural context of health care provision. *Journal of Interprofessional Care, 16*(2), 107–15. https://doi.org/10.1080/13561820220124120

# 13 Teaching community engagement in the Digital Age

## Reflexive pedagogical experiences

*Judy Lawry and Linje Manyozo*

## Introduction

This chapter explores the experiences of teaching and building the Community Engagement Stream at Royal Melbourne Institute of Technology (RMIT). Since 2016 it has been a part of the College of Design and Social Context's Innovation Project (IP Project). Community engagement (CE) is increasingly recognized as a central component of policy and service development. Originally associated with international development, CE has increased in breadth and scope over the past decade. It is now found in fields that including urban development, community revitalization, mining, and construction. Increasing statutory requirements for public institutions to engage communities have generated growing demand for skilled CE practitioners across many industries and sectors. As employment opportunities increase, higher education courses in CE have emerged in a range of disciplines including Urban Planning, Communication, and Political Science. Yet, despite increased demand for CE professionals, to date there is little empirical research that explores the theoretical foundations of CE pedagogy or provides guidance on how academia and industry could collaborate to increase professionalism in CE practice. To address this knowledge gap, the International Association of Public Participation (IAP2) Australasia has embarked on a project to explore methods to advance CE professionalism in Australia and New Zealand.

As part of building this CE Stream, our Master of Communication Program consulted the industry as well as former and current students, and it has participated in the IP Project since 2016. This chapter reflects on those collective experiences. Part of our industry partnerships involved a collaborative pilot study, undertaken by RMIT and IAP2 Australasia, on CE training in Australia and New Zealand. This study reviewed the nature, design, and delivery of postgraduate CE courses and explored how academic-practitioner collaborations could strengthen the CE Stream at RMIT. IAP2 CEO Marion Short emphasized, in an email to the research's participants, that:

> IAP2 Australasia has partnered with RMIT University to review post-graduate courses in community engagement and public participation, offered by universities in Australia and New Zealand. As IAP2 considers the development of a

DOI: 10.4324/9781003282075-16

tertiary-based accreditation process for courses in community engagement, this review will provide an understanding of the elements of a typical tertiary CE course.

(Email communications from Marion Short, CEO
(Chief Executive Officer) of IAP2 Australasia
to Linje Manyozo, May 11, 2020)

This chapter draws on literature on communication and political theory to define terms used in relation to CE. That discussion will then be followed by a description of Australia's and New Zealand's CE regulatory frameworks, and a review of the literature related to CE teaching in higher education. A description of the research method is followed by a presentation of empirical evidence in the form of a thematic analysis undertaken within the framework of communication, social, and political theory. The chapter concludes by identifying key findings related to tertiary-based CE pedagogy.

### What is community engagement?

The concept of *community engagement* (CE) has become increasingly associated with institution-led efforts to partner with local or community groups to improve the quality of social services, strengthen local governance, and, sometimes, to genuinely empower community groups as a way of co-designing socio-economic interventions with them. There is no homogenous conceptualization of CE, a term that is used to encompass different approaches to involving communities in decisions that affect them (Brunton et al., 2017). The practice of engaging communities can occur in a diversity of contexts, from politics to industrial relations to global finance to clinical drug trials (Brunton et al., 2017).

In addition to diversity in contexts, a variety of CE concepts describe the ideology including dialogue, deliberation, community consultation, public participation, and collective decision-making. *Civic engagement* is another term used (Centers for Disease Control and Prevention [CDC], 1997; Ekman & Amnå, 2012). Concerned with communicative dimensions of participation, Rowe and Frewer (2005) proposed the term *public engagement* to describe information flows between participants and organizations during CE activities. In a recent article, Christensen (2019, p. 1) discussed the myriad terms used to describe the practice of public participation, noting that many Australian practitioners describe themselves as 'community engagement practitioners.' In this chapter, the term *CE* is used to describe working collaboratively with communities to address issues that affect them.

The literature provides some evidence of limitations affecting the field of CE, primarily that it is defined and described differently across geographical and disciplinary borders. The diversity of terms potentially confuses or limits professional practice by failing to account for the varied and nuanced CE approaches. In reviewing the literature in community health development, Kilpatrick (2009) described two dominant approaches in CE: the systems and empowerment discourses. The systems discourse exists where pragmatic institutions employ utilitarian models of community involvement and engagement to achieve outputs and outcomes as advocated by specific organizations or government departments. The empowerment discourse entailed a complete ownership of the program agenda by local communities (Kilpatrick, 2009). This does not mean that these two discourses are mutually exclusive. There are moments and occurrences where they work together. There are, however, central concepts that harmonized these two approaches. This has a huge bearing on how the subject ought to be taught in universities if it has to have any bearing on genuine community development.

### Components of community engagement

Divergent opinions regarding the critical components of CE and how it worked were evident in the literature. A common theme was that participatory approaches brought with them issues of power dynamics that had to be addressed if participation was to encourage voice and active listening (Renwick et al., 2020). To address the multi-dimensional nature of CE, many authors have proposed frameworks that differentiate participation levels and associated engagement techniques. Arnstein's (1969) seminal work defined participation as power and was perhaps the most well-known example.

The literature also revealed numerous practice-based CE models, developed by government departments and engagement consultants. CE scholars and practitioners were faced with this array of frameworks and models, with no definitive theory of what engagement 'level' or 'technique' enabled effective CE in each context (Rowe & Frewer, 2005). An audit of CE practice conducted by Victoria's Auditor-General Office (VAGO, 2017) confers with Rowe and Frewer. The report noted an inconsistent approach to participation, a lack of strategic approaches or consistency in frameworks used, and poor monitoring and evaluation (VAGO, 2017). This study built on VAGO's findings by examining how different university disciplines understand CE and whether a cohesive body of knowledge provided a foundation to advance CE engagement pedagogy, professionalism, and practice.

### Legislation, regulation, and policy

Increasing numbers of tertiary-based courses, CE courses have followed increased government legislation for public participation (Christensen, 2019; Johnston, 2010). A review of Australia and New Zealand's regulations for public participation revealed that both countries were members of the Open Government Partnership (2020), an international regulatory body that aimed to empower citizens to become publicly engaged.

The various Australian government states, departments, and sections do provide engagement resources, frameworks, and guidelines for public servants. The framework's prescriptive nature provided little acknowledgement that engagement initiatives might be context-specific, suggesting that CE was understood to be a series of technical skills, rather than a profession where practitioners were required to draw on their knowledge and skills when designing approaches (Bryson et al., 2013).

In addition, Australian Local Government Acts outline in-principle duties for local councils to engage and enable participation with their communities (Christensen, 2019).

In New Zealand, public participation information is provided by the Department of Prime Minister and Cabinet where legislation and departmental contacts are required to enact public participation in decision-making are listed. Additionally, regional councils are governed by one Local Government Act that includes consultation guidelines for local councils. This is because CE is ideally based on local needs.

Professional associations provided an additional form of industry regulation in many professions and sectors by establishing and policing codes of ethics, competencies, and standards (Fitch, 2016). IAP2 Australasia, the largest CE professional association in the region, promoted a code of ethics, hosts conferences, judged annual best practice awards, and conducted training programs for practitioners.

A recent study of Australian engagement practitioners reported that 23.2% of practitioners who had undertaken CE training in the last decade had completed an academia-based CE course as part of a degree, compared to 57.1% having completed an IAP2 practitioner-led training course (Christensen, 2019). Literature reveals academic concern at the increasing numbers of engagement consultants, at the expense of "genuine conversation and authentic engagement" (Carcasson, 2014, p. 2). Challenges encountered by academics and practitioners when attempting to work together is well-documented (Escobar, 2017; Kahane & Lopston, 2017). In the practitioners' world, not everyone agreed with the idea that scholarship could help, with many taking a pragmatic view that we need to do the job so do it, and forget the theory (Escobar, 2017). Academics were interested in what drives professional practice and explaining the paradigms in which practitioners worked.

### Researching community engagement in higher education

Information retrieval using the terms 'community engagement' and 'higher education' on RMIT University's library databases revealed that most community–university engagement literature was American-based. Described as service learning or community-engaged scholarship, the American model involved university students in community outreach programs to provide benefits to local communities, whilst teaching students to become civically and socially responsible citizens (Bowen, 2010; Pruitt et al., 2019). In Australia and New Zealand, universities were encouraged to engage with community needs, regional issues and economic development through research and teaching. Encouragement was manifested in the form of government and university research partnerships to address social issues or public infrastructure challenges, business-university partnerships, and government funding contingent on universities enabling civic engagement by students (Winter et al., 2006).

In Australian universities, community-based learning is predominantly conceived as work-integrated learning (WIL), internships, and student volunteer projects (Onyx, 2008). Likewise, in New Zealand universities, WIL is the most common form of student engagement. Our literature search did not find any information on university-based CE pedagogy in Australasia, or campus-based integration of teaching, research, and CE capacity-building (Carcasson, 2014; Renwick et al., 2020). The current study responded to Christensen's (2019) recommendation for further studies in CE education, by examining Australasian tertiary-based CE courses within a framework of transdisciplinary, cross-disciplinary, and interdisciplinary pedagogical approaches (Smith et al., 2014). Responding to and building on the literature and previous research, this study conducted the first in-depth examination of tertiary-based CE courses.

## Research methods

This study employed a number of research methods and tools to analyse the planning, content, and delivery decisions and pedagogical goals of postgraduate CE course design in Australia and New Zealand. We firstly considered this to be a Participatory Action Research (PAR) intervention as both of us had been involved in the IP Project to transform the Master of Communication. This mixed method approach was exemplified by Bowen (2010), who described the rationale for combining document analysis with other forms of data collection as a means to defend the credibility of sources as well as to reduce particular predispositions. The intention of the data analysis was to combine data from the interviews with textual analysis of related course documents to provide a detailed understanding of what was occurring in postgraduate CE pedagogy across Australasian universities, and to establish whether alignment with future employment prospects was evident.

Purposive sampling was used to identify all Australasian postgraduate CE courses offered in 2020. The research tools involved: (1) document analysis; (2) literature review; interviews with course coordinators offering postgraduate CE courses; and (3) participant observation. Our professional experiences were part of the IP Project implemented by the College of Design and Social Context from 2016. The IP was set up to implement program amendments addressing learning needs of our MA Students who expressed a desire to see specific specializations in the Master of Communication. Ethics approval was obtained, and eight course coordinators agreed to be involved.

## Data collection and analysis

Eight semi-structured interviews were undertaken by videoconferencing following an interview protocol based on the literature review. Otherwise, study documents included publicly published course outlines and reading lists. Textual analysis was selected as the method of document analysis. A matrix was developed in NVIVO 12 to code the data, based on the standard headings of published higher education course guides. Matrix categories included a course description (aims), learning outcomes, learning and teaching activities, resources, and assessments. The two researchers selected one course outline and coded it together, to ensure coding validity and reliability (Robson, 1993). Once data was organized into the categorization matrix, it was further coded into sub-categories using a deductive approach that was derived from theoretical knowledge identified in the literature review (Kyngäs & Kaakinen, 2020). Themes were further explored in semi-structured interviews with course coordinators, including the first author, who coordinated one postgraduate CE course.

## The Innovation Project

For us, the experience of teaching CE at master's level started in 2016, when our College of Design and Social Context started experimenting with the Innovation Project. A group of master's programs in the college were selected and brought together to see how they could work together toward developing more iterative and interactive forms of learning using digital technologies. The aim was to improve the customer experience of students. While this was a noble idea in emphasising the centrality of the student experience as being vital to formulating beneficial and relevant academic experiences, it was also understood from our Freirean background that a student cannot be a customer. A student, for us, is a co-teacher and a co-learner. But we also understand that these are liminal spaces that we have to navigate around in Western educational institutions that have their own established pedagogical approaches and preferences.

For the master's program, the Innovation Project provided an opportunity to revamp our offerings. Currently, the program offers a Master of Communication. It is a one and half year long program that has as key courses: Strategic Communication, Leadership and Storytelling, Global Communication in the Digital Age, Communication Management, Critical Inquiry in Media and Communication, and a Professional Research Project. Around these key courses are electives, such as Communication for Social Change, Stakeholder Communications, Civic and Community Engagement, and many other elective courses that are taught by industry practitioners. Central to the IP Project was the establishment of the Community Engagement Course. Based on a rapid appraisal conducted by our program in 2016–2017, it was clear students wanted to specialize in specific skills and field of communication. Two groups emerged: students wishing to specialize in public relations courses (Organisational Communication or Corporate and Public Relations) and students seeking to specialize in CE.

## Community engagement teaching

In August 2016, we approached the executive management of the school to propose a new NQF Level 8 course option for Community and Citizen Engagement to be offered in the Master of Communication. In that communication, we outlined four factors that justified the proposal. These comprised: (1) rationale for the minor amendment, (2) how the course option would improve the MC248 pedagogy, (3) evidence that the new course option would provide savings of $8,113.50 in the MC248 sessional budget, and (4) marketing opportunities for the School and MC248. This proposal was being made in light of our participation in the Innovation Project.

First, regarding the rationale, we noted that engaging with former and the then current students, industry practitioners and our key stakeholders showed that two main cohorts were emerging: The Corporate Communication and Communication for Social Change (CfSC) streams as separate career paths taken by students who enrolled in MC248. At present, the design of the MC248 scaffolded learning for the Corporate Communication cohort but did not provide the basic-level foundation for the Communication for Social Change cohort. The proposal was that Community and Citizen Engagement be taught as a Level 8 course to provide basic scaffolding for a proposed new Communication for Social Change stream. This, in turn, would provide foundational learning that would be built on in advanced courses COMM 2083 Communication for Social Change, COMM 1107 Stakeholder Relations and Emergent Media and the research courses offered in the final year. Secondly, the question of pedagogy was key in introducing a new way of delivery to students.

The teaching philosophy for this proposed course including Communication for Social Change Course, which we were already co-teaching, would be built on Freire's (1970) critical pedagogy as outlined in *Pedagogy of the oppressed*. We had argued that critical pedagogy would mean we would challenge students to come to class prepared, having completed not less than one reading, in order to ensure they were active participants in the tutorials and lessons. It also meant a prior reading of *Pedagogy of the oppressed* was strongly recommended before a student attended any courses within the CfSC stream. We emphasized that the tutorials would be student-led and would feature panels and debates in which the students were active participants. That the course expected students to bring their personal experiences in the classroom as such they have to be prepared to share their stories and background as a process of contributing to this course. This course was also rooted in group assignments, since horizontal learning was key to the teaching and learning ethos in critical pedagogy.

Thirdly, as a result of consultative meetings with students, industry practitioners and our organizational stakeholders including our industry fellows, a need for specializations was stressed rather than the generalized communication offerings in the program. It was also clear that an increasing number of our students were interested in contributing to social change and were thus interested in acquiring skills and tools to enable them to facilitate that community and societal change.

Fourth, there was a financial rationality to the proposal. It was noted that the teaching of the course would not put pressure on the financial and human resources. This was accomplished by converting an empty shell of a course that was no longer relevant. It was also pointed out that by offering Community and Citizen Engagement, which was going to be taught by ongoing staff, the sessional budget could be reduced by $8,113.50. Within this financial rationale was that Community and Citizen Engagement had a great market. It would be offered across the university as a School of Media and Communication elective. Additionally, the Victorian Government had an increased emphasis on CE in its local government and community development interventions, which meant they would be seeking training providers to produce graduates with CE skills. It was expected that a condensed version of the course could be offered as a 'Master Class' to test the external market.

**The course**

Central to our understanding of CE were a number of critical tools, such as tools and skills for living with and engaging with people. These encompass listening, dialogue, participatory action research, and the CE spectrum, Paulo Freire, and the IAP2. Based on our consultative meetings in 2016, an outline for the course was developed. The course would be divided into five modules to be delivered as an Intensive in a week.

First, regarding the structure, the course was usually delivered in a week as an Intensive with each day covering a module. The Module topics were key concepts and approaches in community and civic engagement; power, politics, and participation; facilitating engagement; planning CE; and industry engagement. The class began by reflecting on one or two significant readings or a video on the topic, where the class would have a debate on key topics and concepts. Students would then be divided into smaller groups to discuss specific concepts to enable them to arrive at their own explanations and examples. These would be discussed later and were shared with other group members. After discussing key concepts, students were sent back into groups to engage in scenario planning to discuss specific projects, which they brought to class or were assigned by the teacher. The last part of the class was where students return to their groups to complete the assigned tasks together.

Second, participatory technology has been key to the delivery. When it was originally launched in 2017, we used Google+ Community. This platform allowed students to organize live chat discussions with group members who were not in class. Later on, as part of the Innovation Project, RMIT's Digital Learning Team (DLT) would train us in online teaching using teams, which had just been introduced at that time. Another online teaching platform, Canvas has participatory learning facilities, but teams offered more solid approaches to participatory learning and teaching as students and educators could engage in real time using Notebook or other facilities, such as the Posts window. The Teams Platform enabled us to manage online and face-to-face with students to start with. We were also able to teach aspects of digital engagement (DE) efficiently. Students have had to learn to think about and design digital forms of CE.

Thirdly, the pedagogical approach has been Freire's critical pedagogy. This was based on the idea that banking education was out of touch with any transformative education. As such we emphasized the increased role of students in the learning process. This has not been without its challenges. Employing critical pedagogy in a Western university where there was an emphasis on students as clients and consumers has meant that we have had to repeatedly explain to students why this course, alongside the Communication for Social Change Course, had to be delivered that way. This was the reason the CfSC (or CE) Stream would be proposed so that it meant meeting the different learning needs of students in this area.

**Partnership with the International Association of Public Participation**

Australia and New Zealand have a well-established tradition of CE in local government. The International Association of Public Participation was an embodiment of that. The Association was established in the 1990s as a member-driven organization sought to consolidate the research, training, networking, and capacity building in all forms of public participation. Regarding training and development, IAP2 published the *Journal of Deliberative Democracy* (formerly *Journal of Public Deliberation*) as well as it offered online courses in these areas: Foundations in Public Participation, Strategies for Dealing with Opposition and Outrage in P2, IAP2 Certificate in Public Participation, and Public Participation for Decision Makers (IAP2, 2007).

IAP2 would be consulted with the aim to partner the MA Communication Program. The consultation with IAP2 involved two strategies. The first entailed engaging with CE specialist organizations and firms to enhance our teaching so that it rises to the level of industry expectations and in this area, the Melbourne-based CE consultancy firm, Capire Consulting has proven critical in strengthening CE practice in our teaching. The second strategy involved bringing in IAP2 to link their online certificate offerings in line with this master's course, and, of course, the CE Stream as we would define the CfSC stream later. This would be done through a collaborative research project between our program and IAP2 to research tertiary offerings in CE in Australia and New Zealand.

For our MA Communication Program, the interest was to link with the two Certificate Courses: The IAP2 Certificate in Engagement and the IAP2 Australasia Advanced Certificate in Engagement. The Certificate course targeted emerging practitioners and was delivered online (with face-to-face interactions available). The aim was to introduce students to and engage them in "the IAP2 concepts of Core Values of Public Participation, Code of Ethics and the IAP2 Spectrum," while examining "key concepts and processes in IAP2 Australasia's Engagement Model" (IAP2 2007). Courses included Engagement Essentials, Engagement Design, Engagement Methods, Conflict in Engagement, Engaging with Influence, Engagement Facilitation, and Engagement Evaluation (IAP2, 2007).

On the other hand, the Advanced Certificate Course targets experienced practitioners "whether at the front-line, design and planning stages; or at the management and leadership level overseeing engagement practice—the Advanced Certificate is designed to stretch you as an individual and as a practitioner" (IAP2, 2007). Courses included Strategies for Complex Engagement, Building an Engaging Organisation, and Strategies for Dealing with Opposition and Outrage (IAP2, 2007).

The outline and objectives of these certificate courses provided the epistemological and empirical framework for developing the content of the Community Engagement Course as well as building a framework for the Community Engagement Stream in the Master of Communication.

## Lessons learned

### Lesson 1: Digital engagement vs community engagement in the digital era

Our pedagogical experiences with students have shown the stark differences between two forms of engagement. This was reflective of agriculture and rural extension practice in general. We learned that in discussing CE in a digital context that there needs to be nuance. DE entails online forms of engagement that do not necessarily have anything to do with community. It is about harnessing the power of hardware and software to reach out to more people—probably because face-to-face opportunities are not available or they are limited. Agriculture and rural extension were able to use DE as a form of mass communications. Critically, DE is about reaching out. It is about extending that reach out to ensure increasing numbers of people are included in the process of participating in saying something or accessing certain social services. DE is more often about ensuring the biggest audience reach.

On the other hand, CE in the digital era entails thinking of how to execute CE using ICTs especially computers and the Internet. How do we apply the face-to-face strategies of CE using online platforms? These strategies that include community cafés, participatory action research, appreciative inquiries, or theater for development? These are important considerations because they remind us that certain CE strategies are not transferable but can be negotiated. In the face-to-face version, participatory mapping largely involves development or community workers liaising with community groups to use draft community maps to identify and describe services as part of community audit.

Yet, in the digital version of CE, as in agriculture and rural extension, the strategies will have to vary but with principles that remain the same: inclusion, interaction, and participation. Participatory mapping, for example, can be replaced with a platform such as Ushahidi, the Swahili word for testimony. It refers to, according to Wikipedia, an "open-source software application" that uses crowdsourcing, as a way to use "user-generated reports to collate map and data." Originally employed after the contested 2007 elections in Kenya, Ushahidi aggregated mobile phone text and email eyewitness testimonies of political violence in poorest neighbourhoods and then recorded them on Google Maps.

Considering the question of whether DE required us to examine our reference points when it comes to engagement in and with the community within the field of media and communications, there are two dominant views regarding how the field approaches the question of engagement. This does shape the way we think about DE. In the face-to-face engagement, as was elaborated in Arnstein's concept of a ladder of citizen engagement (1969), there are issues that stemmed from non-participation, tokenism, and citizen power. Within these three phases are the eight 'ladders' comprising manipulation, therapy, informing, consultation, placation, partnership, delegated power, and citizen control. These eight ladders are divided into three forms of engagement: non-participation, tokenism, and citizen power.

For Arnstein (1969), *non-participation* involves informing, therapy, and manipulation all of which create an illusion of people feeling they have been involved and engaged. *Tokenism* involves a small degree of informing, and more of placation and a little bit of consultation. Key to placation is reporting in documents that citizens were involved or are being involved, and yet within that institutional arrangement, it is organizations that have the control of the narrative. Citizen power seems like the highest form of CE. For Arnstein, *citizen power* enabled citizens to use financial, research, and epistemological resources available to them to exercise a degree of independence in making decisions affecting their lives, families, and communities.

### The public relations approach

In the public relations approach, engagement became more about generating and yielding audience and customer views with the purpose of increasing sales margins for organizations or strengthening their public relations profile. We understand that Grunig and Hunt managed to identify four models of public relations, namely, press agentry / publicity, public information, two-way asymmetrical communication, and two-way symmetrical communication. Central to these approaches and theoretical frameworks was the idea that the organization needed to harness an understanding of its audiences so that the generated communications were widely accepted and understood, and in the process, strengthened the business profile of the institution. Hueffner (2020) observed that "better digital customer engagement leads to more customer data and better customer experiences, which can lead to higher profits."

### The empowerment approaches

We have empowerment approaches to engagement, which shape the way we think about DE. In this case, engagement at the community level especially in agriculture and rural extension. It is about providing the space where local people and groups, such as Farmers' Clubs, can contest the flow of socio-economic and political power. In that framework, the digital is not just a tool. It is embedded in the agriculture and rural extension strategies. In this case, therefore, extension approaches have tended to use mobile phones, Twitter, Skype, laptop, or participatory technology as platforms for implementing participation in practice. But there are certain elements that are critical to this discussion, and Gumucio-Dragon's (2003) "Take Five" piece offers a platform for thinking about CE in the Digital Age within the development context. Gumucio-Dragon (2003) argued that serious questions raised about modernist development need to be asked of ICTs especially in their application to development contexts and that this has consequences for CE.

### Lesson 2: Spread of CE courses

A major finding of the study was a revelation of the number of CE courses on offer at different universities. Whereas CE initially started as an extension strategy of most universities to connect with local communities—to share knowledge, it has continued that tradition. However, it has done so with the growing realization that this is a great opportunity as an academic discipline. What the research revealed is that within the academy there is a consistent understanding of CE as a practice during which people participate in decision-making on issues that affect their lives. In this case, the research left out extensionist CE programs that have nothing to do with training. Table 13.1 provides a view into these offerings:

*Table 13.1* Community engagement training

| Discipline | Community engagement course | Degree | University |
|---|---|---|---|
| Social Science (six courses) | Public Involvement & Community Development | Urban & Environmental Planning | Griffith University |
| | Participatory Planning | Urban Planning | University of Melbourne |
| | Community Engagement | Environmental Policy & Management | Adelaide University |
| | Community Engagement | Graduate Diploma Communication (Local Govt.) | Southern Institute Technology |
| | Community & Civic Engagement | Communication | RMIT University |
| | Communication for Social Change | Communication | RMIT University |
| Political Science (two courses) | Participation, Community Engagement & Public Talk | Public Policy: Policy Communication | Australian National University |
| | Digital Engagement in the Public Sector | E-Government | Victoria University Wellington |
| Science (two courses) | Effective Stakeholder Engagement | Agribusiness | University of Queensland |
| | Community Planning & Participation | Urban & Regional Planning | University of Queensland |
| Business (one course) | Community Consultation & Engagement | Business | Queensland University Technology |
| Engineering (one course) | Community Engagement for the Resources Sector | Responsible Resource Management | University of Queensland |

## Lesson 3: Understanding of community engagement

Consistent with literature reviewed (Renwick et al., 2020), 'participation' and 'community engagement' were key terms that were used interchangeably by respondents, yet recognizing the contradictions in practice. There was a recognition that even the most inclusive of strategies cannot be "inclusive for everyone," after all, practitioners have to "make decisions about who to include and who to exclude" (C5, 2021, Interview in Lawry, 2020). (All the interviews that follow are from Judy Lawry, 2020.) Along with that understanding of these key concepts, the participants also expressed an appreciation for power as a key component of CE in line with Johnston (2010) and Renwick et al. (2020). Interviewee C7 (2021) conceptualized CE as involving an "understanding that we are not equal and that how you relate to the public." These findings showed a consensus regarding four components considered fundamental to CE (Renwick et al., 2020).

These components were participation, paying attention to the inclusion of multiple views, existence of power struggles during decision-making, and community empowerment. There was a general understanding of what CE was all about. More than half the respondents described CE as a practice that empowers community members. For Interviewee C5, practitioners work in CE "because they want to empower communities and because they want better outcomes." This view is supported by C1's conceptualization, who described CE as a strategy of "giving power over, and I'm … interested in how communities take power back" (C1, 2021: Interview in Lawry, 2020).

Despite the variations in and where CE was located as a discipline, respondents seemed to agree about the centrality of CE as a central communicative strategy in "the context of planning as government casting a net and bringing community in conversation … on planning proposals, planning projects" (C1, 2021: Interview). University educators' pedagogy was shaped by this belief, such that CE was a process that brought people into decision-making that affects their lives and their communities, broader than any university discipline boundary. Therefore, teaching CE entails providing a "broad curriculum perspective … for students in terms of helping them think critically about what … community is and, and what engagement is" (C5, 2021: Interview).

## Lesson 4: A cohesive body of knowledge

The study sought to determine whether a cohesive body of knowledge was practiced across disciplines. An important emphasis in most courses is on theories and models of power and participation (90%) and the development of critical analysis and reflection skills. Results concur with literature that knowledge of CE's legislative environment is important (Christensen, 2019; Johnston, 2010), with regulations for public participation taught in 73% of courses. As one interviewee explained, "in order to understand participation, you have to understand the governance system on which it's being conducted" (C1, 2021: Interview). Perhaps missing from most of these courses is the fact that DE or CE in its online platform would look like. It has to be acknowledged that the research was conducted during COVID-19, and even though most universities have been involved in online deliveries (including courses in CE), delivering these courses with digital implementation in mind was something that was being thought out.

Seminal work by Arnstein (1969) on power and participation as well as Jürgen Habermas on communicative action and power were evident in most courses across academic discipline boundaries. Findings revealed that social movement concepts are included in 55% of courses situated across four different disciplines. Deliberative democracy, Indigenous health, stakeholder analysis, and facilitation skills were also taught in more than half of the courses (Table 13.1). These results indicated that a cohesive, transdisciplinary body of knowledge was taught across academic discipline boundaries.

Interviewees described how the *context* of CE influences decisions on what literature to include. Consistent with Smith et al. (2014) description of cross-disciplinary pedagogy, one interviewee describes using the context of mining was a dominant analytical lens to view cross-discipline CE concepts. In another course, concepts from various disciplines were analysed through the lens of public relations. As observed by some respondents (C4, 2021: Interview; C7, 2021: Interview), cross-disciplinary pedagogy was also used in urban planning CE courses where discipline-specific literature was combined with communication and political science literature. Similarly, communication-based CE coordinators described the importance of introducing seminal work on power and participation from urban planning and political science disciplines, while providing a discipline-specific perspective from communication scholars.

**Lesson 5: Centrality of Freire**

The study also examined pedagogical approaches used in CE courses, followed by an analysis of preferences to industry involvement in course design and teaching. As class format and delivery options were influenced by student numbers, interviewees were asked the average size of their CE class (Harfitt, 2013). Classes ranged from 15 to 300 students. One large class enrolled between 130 and 200 students from a range of different master's programs. More commonly, however, class sizes comprised between 15 and 45 students. Harfitt's (2013) description of transformative learning styles corresponded to findings from 45% of respondents, with Freire's transformative pedagogy a commonly reported approach (C5). Likewise, Interviewee C2 emphasized the centrality of Freire in teaching CE as Freire's works center around the collaboration between the student and the teacher. Alongside Freire's pedagogy, another theoretical framework has been Arnstein's Ladder of Participation (C5, 2021).

Most CE classes have largely been delivered as combined lecture / workshops where student-led role-plays and discussion were used with case studies. Attempts to weave theory, practice, and self-reflection throughout a course were also reported. As observed by C5 (2021: Interview), for example,

> theory kind of does come the whole way through in the sense that, throughout they've got readings that are … a mix of … more practical case study type stuff. And … also … more abstract readings. They do a reflective learning journal throughout.

In these pedagogical strategies, lectorials are employed as interactive lectures and discussions to introduce theoretical concepts in a mixed undergraduate-postgraduate course offering. Additional extension seminars for postgraduate students were offered each week to provide more theoretical discussion in terms of community participation frameworks and why community participation in planning is important. Most of these learning opportunities and facilities provided opportunities for all students to interact and network with industry members.

## Conclusion

Teaching CE always comprises lectures, industry sessions, tutorials, and assignments. Lectures can be delivered both online and face-to-face, which means moving these interactions to the digital platform would not be challenging. Postgraduate CE courses are situated across a range of academic disciplines, indicating that universities responded to the need for CE expertise in a variety of employment contexts. Delivering CE digitally entails the need to rethink the delivery mechanisms: How do we challenge the students as we would in face-to-face situations?; How would scenario planning be implemented?; And, is it just by watching YouTube videos?

Second, a cohesive, transdisciplinary body of knowledge was taught across discipline areas. Critical analysis and reflection, power and participation, CE governance, diversity, and social movements are fundamental concepts in CE pedagogy, regardless of academic discipline or engagement context. These findings provide a foundation for standardization in CE pedagogy and offer valuable insights to increase CE professionalism (Christensen, 2019; Rowe & Frewer, 2005).

Third, findings confirm a disjuncture between CE teaching in universities and practice occurring on the ground, with empirical evidence supporting claims in the literature that academia and practitioners face challenges when attempting to work together (Escobar, 2017; Kahane & Lopston, 2017). This has an impact on the way we think about agriculture and rural extension pedagogy and practice. There was no evidence of academic-industry collaboration to align course content with CE emerging employment trends, and barriers to expanding industry-academia collaboration were identified. An important finding was that practice-based CE models were critically analysed within a framework of theoretical concepts in higher education courses. Combining this academic, conceptual interrogation with practice-based techniques highlighted a difference between university CE courses and skill-based CE training in developing CE professionalism. Two opportunities that could further CE course alignment with employment trends and advance CE theory and practice, emerged from the study. A proposed IAP2-led micro-credential module for inclusion in higher education CE courses, and an industry-academia 'community of practice' have potential to reduce barriers to CE industry-academic collaboration, and to bridge the current divide between skills-based training and university-based education in CE.

## References

Arnstein, S. (1969). A ladder of citizen participation. *Journal of the American Institute of Planners, 35*(4), 216–24. https://doi.org/10.1080/01944366908977225.
Bowen, G. A. (2010). Exploring civic engagement in higher education: An international context. *Asian Journal of Research and Synergy, 2*(2), 1–8.
Brunton, G., Thomas, J., O'Mara-Eves, A., Jamal, F., Oliver, S., & Kavanagh, J. (2017). Narratives of community engagement: A systematic review-derived conceptual framework

for public health interventions. *BMC Public Health, 17*(1), 944. https://doi.org/10.1186/s12889-017-4958-4

Bryson, J., Quick, K., Slotterback, C., & Crosby, B. (2013). Designing public participation processes. *Public Administration Review, 73*(1), 23–34.

Carcasson, M. D. (2014). The critical role of local centres and institutes in advancing deliberative democracy. *Journal of Public Deliberation, 10*(1), Article no. 11. https://doi.org/10.16997/jdd.199

Centers for Disease Control and Prevention (CDC) (1997). Screening young children for lead poisoning: Guidance for state and local public health officials. Atlanta: Department of Health and Human Services. https://www.cdc.gov/nceh/lead/publications/screening.htm

Christensen, H. (2019). Participatory and deliberative practitioners in Australia: How work context creates different types of practitioners. *Journal of Public Deliberation, 15*(3), 5.

Ekman, J., & Amnå, E. (2012). Political participation and civic engagement: Towards a new typology. *Human Affairs, 22*(3), 283–300. https://doi.org/10.2478/s13374-012-0024-1

Escobar, O. (2017). Making it official: Participation professionals and the challenge of institutionalising deliberative democracy. In L. Bherer, M. Gauthier, & L. Simard (Eds.), *The professionalisation of public participation* (pp. 141–64). London: Routledge.

Fitch, K. (2016). Professionalising public relations: History, gender and education. London: Palgrave McMillan.

Freire, P (1970) Pedagogy of the oppressed. New York: Continuum.

Gumucio-Dragon, A. (2003). Take five: A handful of essentials for ICTs in development. Paper presented at the Food and Agriculture Organisation (FAO) *International Workshop on Information and Communication Technologies Servicing Rural Radio: New Contents, New Partnerships*. https://www.fao.org/3/Y4721E/y4721e05.htm#TopOfPage

Harfitt, G. J. (2013). Why 'small' can be better: An exploration of the relationships between class size and pedagogical practices. *Research Papers in Education, 28*(3), 330–45. https://doi.org/10.1080/02671522.2011.653389

Hueffner, E. (2020). What is digital engagement? + Customer engagement strategies. 2020. https://www.zendesk.com/au/blog/digital-customer-engagement/

IAP2 (2007). IAP2's Public Participation Spectrum. https://www.iap2.org.au/resources/spectrum/

Johnston, K. A. (2010). Community engagement: Exploring a relational approach to consultation and collaborative practice in Australia. *Journal of Promotion Management, 16*(1–2), 217–34. https://doi.org/10.1080/10496490903578550

Kahane, D., & Lopston, K. (2017). Negotiating professional boundaries: Learning from collaboration between academics and deliberation practitioners. In L. Bherer, M. Gauthier, & L. Simard (Eds.), *The professionalisation of public participation* (pp. 165–88). Routledge.

Kilpatrick, S. (2009). Multi-level rural community engagement in health. *Australian Journal of Rural Health, 17*(1), 39–44. https://doi.org/10.1111/j.1440-1584.2008.01035.x

Kyngäs, H., & Kaakinen, P. (2020). Deductive content analysis. In H. Kyngäs, K. Mikkonen, & M. Kääriäinen (Eds.), *The application of content analysis in nursing science research* (pp. 23–30). Springer.

Lawry, J. (2020). *Collected transcripts from the interviews with CE practitioners, Researchers and educators: A review of community engagement pedagogy in higher education in Australia and New Zealand*. [RMIT University, Unpublished.]

Onyx, J. (2008). University-community engagement: What does it mean? *Gateways: International Journal of Community Research and Engagement, 1*, 90–106. https://digitalcommons.unomaha.edu/slcepartnerships/41

Pruitt, S. T., McLean, J. E., & Susnara, D. M. (2019). Building a conceptual framework for community-engaged scholarship. [Paper presentation.] 20th annual engagement scholarship consortium. Deepening our roots advancing community engagement in higher education. Denver, Colorado, USA. October.

Renwick, K., Selkrig, M., Manathunga, C., & Keamy, R. K. (2020). Community engagement is…revisiting Boyer's model of scholarship. *Higher Education Research & Development, 39*(6), 1232–46. https://doi.org/10.1080/07294360.2020.1712680

Robson, C. (1993). *Real world research. A resource for social scientists and practitioner–researchers.* Blackwell Publishers. https://doi.org/10.1177/144078339403000311

Rowe, G., & Frewer, L. J. (2005). A typology of public engagement mechanisms. *Science, Technology and Human Values, 30*(2), 251–90. https://www.sparc.bc.ca/wp-content/uploads/2020/11/a-typology-of-public-engagement-mechanisms.pdf

Smith, D., Tiwari, R., & Lommerse, M. (2014). Navigating community engagement. In R. Tiwari, M. Lommerse, & D. Smith (Eds.), *M² models and methodologies for community engagement* (pp. 1–24). New York: Springer.

VAGO (2017). *Public participation and community engagement: Local government sector.* Victorian Government Printer. https://www.parliament.vic.gov.au/file_uploads/VAGO_PP-Local-Gov_qQN5QyHk.pdf

Winter, A., Wiseman, J., & Muirhead, B. (2006). University-community community engagement in Australia: Practice, policy and public good. *Education, Citizenship and Social Justice, 1*(3), 211–29. https://doi.org/10.1177/17461979060646

# 14 Masculinity and participation in China

## Exploring ideals and practices of development in a Heyang village

*Byron Hauck*

## Introduction

I used a transcultural political economy of communication framework to specifically address how cultural practices, such as idealized characteristics of masculinity transform and are negotiated, in response to prevailing political economic conditions (Chakravartty & Zhao, 2008). With reference to stories shared by three Chinese men regarding their understandings of participation and development in their home village, this chapter investigates masculinity and participation in development in China. These perspectives provide a contrast to how participation and development are idealized moving from individual ability to finding mutuality in collective dependence. A key aspect of how this participation occurs is in the shared-time temporality of using time keeping to collectivize efforts aimed at addressing the well-being of others.

## Masculinity in China

Gender is performative (Butler, 1999). It is about individuals trying to act with reference to an ideal (Butler, 1999). For masculinities the ideal is also the hegemonic form of patriarchal systems (Connell, 2020). While masculinity studies became a rising field in the mid-1990s, such research primarily focused in on Western gender performances leading Connell to call for more global understandings of the construction and performance of masculinity (cited in Louie, 2014). Since 2010 scholarly work on Chinese masculinities has grown (Louie, 2014). However, this research has primarily focused on urban masculinity with rural men mostly referred to as a sounding board of negative qualities against which urban ideals are constructed.

One of the most dominate frameworks of Chinese masculinity is Louie's (2002) *wen* / *wu* bifurcation. In this hierarchical division *wen* is related to mental abilities. While traditionally associated with the cultural capital of passing the civil service examination *wen* is now expressed through university education and the prowess required for financial success (Louie, 2014). *Wu*, in contrast, relates to martial skills with the ability to commit violence and the restraint to know how and when to deploy violence. Louie claims that while the most ideal, and thereby hegemonic,

DOI: 10.4324/9781003282075-17

masculinity should display a mixture of these qualities, *wen* dominates over *wu*, or rather, men who demonstrate qualities of a refined culture dominate over those who are rough around the edges.

Lin (2013) deployed Louie's framework to describe peasant migrant workers as displaying *wu* characteristics citing government documents that label them as "an army." Lin claimed that while valued under Mao Zedong's leadership a once hegemonic *wu* masculinity is no longer desirable but rather indicative of the contrasts of tradition against a *wen* modernity. Lin then focused his analysis to show that male peasant workers were not hegemonically *wu* but must negotiate their gendered and othered status through the class conditioned contexts of China's project to modernize.

Liu gives a nod to the *wen* / *wu* distinction but claims that a new form of hegemonic masculinity is at work. This he encapsulated under the term *chenggong* (success; outstanding accomplishment) and this was associated with younger men, including rural-urban migrants (2020, p. 115). He contrasted this against the concept of *guorizi* (literally, passing the days; getting by in a livelihood; making ends meet) that he ascribed to older men, especially those who still lived in villages. Liu asserted that the ideal of *guorizi* is related to disenchantment with past Communist propaganda that one could contribute to the nation and be rewarded for self-sacrifice (2020, p. 125). For both Liu and Lin, socialist masculinities are no longer a source of dignity, or a form of belonging, that is part of contemporary China.

Liu's *chenggong* framework helped to highlight the cultural transformations in China from its traditional society through socialism and into the contemporary era of China's engagement with global capitalism. It is not, however, a holistic reworking of Louie's discussion of *wen* masculinity, which is also recognized as having adapted from public service to financial success. From this continuity we found that there was epistemic prioritization of dominate political economic institutions over the formation of social roles such as masculinity. Capitalism has colonized cultural forms by eschewing past social ethos that do not relate to its regime of accumulation.

## Theories of participation

The ideal, or presumed, agent of participation theory, when discussed as a universalized abstract and when not particularized to perspectives such as gender, is masculine (Smith, 1990). The universalized actor is often framed as a nonracialized middle-class male. This hegemonic actor is very similar to the theorized ideal Chinese masculinity. We can also abstract this universalized agent from Carpentier's theorization of participation.

Carpentier recognized that the entomological root of participation has "a strong emphasis on 'taking part,' which implies the presence of more than one actor" (2014, p. 1002). He proposed that participation is valued because of the belief that one's actions will have an impact on what he calls, "the make ability of the social, or in other words, the belief that individual agency... 'truly' matter[s]" when "actions ... reach beyond the individual level" (2014, p. 1007). Carpentier believed that "This stretching beyond individual control [however] is exactly the characteristic of structure that frustrates" the concepts of "freedom and agency," which inform the normative good of participation (2014, p. 1006). In explaining this concern, Carpentier (2014) sided with Giddens in recognizing that "structure is the counter-weight of agency" (p. 1006). He observed that within this context of people coming together there will be power imbalances that would be "addressed (and limited) through the participatory process itself" promoting further equality (2014, p. 1002). However, the mechanisms by which more opportunities to participate lend their selves to more equality were not described.

In these commitments to an antagonistic society of individuals whose agency is limited by the participation of others, Carpentier laid down ontological assumptions that corresponded to a world of independently abled men. The need to depend on others as well as the need to take actions or make decisions was minimized. The ideal actor is one who was able to take advantage of opportunities to participate. In such forums, they participated to such a degree that they can reproduce the material and social conditions to participate again.

A different take on participation comes from Dean (2012). Dean critiques the liberal celebration of "participation" as intertwined with capitalism because the emancipatory potential of participation is pacified through affective appeals to one's individuality. She argued that digital networks feed off of our human desire for collectivity, claiming:

> The very appeal, the affective charge, of the spectacle [of digital communications] is its mass quality, the way it makes us feel connected to a larger 'we' to which we belong.
>
> (2012, p. 151)

In contrast to the individualizing appeals of participation in capitalist contexts Dean developed the term "comrade" as a form of political belonging:

> The term comrade indexes a political relation, a set of expectations for action toward a common goal. It highlights the sameness of those on the same side .... Comradeship binds action, and in this binding, this solidarity, it collectivizes and directs action in light of a shared vision for the future.
>
> (Dean, 2019, p. 2)

In this framing of participation, there is an emphasis on recognized mutuality in another as co-existing beings having intentionality. Participation is not just about impacting a wider society but also about finding belonging in collectivity. Where this affective desire is at once promised by digital networks and their objective powers of connectivity, it is revoked by the privatization of the knowledge that arose from participation in them (Zuboff et al., 2019).

Where the surveillance of digital companies is, at best, indifferent to individualized users simply seeking data points, and was interested in, at worst, keeping users to keep behaviors transparent and recordable (Zuboff et al., 2019). We might think it a better facilitator for our participatory actions than the state, especially a Communist state. Digital companies ask no obligations from us except to assert our agency, to be like the idealized male who was willing to act without considering what such action depend upon, but merely concerned that actions produce. The autonomy expressed in ideals of participation then are free from obligation, this is the participation realized by the mandarins of our current age, universalized and normatively projected as possible for all.

China's historic engagement with Communism and mechanisms to realize a collective "we" reverses these assumptions. Leaders should not be free, autonomous, and valued according to the individualized personalities they bring forth. Rather, they are asked to recognize the dependent nature of the masses. They are asked to participate in the disadvantage of dependance in order to help bring about the sharing of the benefits of participation in a modern industrial society. The point of participation is not simply to make a place for one's individualized self but to work toward creating comradeship to emancipate a community from the conditions that marginalize it. In Heyang village, understandings and conceptions of empowered participation do not fetishize a digital solution but point us to the importance of social structures.

## Development in a Heyang village

My research in a Heyang village took place during multiple visits between 2015 and 2019. I was able to conduct research in Heyang because of Yuezhi Zhao's Heyang Institute for Rural Studies was set up to enable urban and international scholars to conduct field work in rural China (Zhao, 2017). My field work consisted of six months of ethnographic research, participating in daily labor such as small plastic manufacturing with women and village inspections with the male fire brigade. Two rounds of focus group interviews totaling 27 groups with 172 participants, and 15 one-on-one interviews were also conducted. Participants in the focus groups were chosen to represent a large range of age, gender, class, and education levels in the village including ordinary residents, people who work in the village (be it employees of the state-owned Tourism Management Committee, or at the school), and local leaders. The participants cited in this chapter represented a small selection of this research but helped to provide a focus on core issues of development and participation. Importantly, being men who had different relations to local government they helped provide insight on the place and function of masculinity in the processes and evaluations of participation in development.

Heyang is a village in the central mountains of China's Zhejiang province. It represented some of the development struggles of rural China in that its local leaders had plans for how it could be integrated into nationally supported programs to attract wealth from urban markets. This developmental path was heritage tourism based on the existence of numerous family homes and ancestral halls preserved since the Ming Dynasty (1368–1644). The plans for this developmental path formed under the local party secretary and mayor in power during the late 1990s. Their decision led to the formation of a state enterprise for tourism and a management committee tasked with curating the tourism potential of Heyang (Hauck, 2020). As such the governance of this village was caught between the negotiations of those responsible for preserving architecture and building new facilities and the village council responsible for general governance. Major decisions regarding building projects and changes to the village's material structure were decided through a joint committee.

The impact that tourist development had on resident villagers is largely experienced in terms of housing. The central core of Heyang is where most of the preserved structures are located, and it had a special designation limiting the construction and renovation of buildings to set historical standards. Some wealthy residents accessed small grants and gained permission to rebuild their homes. However, many who lacked the wealth to buy the materials and build to code were left with leaky roofs and poor living conditions. Caught in the capitalist realism of the moment, money was agreed to be the main factor impacting individuals' and families' living standards. Informed by the past experience of China's Communist revolution, however, visions of what socially just development could look like were very socialist in nature and related to state society relations under Mao.

Three men were selected and their responses from focus groups were edited into short vignettes to highlight how their social position and personal experiences informed their understanding and engagement with Heyang's development. These men were given pseudonyms to protect their identities.

### Xin's perspective from the top

Xin was a 40-year-old with a middle school education. He ran his own business based on his farm work and his wife had a clothing store in the local township. When he was young, he left Heyang to help his cousin run a business in Shanghai but returned after finding that the cost of living was too high for him there. He was one of the newest village council members.

In reviewing recent development projects, he highlighted that there is a major contrast between projects developed and carried out locally and those instituted by higher levels of government. One recent project was the construction of new sewer systems that were carried out at the county level:

> These [new sewage structures] only cause damage, they're of no use. This is because they didn't even notify our village when they came to do the initial check and inspection of the project. If they actually sent someone to take a second look, it would be clear that it wouldn't pass inspection at all. Was this really done in the name of benefiting the lives of villagers and common folk? I think not.

In contrast, he has a high praise for the work carried out by the village council:

> I think our current cohort is doing quite well. Why? Because when we are doing things for the village, its visible [people can see us], we act fairly and we are attentive ... [But] when it comes to something as simple as tearing

down pig pens, the person in question's interest will be the most directly impacted .... However, all we can do is balance this bowl of water as best as we can .... But to some, they might start comparing whose home is bigger, better, etc. When we encounter these situations, we typically do try to let villagers get a bit of the upper hand. Whatever we want to do in the village, we still have to yield to villagers' interests a bit more.

From these accounts we see that there was a desire to put people first. There was the mass line understanding that leaders should be seen to be involved in seeing projects were working and attentive to the material politics of the impacts that decisions can have on individual families.

This touches on Xin's ideas of development, "We actually have an opportunity here, why? Because right now it's all about urbanization, if we can develop into a place that the city aspires towards, then that would be pretty good."

This notion of development as integrating with urban markets was the ideal toward which many rural development projects aim. Here again Xin saw the path forward according to a *chengong* financial understanding of success related to processes of urbanization.

But demonstrating a high culture, big picture *wen* understanding of the success of village work, Xin does not boast about individual leadership but emphasizes the importance of unity:

Our village is actually better than the average village, why? Our village is still relatively cohesive, we're all walking in the same direction, all working towards the village's development. Us village cadres, party secretary, and old cadres, including the party member representatives, we've all come to gradually recognize that the only way is forward.

Xin emphasized this unity as part of the village council's methods for development:

The entire year last year we held numerous meetings. For instance, in the past, it wasn't likely that we'd ever really hold the big Village Representative Assembly more than once or twice. But now it's different. Now we have several in one year alone. This is because they sincerely want to make an effort for the village's development. Even though we have no money, but we're trying our best to make significant changes to our village with the little we have.

This emphasis on unity can speak to the *wen* of a ruler being responsible for producing harmony in society. But where the traditional Confucian ideal was for a strong leader to make decisions and have them be carried out Xin's ideal differs. Not only is there the sense that leadership works best when there is regular dialogue but also that depending on the organization of local people can achieve results—even when there was little money. This stance then breaks free from both *wen* and chengong understandings of masculinity offering up a desire for neither the "tradition" nor the neo-liberal "modern," but an openness to the agency of those involved to confront the present by sharing understandings of conditions and abilities in dialogic meetings.

As united as Xin claimed the village to be, he had a very stark appraisal of the need to develop claiming:

> It's onwards or die. We really feel this way now. Because our skills, financial resources, and all other aspects are limited. What we can do is limited. But at least for now, we are still going down a positive developmental path. Otherwise, we'd just disappear.

What Xin meant by disappearing was the notion that village life would no longer be economically sustainable for its residents. He compared Heyang to some neighboring villages:

> There aren't even any more people left [in those other villages]. For instance, a village of 500 people, if you visit on a regular day, you'll just see a couple dozen old men and old ladies, what are you going to use to revitalize it? They have nothing to eat at home.

These appraisals reflect the precariousness that informed Xin's marginalized male subjectivity. Here the struggle to still desire market success was not just about success but the means of having ends meet in capitalist conditions. While individual financial success may be a hegemonic form of masculinity, the failure to achieve this ideal was not to become passive, but to become more concerned about collectivity and the need for a material base for marginal people to organize from.

### Ding's perspective from the outside

Ding was a man in his late 50s who was part of the Lotus Cooperative in Heyang. Never seeing a future in farming, he claimed that he was one of a few who would venture out of the village before migrant labor was legalized in part with China's 1978 reforms. He returned to Heyang full time in 2007 and while he was a member of the farming co-op, he claimed that most of his work was in local factories to make money. He was frustrated with the governance of his village. According to him this was because while Heyang had a main plan to develop tourism, the plans for how to accomplish this changed with different directors of the tourism management company.

Ding was a member of the Red Guard in his youth. Recollecting this time he says, "Back then, we'd all have to participate in some way when something happened in the village. Basically, whenever there was a large assembly held, we'd all have to go." When I inquired about how decisions for the village or the Red Guards were decided he remembered:

> It came down level by level. It was always passed down through the Party Committee. The township, the village, the commune to the brigades, and from the brigades to our respective production teams. That's how the directives were sent down. They were sent down to hold meetings.

In these recollections, we can appreciate his experience of socialism. In them we can see a strong emphasis for the Chinese Communist Party's vanguard social position sending instructions top down. The emancipatory aspect of these politics, however, was the dialogue these meetings enabled. They gave leaders an appreciation of the contexts that informed peasant perspectives.

Ding remembered these as times when leaders both relied on their fellow residents to carry out projects and would share in the work:

> Back then, if there was a meeting, that meant something was happening. Or it was notifying us of a new construction site / project, allocating the work to different brigades. So we'd work alongside them [the cadres] every day. It isn't the same now, we don't do this anymore.

When asked about the difference he saw in relations to the village today, Ding continued, "The leaders have become more estranged, it's less democratic now. Leaders take full command, whatever they say goes. Villagers [are treated as if we] don't know anything."

In these accounts of his past and the present we can see how Ding's historical experiences became values by which he appraised his current conditions. Rather than being disenchanted with the socialist message as Liu (2020) assigns in the term *guorizi*, we see that Socialism remained a vital aspect of his political subjectivity. One that he used to legitimate his desire to be involved in the village's development.

When asked about his thoughts on the heritage tourism plans, he continued:

> It's not what I would envision. It has limited Heyang villages and its villagers' development. In reality, if tourism took off properly and if you could develop from there, then yes, villagers would benefit. When the village just started doing tourism, medical insurance was paid for us by the village committee, villagers didn't have to pay anything. But [now] villagers have to pay. This is because they say they aren't making any money, but it's all just a waste.

His evaluation of why the tourism plans were not leading to more success and collectively shared wealth, he complained that:

> The director has been changed, a new term, your plans were your plans, but now it's his. Once these blueprints are complete, he's no longer director, and now it's his successor. So he starts drafting his own plans.

Ding was asked if he had always felt out of the loop on this development. He replied "no" and discussed the formation of the Lotus Cooperative:

> The former director spearheaded it. We wanted to plant lotus flowers. Have a field of lotus for people to see. It'd be the pearl of Heyang. As it was a joint venture we'd work together [with the management committee]. So everyone

who agreed contributed 10,000 yuan and we had a collective fund. But now we're about to get to 0, we have no money. This director left and his successor doesn't care about us. I think it's coming to its end.

Ding delivered this last reflection on the potential demise of the cooperative with a bitter laugh, and reflected:

Yes our quality of life has improved, but this has only been the result of individual efforts. We can fill our bellies, we can dress warmly, if we can keep going then we will work another several years. But if we can't, then we'll just retire at home.

In his final appraisal, Ding was less pessimistic about development and failure than Xin was. This may be part of a *guorizi* mentality. However, the material grounding enabled him the privilege to be less concerned was rooted in the historical legacy of having rights to use collectively owned land (Lin, 2013). Ding, like many other peasants in China, looked to family plots as a means to ensure sustainability through one's retirement.

## Hong's perspective from the bottom

Hong was a farmer in his 60s. Like Ding he was a Red Guard in his youth. Unlike Ding, Hong was granted permission to leave Heyang in his youth for school. However, he was one of the first sent down youth in the late 1960s. He ventured out of the village after 1985, doing various jobs from construction and masonry to selling clothing. Unpaid wages and failing health made him give up life as a migrant worker and he has retired back to the village where he collected an old-age pension. While he came off gruffer and more plainspoken than Xin or Ding, he was also passionate about the collective well-being of Heyang. He has served as a leader of one of Heyang's 32 work teams both before and after China's Reform and Opening. In his role as a village representative in the early 1980s, he worked to ensure costs were kept in control. While he has had some authority in the village, Hong's story represented the toughness of the *wu* masculinity argued to have been valued during the Cultural Revolution.

Exemplifying his character was an experience he had with his son:

Just last year, my son said: 'Ha, you acted as production team captain for over 30 years, yet you haven't even received a single morsel of land or room. Others get allocated something even just after one year as leader.' He was saying how pitiful I am. I hit him, I said, 'Say one more word about this and I'll beat you to death.'

He described his work in the early 2000s as a team leader:

One time—I remember this very clearly—I was tearing down that public washroom. They [the village cadres] told every production brigade leader to take part in this, and they'd receive 30 kuai [colloquial term for yuan] per day. I brought up the fact that I made 200 kuai per day when I worked manual labor jobs outside of the village, how can they say 30 kuai? And for such a dirty and tiring job. Already so much money is wasted by [the village committee], wouldn't it be better to just spend it by paying us locals a bit more for our labour? In the end it was raised to 80, but they didn't call me back to do any more jobs after that.

From this anecdote, we can see that Hong presented himself as concerned with collective over personal benefit, while not being very politically savvy in how he accomplished that goal. Indeed, this is where we can see some correspondence with Louie's insistence that *wen* masculinity dominates over the *wu*. Hong's lack of tact in his fight to obtain just wages showed little in the way of *wen*, and his brashness only helped to reaffirm his subordination after the fact.

Hong discussed the development of Heyang's heritage buildings noting:

Everything needs money. The biggest mistake, there's that really tall building next to the *bashimen*. When CEOs / bigshots from Beijing come here, they're always saying how this [or that needs to be done]. But where does that money come from?

Hong summarized Heyang's heritage tourism development asserting:

Life for us in Heyang is very bitter. We all say it's the "bitter" dwellings. We don't have a penny to our names, neither does the local state.

Hong finished his account by saying, "I'm not dissatisfied, I'm not resentful. I'm just telling the truth; reality. I tell you, this is just the way society is."

Hong's note about 'bitter dwellings' speaks not just to the lack of collective wealth but also to the renovation restrictions on buildings in the designated heritage protection area. Unable to afford historically "authentic" materials, many residents were unable to do repairs to their homes with leaky roofs being one of the main concerns for residents. This ability to identify injustice and to have a subject position to voice it (and not merely act with violence) was a key quality that Mao read into peasant subjectivities as the means to inform the Communist Party vanguard of the real situation of the people.

## Rethinking development from participation that matters

All three of the participants displayed marginalized masculinities struggling to compete and find belonging in China's current urban and market-orientated development. Likewise, all three placed value in a few key aspects of how they framed such marginalization and aspirations, namely, the importance of locals defining projects and being relied upon to realize development efforts. This was no simple coincidence. It speaks about the legacy of China's mass line, a concept that guided state-society relations and communication during the nation's experiments with high socialism under Mao's leadership. Summed up in the paraphrase "from the masses; to the masses; everything depends on the masses," the mass line was a means to keep the Communist Party connected to the people. It was a mechanism by which the Party and various classes of people could learn from one another through shared experiences and discursively developed goals aimed at addressing shared mutuality.

The mass line is currently not practiced to its ideal in the Heyang village. It existed as terminology by local leaders, such as the director of the tourism management committee, but this was mostly in name only and with the occasional village meeting about key decisions.

The mass line was revived during the Chinese Communist Party's response to the COVID-19 pandemic. This has been both in terms of technology to know what was happening with the people, but also the *fengqiao* experience (Cai & Yan, 2021). Tarnished in Western academia for symbolizing the totalitarianism of a Communist state (Bandurski, 2013), and praised in Chinese literature as a means to reduce the state's reliance on violence to realize its social objectives (Yan, 2019), the *fengqiao* experience is a means by which the national framework of mass line relations could be realized in a localized geography, with Party members engaging the people. While technology continues to help fulfill the mass line's communicative function of learning "from the people" and propagating knowledge "to the people," such communication is meaningless until it had put leaders and people into direct contact. In doing so, people do not act in conditions outside of their control but belong to the mechanisms through which lived conditions are realized through mutually dependent participatory action.

Development that hid behind a forefront of technological solutions puts continued pressure on the marginalized in the logics of accumulation that have enabled the globalization of capitalism and its current digital infrastructures. Participation does not occur through technology but through social mechanisms. The men who were interviewed realized their participatory potentiality through social categories of gender describing empowerment in terms of how masculinity was transformed by Communism and the historic practice of the mass line.

These responses were not reactionary and dependent on the dominate social form, but were realized as a synthesis of dialectical reasoning. Masculinity under capitalism reflected so much of traditional masculinity that Louie continued to use the term *wen* to depict the ideal, independent owners and users of the products of their actions. What has changed was merely what social accomplishments enabled a man to find that sovereignty to act. This ideal was contrasted against a notion of dependance, the duty to recognize how others enable one to act rather than using them as a means to an end, and to recognize them as ends—a caring ethic of feminism (Liao, 2006). The synthesis of these is not a feminized male, nor an unproductive *guorizi* masculinity, but rather a commitment to comradery, mutuality in coexistence where the benefits of participation were socialist as the value came from "being shared" (Buck-Morss, 2010, p. 70). This enabled more, not less, opportunity for others to act in turn.

## Conclusion

What can be derived from the above practices, this is not development as it is usually defined in the literature, nor the development aims of national policies by the Chinese Communist Party. Collectively, these have worked to shape the conditions needed to prioritize the urban-orientated heritage tourism industry in Heyang. There was an understanding of development offered that spoke to the underlying concerns of marginalization expressed by Xin, Ding, and Hong: to recognize vulnerability and working toward sustaining the village through addressing the basic needs of its residents through their own efforts.

Digital platforms need not be separate from these efforts. However, they were useful only in so much that they instilled the importance of localized contact between people privileged and marginalized in society. The current condition of digital platforms to privatize the product of our efforts to engage the promise of a collective "we" makes this difficult. It served to distract us from who in our society realized the ideals of participation. Rather than universalizing the ideals by which the socially privileged were able to act, we must refocus theoretical efforts to understand how the conditions to participate in society rest in recognizing the dependence we have on each other and the responsibilities to care for one another that this recognition entails.

## References

Bandurski, D. (2013). Xi Jinping playing with fire. *China Media Project*. https://chinamediaproject.org/2013/10/17/xi-jinping-playing-with-fire/

Buck-Morss, S. (2010). The second time as farce… historical pragmatics and the untimely present. In C. Douzinas & S. Zizek (Eds.), *The idea of communism* (pp. 67–80). Verso.

Butler, J. (1999). *Gender trouble: Feminism and the subversion of identity* (2nd ed.). Routledge.

Cai, M., & Yan, Z. (2021). *Life first: The strategy for the Chinese police under normalization stage of prevention and control of COVID-19*. In M. Z. Bin & N. A. Ghaffar (Eds.), *Proceedings of the 2021 4th International Conference on Humanities Education and Social Sciences (ICHESS 2021)*, 615 (pp. 2047–50). Curran Associates. https://doi.org/10.2991/assehr.k.211220.351

Carpentier, N. (2014). 'Fuck the clowns from grease!!' Fantasies of participation and agency in the YouTube comments on a Cypriot problem documentary. *Information Communication and Society*, *17*(8), 1001–16. https://doi.org/10.1080/1369118X.2013.875582

Chakravartty, P., & Zhao, Y. (2008). Introduction: Toward a transcultural political economy of global communications. In P. Chakravartty & Y. Zhao (Eds.), *Global communications: Towards a transnational political economy* (pp. 1–22). Rowman & Littlefield.

Connell, R. W. (2020). *Masculinities* (2nd ed.). Routledge.

Dean, J. (2012). *The Communist horizon*. Verso.

Dean, J. (2019). *Comrade: An essay on political belonging*. Verso.

Hauck, B. (2020). The shared time of the mass line: Economics, politics and participation in a Chinese village. *Javnost – The Public*, *27*(2), 186–99. https://doi.org/10.1080/13183222.2020.1727273

Liao, H. A. (2006). Toward an epistemology of participatory communication: A feminist perspective. *Howard Journal of Communications, 17*(2), 101–18. https://doi.org/10.1080/10646170600656854

Lin, X. (2013). *Gender, modernity and male migrant workers in China. Gender, modernity and male migrant workers in China.* Routledge. https://doi.org/10.4324/9780203590904

Liu, F. (2020). *Modernization as lived experiences. Modernization as lived experiences.* Routledge. https://doi.org/10.4324/9781315441245

Louie, K. (2002). *Theorising Chinese masculinity: Society and gender in China.* Cambridge University Press.

Louie, K. (2014). Chinese masculinity studies in the twenty-first century: Westernizing, Easternizing and globalizing Wen and Wu. *Norma: International Journal for Masculinity Studies, 9*(1), 18–29. https://doi.org/10.1080/18902138.2014.892283

Smith, D. (1990). *The conceptual practices of power.* Northeastern University Press.

Yan, C. (2019). Some reflections on China's grassroots social governance: Taking "Feng Qiao experience" as the object. *Asia-Pacific Journal of Humanities and Social Sciences, 1*(1), 1–12.

Zhao, Y. (2017). Global to village: Grounding communication research in rural China. *International Journal of Communication, 11*, 4396–422. https://ijoc.org/index.php/ijoc/article/viewFile/8265/2173

Zuboff, S., Moellers, N., Wood, D., & Lyon, D. (2019). Surveillance capitalism: An interview with Shoshana Zuboff. *Surveillance & Society, 17*(1/2), 257–66. https://doi.org/10.24908/ss.v17i1/2.13238

## 15 Participatory practices and lessons from Scientific Animations Without Borders and a WhatsApp network in a post-COVID age

### The case of video animations for rural agriculture

*Anne Namatsi Lutomia, Julia Bello-Bravo,*
*Noel Iminza Lutomia, James Kamuye Kataru,*
*John W. Medendorp and Barry R. Pittendrigh*

### Introduction

COVID-19 made the challenging problem of delivering life-changing educational information to people globally more difficult as it affected the ability of people at all levels of society to move freely. International development has traditionally preferred on-the-ground, in-person engagement for information dissemination. This approach needed to be altered when COVID-19 brought international travel to a halt. The pandemic brought extraordinary technological, pedagogical, and logistical difficulties. It also introduced previously unanticipated impacts, both positive and negative.

The repercussions were vast in the domain of education. The makeshift solutions cobbled together during the emergency proved ineffective and often exacerbated existing inequalities in education, especially as a result of unequal access to technology and networks (Haelermans et al., 2022; Parker et al., 2020; Reimers, 2022).

As the COVID-19 crisis dragged on, grassroots and non-formal sectors emerged as more reliable and effective access channels for receiving information (Pollett & Rivers, 2020). Rather than relying on educational institutions' digital infrastructures, or lack of them, for delivering content to adult or youth learners, various social platforms dramatically increased in use as communication channels, with software applications like WhatsApp and Zoom moving from somewhat infrequently utilized technologies to almost mandatory ones (Mu'awanah et al., 2021). Although the process has been uneven and uncomfortable, it was fortunate that the sudden, sheer volume of digital communication traffic could identify effective channels for people to interact during periods of movement restrictions, curfews, and isolation (Nchanji & Lutomia, 2021; UN Migration Agency, 2020).

DOI: 10.4324/9781003282075-18

*WhatsApp*

The reasons for forming WhatsApp group chats varied. They ranged from formal to informal groups for information dissemination, staying in touch, and learning from each other. The only requirements for forming a WhatsApp group were: having a purpose that was appealing to members; having a mobile device that could access WhatsApp; and having sufficient funds to buy subscriptions, bundles, or credit for online access. Notably, WhatsApp members should have also been able to charge their phones to send and receive messages. The network also required an administrator or leader who moderated discussions, invited and removed members, and ensured that members stayed informed.

Before COVID-19, patterns of international collaboration were well-documented and understood, both in their affordances and hindrances (Lutomia, 2019; Madela, 2020). With the advent of COVID-19, those familiar patterns were disrupted or halted while new ones emerged. This situation offered, or forced, a unique opportunity for international collaboration to occur virtually due to lockdowns and travel bans. Therefore, those in the West with projects in the Global South or other parts of the world engaged in collaboration virtually. The WhatsApp network group of farmers in rural Kenya is one such instance.

In this chapter, we will explore and describe the experiences and tools related to digital participation used by rural smallholder farmers, the network manager, and overall project leadership in that situation. Kenya had the highest monthly WhatsApp usage compared to the rest of the world—an astounding 97% of its Internet users, according to the Global Web Index's 2020 *Social Media User Trends* report (Bayhack, 2020). We will focus on new practices, experiences, and tools that arose in Kenya through a WhatsApp network for distributing food security videos. A knowledge management group, Kataru Concepts, extended an educational network for food security and COVID-19 mitigation to about 44 out of 47 counties in Kenya, with some areas being more active than others.

### About WhatsApp

WhatsApp allows users to share real-time digital information, such as photos, videos, messages, and other documents (Acton & Koum, 2014; Ahad & Lim, 2014). It now has more than 500 million users worldwide. Research on WhatsApp affordances included its impact on individuals and groups (Church & De Oliveira, 2013; Devi & Tevera, 2014; Soliman & Salem, 2014) and its impacts on student learning and performance (Bere, 2012; Yeboah & Ewur, 2014). Nevertheless, the practices and experiences of people who used WhatsApp in rural African communities are less well understood.

### Problem statement

Our study aimed to build on the knowledge of WhatsApp group use among farmers in Kenya, highlighting their incentives to be part of the group and their interest in accessing, sharing, and learning. This chapter examines the WhatsApp group participation among participants, specifically young people interested in or already practicing farming in various counties in Kenya.

### SAWBO and digital learning in Africa

Mobile penetration in Africa, especially among young subscribers, has grown exponentially in the last five years. Mobile phones, with their capacity for Internet access, chat, text, call, messages, music, video, and applications, are prevalent worldwide and have, since 2017, become the primary access device for digital information globally (Bello-Bravo et al., 2021a). In recent years, WhatsApp individuals and groups sharing real-time messages, information, and videos have increased organically without any regulated control.

Founded in 2011, the mission of Scientific Animations Without Borders (SAWBO) has been to create educational materials in the form of animations for low-literacy learners in local languages. The SAWBO library contains over 140 animation topics and over 240 language variants, including nearly 200 distinct languages. More than 50 million people have accessed these videos in more than 120 countries. They focus on agriculture, health, women's empowerment, peace-building, and climate change. These videos can be accessed through mobile phones, tablets, laptops, desktops, televisions, and radios and can be communicated to larger groups with projectors—all free of charge. Translation of the videos has been implemented by volunteers who translate, record, voice overlay, share, deploy, and train others. SAWBO draws from scientific and Indigenous knowledge to reach those marginalized from accessing information due to geographical distance or illiteracy.

The SAWBO Responsive, Adaptive, Participatory Information Dissemination (*RAPID*) Project was funded by the United States Agency for International Development (USAID) to improve food security in four COVID-19-affected countries (Bangladesh, Ghana, Kenya, Nigeria). This chapter focuses on new practices, experiences, and tools that arose in Kenya through a WhatsApp network for distributing food security videos.

Communities can increase their food security by increasing their ability to create preventative and remedial solutions for the stresses being experienced, leading to agricultural resilience. Agricultural resilience equips farmers to absorb and recover from shocks and stresses to their agricultural production and livelihoods.

### Theoretical approach

Adult learning (or andragogy) theories differ from pedagogy theories that inform children's learning. Adult learning theories suggest that the end user should be prioritized while building training programs or tools, such as video animations. While it is necessary to connect the curriculum to the lived realities of learners (Chaiklin, 2003; Vygotsky, 1978), adult learners are typically already motivated to learn the material (Knowles et al., 2012). It is imperative to present the information in easily accessed and culturally competent ways (Bello-Bravo et al., 2021b).

Andragogy leverages adults' self-motivated learning. Adults appreciate the value of education and generally have a specific goal in mind when pursuing an educational topic. Adults also have existing funds of knowledge that allow them to understand new concepts quickly, whereas children often encounter a subject with limited or no prior knowledge or experience (González et al., 2005; Moll et al., 1992). Adults are more likely to be self-directed to solve problems and learn new concepts but will also stop participating if educational materials seem irrelevant, culturally unfeasible, or non-authoritative. Children require more guidance and direction but are less likely to balk at the material, especially at younger ages. For older students, the compulsory nature of primary education and the mandate to do well on tests can motivate students toward the material, even when an instructor is incompetent. Nonetheless, *how* an educational subject is presented can be more important than *what* is presented (MacLean, 1962; McLuhan, 1964).

For farmers in Kenya, the socio-economic effects of COVID-19, including lockdowns, movement restrictions, and prohibitions on marketing crops, had already motivated farmers to be interested in food security and COVID-19 mitigation educational materials. The main goal was to make such materials accessible without requiring face-to-face interaction through digital means (Ribot & Peluso, 2003).

In this chapter, we consider how smallholder farmers acquired a practice centered on the pursuit of learning using animations supplied through WhatsApp. The term *practice* herein implies how individuals do work, the knowledge they draw on, what they evaluate as correct to do, and what they value.

## Literature review

### *Agriculture in Kenya pre-pandemic and during the pandemic*

Agricultural extension services were vital in Kenya because its economy relies on agriculture for development. Additionally, 70% of Kenyans live in rural areas and participate in agricultural activities. Varying extension models continued to be employed in the dissemination of information. These included field days, agricultural shows, face-to-face extension, on-farm demonstrations, farmer teachers, and mass media such as radio, local public gatherings, printed matter, and Farmer Field Schools (FFS). FFS relied on a group of farmers trained to train others to disseminate information to enhance their skills. These schools were also a space for transformational change that included "personal transformation, changes in gender roles and relations, customs and traditions, and community relations, and an increase in household economic development" (Duveskog et al., 2011, p. 1529). Since its independence in 1963, Kenya has embraced agricultural extension by delivering information and inputs to farmers, human resource development, and services connecting researchers to farmers (Mwombe et al., 2014). The fee-for-service model also contained farmers' groups contracted public and private extension agents to provide information and services.

In 2013, Kenya's system of governance was de-centralized. As a result, county governments took on more responsibility in various sectors. County governments were charged with providing demand-driven extension services within the agricultural sector. The challenge that county-level extension faces is that although expenditures on agriculture by the national government were increased, there had been a decline in total budget expenditures. Agricultural expenditures had been low in some county governments' budgets (Njagi & Kombo, 2014). Another problem is the low ratio of extension staff to farmers: 1:5000. This led to the government developing a national policy that recognizes this need and offers an increased role for private sector extension service provision, with the government focusing on its regulatory roles.

Presently agricultural extension is tied to the achievement of Vision 2030 as the agricultural sector is expected to drive the economy to a projected annual growth of approximately 10% (Vision 2030, 2007). In the agriculture sector, the Vision 2030 document points to a transformation from the current practices to commercially oriented agriculture. Kenya plans to undertake an institutional reform driven by high performance to trigger growth in private sector agriculture (Ndung'u et al., 2011). For the social pillars of sustainability, Kenya has set out to increase labor and employment and recognize the importance of science, technology, and innovation.

In line with agricultural extension services, Aker and Mbiti (2010) noted that the proliferation of mobile phones in developing countries had introduced cheaper, accessible, and easy-to-use technologies. They noted that mobile phones could be used in the agricultural sector in extension services, data collection, and e-learning. Kenya has also taken advantage of its information and communication technologies (ICTs) platform innovation and applied it to agriculture. The spillover effects from innovations like M-PESA, a mobile phone banking system where users can deposit, send, and withdraw money, can also be seen in Kilimo Salama. This micro-insurance program allowed farmers to insure their crops using phones (Irungu et al., 2015). Other communication modes farmers included were radio programs and television programs. Kenyans interested in agriculture also shared information on other platforms, including WhatsApp, Twitter, and Facebook. Mobile phones facilitated agricultural information on market prices, weather, transport, and agricultural techniques via voice, short message services (SMS), and the Internet (Aker, 2010). This shift has impacted extension services provision. Currently, they are driven by the demand for effective and appropriate extension services, dwindling government budgets, and advances in telecommunication technology worldwide (Omotayo & Isiaka, 2005). The growing number of young farmers and a shift toward agriculture by those employed but seeking to supplement their salaries, returnees from abroad, or agricultural entrepreneurs also influenced how agricultural information, such as market prices, weather, transport, practices, and meetings to attend, are shared.

ICTs provided the potential to increase agricultural productivity through the dissemination and communication of information and knowledge to rural agricultural communities (Munyua & Stilwell, 2012). Kenyan smallholder farmers accessed agricultural information via mobile phones, radios, television, and the

Internet. Promoting ICTs in agriculture has mainly focused on their adoption and use by extension officers. This was seen, for example, in employing participatory approaches that involved the end users of technologies around creating and transferring such technologies. Some ICTs enabled horizontal communication. This allowed farmers to share information among themselves.

Reducing the cost of mobile phones and connectivity gains have increased phone ownership and Internet access (Wyche & Olson, 2018). In the agricultural sector, however, socio-economic barriers like gender, age, and education can affect access and the use of technology in extension service delivery. Tata and McNamara (2016) cited women's limited access to higher levels of technology, women's linguistic print non-literacy, social attitudes that prevented women from using technology, and women's lack of control over or ownership of technologies as factors affecting access and use.

### Agriculture in Kenya and COVID-19

Agriculture in Kenya was an input-intensive sector. For most smallholder farmers, key agricultural inputs included fertilizers, plant protection products (pesticides, fungicides, herbicides), seeds, fuel, and labor. Disruptions in the supply and availability of these inputs likely result in output reductions. The COVID-19 crisis led to lockdowns and curfews, increasing prices, and decreasing access to farm inputs. Between March and November 2020, Kenya applied lockdown and quarantine measures to combat the COVID-19 pandemic. During this period, there was limited production and supply of plant protection products. This affected crop protection activities, especially in rural areas that depended on transport to receive farm inputs. The domestic and international supply chain disruptions to pesticides caused by COVID-19 resulted in reduced outputs (Brewin, 2020; Schmidhuber et al., 2020). East African transportation costs increased three-fold while delays occurred due to fewer inbound flights (Gebre & Herbling, 2020).

Ways in which people organize themselves to learn about topics of mutual interest have been understood through the lens of communities of practice. The farmers discussed in this chapter who organized themselves in their WhatsApp group to learn from each other and to share information using video animations can be understood as a virtual *community of practice.*

Communities of practice denote a group or network of connections among people who share a concern or a passion for something they do to better their knowledge and skills as they work on something. Wenger (2001) observed that communities of practice "address the informal and tacit aspects of knowledge creation and sharing, as well as the more explicit aspects" (p. 3). Membership in a domain of interest is tied to a shared competence. These members engaged in relationship-building, activities, and discussions to help share information as they learned. According to Wenger (2001), communities of practice typically address non-formal and implicit knowledge creation and sharing features. A community of practice represents shared histories of learning, where participants make meaning, sense, and understanding of their work. Through mutual engagement and interrelated forms of participation, members of a community of practice negotiate the meaning of their practices and form an identity of who they are (Lave & Wenger, 1991).

Lave and Wenger (1991) asserted that learning provided a means of becoming an insider and acquiring knowledge and skills that lead to becoming a practitioner. Learning languages, technical skills, and cultural knowledge occurs through increasing participation. The learning process encompasses conversation, storytelling, experimenting with new approaches, and commenting on each other's solutions. Learning emerges through the process of working or practicing. One's craft, knowledge, and sense of self are gained and draw meaning from the community's activities. What emerges from such learning is the social production of knowledge that helped solve problems. Taken together, WhatsApp groups intended for virtual communities of practice provided affordances of support from other members to access and share learning resources and services. By doing so, they connected and included those otherwise on the margins of communities (Abiodun et al., 2020).

## Methods

### Context

We examined the crossroads between educational content in videos and communication through WhatsApp groups in Kenya that started during COVID-19 through a USAID-funded program, dubbed SAWBO *RAPID*, to secure agricultural and food resilience during the pandemic. The SAWBO Kenya WhatsApp network members drew from 44 out of 47 counties in Kenya, with more activities in Kakamega County. Members of the group shared and disseminated information on agricultural and health practices. They shared their agricultural problems and provided each other with possible solutions through text and photos.

## Case study: SAWBO Kenya WhatsApp group

### Participants and data collection

This qualitative case study of practices focused on experiences and tools. It emerged from a WhatsApp network that electronically distributes educational materials about food security and COVID-19 mitigation. It combined documentary analysis, indepth remote interviews, and autoethnographic methods to develop an understanding of the experiences, practices, and tools that emerged due to COVID-19 restrictions.

Participants were sampled for their direct involvement in digital changes to educational efforts in international development due to COVID-19 constraints. This purposeful sampling included the knowledge manager of the WhatsApp network in Kenya as a unique and critical player in the phenomenon under examination (Ngulube, 2015). This participant also had extensive access to the thoughts, feelings, and experiences of 230 farmer participants and users of the WhatsApp network and was able to uncover critical data about the phenomenon (Ngulube, 2015). While this purposeful sampling afforded access to the perhaps only source of direct data concerning the central phenomenon, i.e., changes to practices, experiences, and tools prompted by educational efforts during COVID-19, this was also a limitation as the uniqueness of the participant's access makes generalizing to other circumstances problematic.

Documentary evidence included primary source emails and communications by project management; this included instructional webinars to orient distributors of the educational materials and published interviews with three network participant farmers. Also, autoethnographic self-reflections, i.e., one author's expertise, experiences, and tools, are drawn on to document daily participatory practices by the project knowledge manager in Kenya throughout the SAWBO *RAPID* project (Figure 15.1).

## SAWBO WhatsApp group network participatory practice

### *Data analysis*

All data collected from documentary evidence and autoethnographic self-reflection were coded by three researchers with conferencing to resolve any discrepancies in

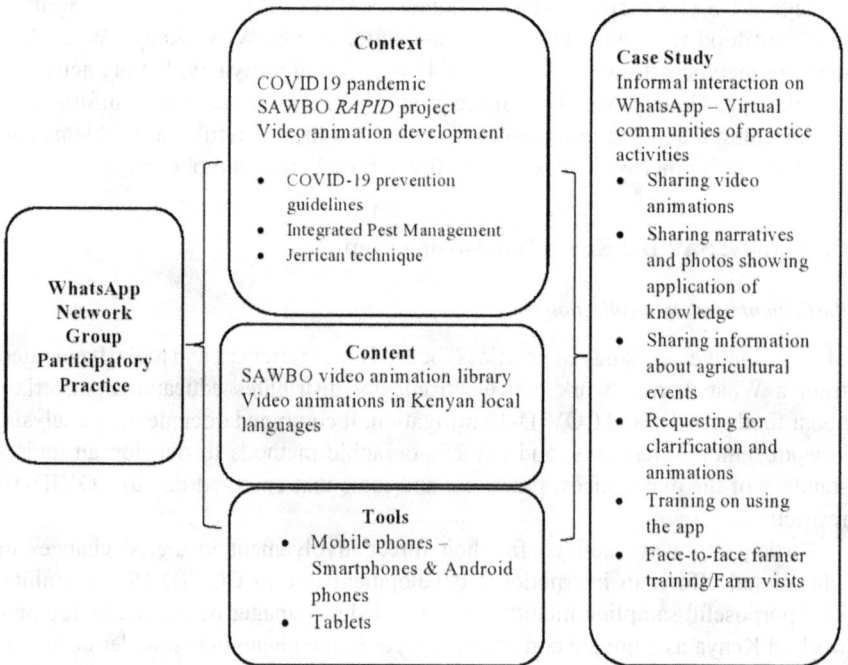

*Figure 15.1* SAWBO WhatsApp group network participatory practice.

coding until an inter-rater agreement of 100% was achieved (Gwet, 2014). Codes were then grouped by two researchers for the emergence of themes, with discussions again to explore discrepancies in categories and to achieve full agreement. Member-checking was utilized to ensure that codes and themes matched participant descriptions, especially for the autoethnographic content (Creswell & Poth, 2016), while triangulation was applied, where possible, to interview and obtain autoethnographic and documentary evidence (Jick, 1983; Olsen, 2004).

## Findings

### *SAWBO Kenya WhatsApp group*

Network manager James Kamuye Kataru developed expertise in developing and managing emergent communities of practice. The blog he references captures his experiences of disseminating videos, sharing information, and visiting farmers (see https://kataruconcepts.com/). Kataru writes:

The WhatsApp network built across Kenya is active in 44 out of 47 counties, with volunteer networks disseminating SAWBO content in their communities. For easy management, the network comprises ten regions, each headed by a regional volunteer, 44 county coordinators working under the regional volunteers, and network members working in their respective counties. This helps the network factor in the communities' dynamics and diversity. The remaining three counties have not worked out due to poor Internet connectivity and the nomadic lifestyle.

The main WhatsApp group has 230 leaders from 10 regions and 10 partners across Africa and the United States. The main group instructs the membership on what videos to share for the day / season, organizes training sessions, deliberates on essential issues that need escalating to regional teams by their regional volunteers and passes important information concerning the network. There are also two more WhatsApp groups, one that was used to train new members and a second for willing members' welfare matters.

During the just concluded Anglican Development Services (ADS) organized farmers' field day and exhibitions in western Kenyan counties, close to 5,000 new members were recruited. We had been identifying people on WhatsApp and adding them to our training group. After an induction session, they will be added to the main group and have regional leaders add them to their respective WhatsApp groups. Those not on WhatsApp shall be encouraged to form groups, and the volunteer teams will then train them to disseminate SAWBO content from their villages. They will be trained on how to share videos using Bluetooth.

The availability of SAWBO videos in local languages helped increase content uptake in the communities, as witnessed in the farmers' exhibition / field days. The participating farmers who watched the video animations confirmed they had read about SAWBO videos from our blogs, watched them on KTNFARMERS television station, and found them on YouTube. Most of the farmers were elated to finally meet 'the SAWBO team' in-person, who were ready to serve them in their local languages and dialects.

The network has not stopped at the three primary training and welfare groups. Regional volunteers are encouraged to form regional WhatsApp groups and add members from their regions with Western, Nyanza, and Central Kenya regions leading to the formation of new groups and network expansion. This expansion is meant to accommodate as many members as possible and have them access SAWBO-animated videos relevant to their applications and in their local languages. These regional teams were also meant to increase the translation and availability of SAWBO videos in local languages and dialects. Members will be encouraged to discuss matters of interest in their local languages to understand the content easily.

The COVID-19 pandemic increased the need to understand what was happening in our communities, the signs and symptoms, how to prevent it, and in case of infections, how to fight it. This coincided with the SAWBO's release of several videos for fighting the pandemic, which became popular within the network and communities accessing the content. Visiting farmers, other groups, and individuals have increased the confidence members have in the network and SAWBO content. Excited farmers state how some videos they've watched helped improve seed quality, crop health, and yield. Farmers on the WhatsApp network prefer in-person visits after WhatsApp engagements. In-person training sessions are popular, especially if done on farmers' farms. Such sessions make the whole dissemination process believable. They act as bonding sessions too.

### Experiences of farmers

Three farmers, Rachel, Janet, and Geoffrey, are community members who benefited from information shared across a WhatsApp network covering all 44 of 47 counties in Kenya:

> Rachel Olukhanda: Before the COVID-19 pandemic, she was a clothing businesswoman in Bungoma, Kenya. Then the government of Kenya closed its borders and locked down the country. She moved to the village and started practicing agriculture on her small plot of land. After watching the SAWBO animation on raised bed farming on KTNFarm television station in Swahili and English, she was inspired to practice farming. Her yield at the end of the 2020 planting season was enough to feed her family and sell the remainder to neighbors. She reported that she has watched and used a variety of SAWBO video animations and shared them with her neighbours, as well as helped them develop their farms using the raised bed approach. She added that farming had become an essential source of income and food during COVID-19. She was later approached by the Knowledge Manager and joined the WhatsApp group. Olukhanda is now a member of the SAWBO Kenya Network WhatsApp group.
>
> Janet Adika: In June 2021, at Bulechia Village, Kakamega County, Kenya, she faced the problem of safely storing her postharvest grains for consumption and seed-planting during the next growing season. The solution to her problem came on her cell phone as an animated Swahili video detailing the

steps for hermetically storing beans in jerricans. The animation was downloaded from the Scientific Animations Without Borders (SAWBO) website by a trainer, put on a flash disk, and then played on a TV screen for a group of interested learners. Videos were also shared between phones using Bluetooth, which allowed people to practice safer social distancing in the village.

Geoffrey Anzwenu: In Kakamega County, Kenya, he also watched a SAWBO video on his mobile phone on how to make compost. Geoffrey put into practice the knowledge he received from the video. That learning has reduced input costs and increased soil health in his fields, empowering him to grow food better and improve food security for himself and his family.

## Discussion

The proliferation of mobile phones and the low cost of accessing WhatsApp has increased the use of Mobile Social Networking (MSN) technologies like WhatsApp as mediums of education, particularly as spaces for learning in agriculture and other areas such as foreign language learning. This potential to engage participants online can be formal and informal, focusing on different aspects of lessons.

Learning in these WhatsApp groups is driven by meaningful interactions derived from mediation by the teachers, facilitators, or the knowledge manager. These mediations offer sources of knowledge and feedback and ensure the participants' focus on the learned content (Naghdipour & Manca, 2023). According to Aburezeq and Ishtaiwa (2013, p. 165), there are three levels of interactions that are developed in WhatsApp learning: student-content interaction (54%), student-student interaction (71%), and student-instructor interaction (42%) (p. 165). Their study found that WhatsApp offered "a virtual time and space for sharing ideas, feelings, and exchange of knowledge and information" (p. 165).

WhatsApp group learning is germane to adult learning activities like self-directed and interactive learning because the learner can access the information depending on one's choice. The learner can also engage with others in the group. The use of WhatsApp equips learners with communication and technology literacy skills. However, earlier adult learning theories do not address online processes of student / instructor or student / student interaction mainly because they were developed when online learning was not as ubiquitous and well understood as it is now (Frey & Alman, 2003). Current adult learning theories center on the learner, context, physical space, and spatiality as drivers and inhibitors of learning (Merriam, 2008).

Keogh (2017) noted that the interactions between students form a semblance of a community of practice where they learn from each other, share information, and co-construct knowledge. Further, Keogh (2017) observed that the dialogue leading to the interactions in the WhatsApp group was not organic. Most of the discussions required nudging or were initiated by the teacher. In the case of the SAWBO WhatsApp group, the knowledge manager guided, taught, and redirected the discussion when members went off-topic and provided topics for discussion.

The challenges of using WhatsApp included access to WiFi, affordability and the cost of maintaining the mobile phone, and access to electricity for charging.

Secondly, despite setting up and agreeing on the housekeeping rules, some participants shared irrelevant posts with incorrect information that required deleting, redirecting the group, and providing the correct information by the knowledge manager-leader. Lastly, some members joined the group and never participated.

### Lessons learned

Relationships with partnerships buttressed with active communication, delegation, and high trust levels drove the project's various activities. Due to lockdowns and travel bans, the SAWBO team pursued techniques and approaches that allowed for exclusively remote interactions while also empowering local leadership to take innovative approaches to fill the gap left by a lack of extension activity.

The use of the WhatsApp network worked well in Kenya. Kenya is particularly suited to the WhatsApp approach because of a current reliance on WhatsApp groups to share information, organize communities, and socialize (Macharia et al., 2021). Kenyans who generally use WhatsApp groups include alum associations, community organizing, fundraising for different purposes and friendship groups. They are, therefore, well-versed in engaging on WhatsApp. Lowe (1986) surveyed African countries that had more readily embraced modernized farming technologies than others. The ready adoption of WhatsApp in Kenya may reflect the distinction Arnon (1981) noted between communities that adopted new techniques quickly and required little to no persuasion compared to others who resisted technological change. In their study of young volunteer health workers, Farmer et al. (2016) found that WhatsApp was efficient and inexpensive, allowed for information exchange, problem-solving, and coordination, and could effectively be used to communicate and build capacity.

The WhatsApp approach, using social media platforms as secondary pathways for disseminating expert knowledge, is a social benefit for marginalized groups. Due to its informal, horizontal information-sharing system, it is especially effective for reaching "last mile" communities that would otherwise be overlooked or ignored by formal learning systems. The democratizing aspect of social media allowed for existing content to be pushed out into areas in crisis within days in a highly scalable manner. Virtual networks can be used to (1) understand problems globally, (2) develop culturally and linguistically appropriate responses, (3) deploy content virtually, and (4) establish partnerships whereby the local leaders and participants could develop and deploy countrywide networks to get content to those who most needed it. This can all be implemented without travel, permissions, or bureaucratized hierarchies. This strategy removed the high carbon footprint costs of international development work that required significant international, regional, or local travel. It empowered local populations to take control of their own knowledge acquisition.

COVID-19 provided a paradigm shift in SAWBO's international program management. It provided an impetus for speeding up SAWBO *RAPID* project activities because no travelling was required. All team members and consultants worked virtually. This led to robust collaborations in the development and distribution of content. All teams moved into virtual space and found new ways to collaborate, relying heavily on local leadership to get the job done. Online social networks and teams worked exceptionally well together, and innovative ways of accomplishing the same tasks cheaply and effectively emerged from that collaboration.

Prior relationships made this project possible. However, the real breakthrough was the growth of horizontal networks where leadership was not appointed or anointed by some outside entity but emerged organically from the communities themselves, where trust is intrinsic rather than coerced or leveraged. The pandemic revealed the need for investments in long-term and robust relationships in-country, while cultivating and empowering local leadership. We did not create these networks but rather identified existing networks and tapped into them, reached out to existing groups (e.g., television in Nigeria), and developed a virtual pull strategy to understand the problems so we could make content that is responsive to the needs as locally identified rather than as prescribed by others from outside the context. At the project's beginning, we began with phone interviews in the four countries where SAWBO *RAPID* was working to understand better the locally defined challenges with the COVID-19 lockdowns. Based on their responses, we created COVID-19-related videos to mitigate the secondary effects of the pandemic.

### *Lessons by the SAWBO network leader*

The knowledge manager and project leader in Kenya was responsible for implementing the project on the ground and communicating with SAWBO and the WhatsApp network. When asked to share the lessons learned, he cited that working remotely gave impetus to communicating project goals, learning, and implementing the project online. Webinars were used for training and introducing the program to organizations in different countries, including Kenya. WhatsApp chats and emails were used for project communication, follow-up, and relaying information about the project. These approaches encouraged openness to ask questions and seek clarification.

The virtual network forced the SAWBO *RAPID* team to rely on those on the ground for information about the project since travel was not easy in Kenya during the COVID-19 crisis. Having to rely on our knowledge manager to provide information about the project, who, in turn, had to rely on his network to provide him with information, reoriented the project's leadership to a grassroots level, and opened the initiative to those operating closest to the problem. This devolution of leadership set in motion a reversal of traditional models and empowered even the most marginalized communities. Although international projects have always relied on those on the ground to share some of the information, the paradigmatic shift in international project work represented by this development of virtual networks meant that initiative and authority had moved down to the local level. Last-mile communities no longer needed to wait for a visit from some distant expert to actualize solutions.

Another interesting lesson learned from this project was the management of time differences and technology for remote work. Due to the difference in time zones (seven to eight hours), we had to coordinate our time to meet, which involved staying up late, getting up early, or working over weekends to accommodate differing schedules. This breakdown of the office "walls" allowed for continual communication. Informal conversations could occur anytime to address urgencies, and solutions can be brought in from anywhere.

Providing feedback comprises another lesson learned. The well-nurtured relationship and commitment-support by SAWBO and the network leader or knowledge manager allowed communication and feedback to flow freely. Email, Zoom meetings, and WhatsApp were the modes of communication. Both parties valued timely feedback. The communication and feedback allowed for collaboration in planning, resolving problems, and sharing experiences from the field. Specifically, Trello provided blog feedback demonstrating the team's experience in the field and on WhatsApp. It is plausible that feedback-giving provided affordances for capacity development that showed how team building and recognition of work well-done becomes a culture that is shaped and propagated through the network.

## Conclusions and recommendations

The experiences of farmers and cases that captured and gave voice to the rural-based farmers who used a WhatsApp group to learn, share, and stay in touch with each other provided insights into how the networks formed and grew. The creation and use of the WhatsApp tool connected with the WhatsApp group and facilitated easy downloading of animations that members could use directly.

Future research about this WhatsApp group could include an understanding of their practise and factors that influence participation, along with an analysis of the discourses of virtual communities of practise. Other research included empirical studies around members' knowledge, attitudes, and practices about the WhatsApp platform.

Emerging WhatsApp group networks offered low-cost, low-tech, low-threshold forms of working with farmers in rural Kenya during COVID-19. The lesson for rural development participation is that the use of the WhatsApp platform required an administrator who was an expert in the purpose of the group to ensure the quality of advice and that the discussion was positive and constructive. The administrator catalyzed the group interaction and ensured they received the requested assistance. This emerging form of international development offered many logistical advantages for international development practitioners and unlocked the unlimited potential of localized leadership for local solutions.

## Acknowledgments

This chapter was supported by United States Agency for International Development (USAID) under the Scientific Animations Without Borders (SAWBO) Responsive, Adaptive, Participatory Information Dissemination (*RAPID*) Scaling Program. SAWBO *RAPID* is funded through a grant from Feed the Future, the US government's global hunger and food security initiative, through agreement no. 7200AA20LA00002. The opinions expressed herein are those of the author and do not necessarily reflect the views of USAID or the US government.

# References

Abiodun, R., Daniels, F., Pimmer, C., & Chipps, J. (2020). A WhatsApp community of practice to support new graduate nurses in South Africa. *Nurse Education in Practice, 46*, 102826. https://doi.org/10.1016/j.nepr.2020.102826

Aburezeq, I. M., & Ishtaiwa, F. F. (2013). The impact of WhatsApp on interaction in an Arabic language teaching course. *International Journal of Arts & Sciences, 6*(3), 165–80. https://www.universitypublications.net/ijas/0603/pdf/F3N281.pdf

Acton, B., & Koum, J. (2014). WhatsApp blog. [Blog.] http://blog.whatsapp.com/

Ahad, A. D., & Lim, S. M. A. (2014). Convenience or nuisance?: The 'WhatsApp' dilemma. *Procedia-Social and Behavioral Sciences, 155*, 189–96. https://doi.org/10.1016/j.sbspro.2014.10.278

Aker, J. C. (2010). Dial "A" for agriculture: Using information and communication technologies for agricultural extension in developing countries (Working Paper no. 269). Center for Global Development.

Aker, J. C., & Mbiti, I. M. (2010). Mobile phones and economic development in Africa. *The Journal of Economic Perspectives, 24*(3), 207–32. https://doi.org/10.1257/jep.24.3.207

Arnon, I. (1981). *Modernization of agriculture in developing countries: Resources, potentials and problems.* Wiley.

Bayhack, J. (2020). 97% of internet users in Kenya use WhatsApp; here's how businesses can reach them. https://www.cnbcafrica.com/2021/97-of-internet-users-in-kenya-use-whatsapp-heres-how-businesses-can-reach-them-by-james-bayhack/

Bello-Bravo, J., Brooks, I., Lutomia, A. N., Bohonos, J. W., Medendorp, J., & Pittendrigh, B. R. (2021a). Breaking out: The turning point in learning using mobile technology. *Heliyon, 7*(3), e06595. https://doi.org/10.1016/j.heliyon.2021.e06595

Bello-Bravo, J., Medendorp, J., Lutomia, A. N., Reeves, N. P., Bohonos, J. W., Gardner, R., & Pittendrigh, B. (2021b). A case study of social-technical systems approaches and education video animations (health care/international). In M. A. Bond, S. R. Tamim, S. J. Blevins, & B. R. Sockman (Eds.), *Systems thinking for instructional designers: Catalyzing organizational change* (pp. 159–69). Routledge.

Bere, A. (2012). A comparative study of student experiences of ubiquitous learning via mobile devices and learner management systems at a South African university. *Paper presented at the Proceedings of the 14th Annual Conference on World Wide Web Applications*, Chiba, Japan.

Brewin, D. G. (2020). The impact of COVID-19 on the grains and oilseeds sector. *Canadian Journal of Agricultural Economics/Revue canadienne d'agroeconomie, 68*(2), 185–88. https://doi.org/10.1111/cjag.12239

Chaiklin, S. (2003). The zone of proximal development in Vygotsky's analysis of learning and instruction. In A. Kozulin, B. Gindis, V. S. Ageyev, & S. C. Miller (Eds.), *Vygotsky's educational theory in cultural context* (pp. 39–64). Cambridge University Press.

Church, K., & De Oliveira, R. (2013). What's up with WhatsApp? Comparing mobile instant messaging behaviors with traditional SMS. Paper presented at the *Proceedings of the 15th International Conference on Human-computer Interaction with Mobile Devices and Services*. Munich, Germany.

Creswell, J. W., & Poth, C. N. (2016). *Qualitative inquiry and research design: Choosing among five approaches* (4th ed.). Sage.

Devi, S., & Tevera, M. S. (2014). Use of social networking site in the University of Swaziland by the health science student: A case study. *Journal of Information Management, 1*(1), 19–26.

Duveskog, D., Friis-Hansen, E., & Taylor, E. W. (2011). Farmer field schools in rural Kenya: A transformative learning experience. *Journal of Development Studies*, *47*(10), 1529–44. https://doi.org/10.1080/00220388.2011.561328

Farmer, M. Y., Liu, A., & Dotson, M. (2016). Mobile phone applications (WhatsApp) facilitate communication among student health volunteers in Kenya. *Journal of Adolescent Health*, *58*(2), S54–55. https://doi.org/10.1016/j.jadohealth.2015.10.121

Frey, B. A., & Alman, S. W. (2003). Applying adult learning theory to the online classroom. *New Horizons in Adult Education and Human Resource Development*, *17*(1), 4–12. https://doi.org/10.1002/nha3.10155

Gebre, S., & Herbling, D. (2020, March 22). Coronavirus slowing desert locust response amid new swarms. *Bloomberg*. https://www.bloomberg.com/news/articles/2020-03-22/coronavirus-slowing-desert-locust-response-in-east-africa#xj4y7vzkg

González, N., Moll, L. C., & Amanti, C. (2005). *Funds of knowledge: Theorizing practices in households, communities, and classrooms*. Lawrence Erlbaum Associates.

Gwet, K. L. (2014). *Handbook of inter-rater reliability: The definitive guide to measuring the extent of agreement among raters* (4th ed.). Advanced Analytics, LLC.

Haelermans, C., Korthals, R., Jacobs, M., de Leeuw, S., Vermeulen, S., van Vugt, L., Aarts, B., Prokic-Breuer, T., van der Velden, R., & van Wetten, S. (2022). Sharp increase in inequality in education in times of the COVID-19-pandemic. *PLoS One*, *17*(2), e0261114. https://doi.org/10.1371/journal.pone.0261114

Irungu, K., Mbugua, D., & Muia, J. (2015). Information and Communication Technologies (ICTs) attract youth into profitable agriculture in Kenya. *East African Agricultural and Forestry Journal*, *81*(1), 24–33. https://doi.org/10.1080/00128325.2015.1040645

Jick, T. D. (1983). Mixing qualitative and quantitative methods: Triangulation in action. In J. VanMaanen (Ed.), *Qualitative methodology* (pp. 135–48). Sage.

Keogh, C. (2017). Using WhatsApp to create a space of language and content for students of international relations. *Latin American Journal of Content & Language Integrated Learning*, *10*(1), 75–104. https://doi.org/10.5294/laclil.2017.10.1.4

Knowles, M. S., Holton, E. F., & Swanson, R. A. (2012). *The adult learner* (7th ed.). Routledge.

Lave, J., & Wenger, E. (1991). *Situated learning: Legitimate peripheral participation*. Cambridge University Press.

Lowe, R. (1986). *Agricultural revolution in Africa*. Macmillan.

Lutomia, A. N. (2019). A case study of successes and challenges in a scientific collaboration program based in the United States and Benin. [Ph.D. Dissertation. University of Illinois Urbana-Champaign.]

Macharia, S. M., Obuya, J., & Mulwo, A. (2021). Communication of social messages in task-oriented instant messaging groups: A netnographic study of farmers' WhatsApp Groups in Kenya. *Journal of African Interdisciplinary Studies*, *5*(10), 184–207. https://doi.org/10.2022/ajest.v6i4.735

MacLean, A. (1962). The method is the message. [Pamphlett.] http://www.pacificuu.org/publ/univ/education/maclean_method_message.html

Madela, L. M. (2020). Perspectives on South-North institutional collaboration/partnership research in higher education. [Ph.D. Dissertation. University of Illinois Urbana-Champaign.]

McLuhan, M. (1964). *Understanding media: The extensions of man*. McGraw-Hill.

Merriam, S. B. (2008). Adult learning theory for the twenty-first century. *New Directions for Adult and Continuing Education - Special Issue: Third Update on Adult Learning Theoryducation*, *2008*(119), 93–98. https://doi.org/10.1002/ace.309

Moll, L. C., Amanti, C., Neff, D., & Gonzalez, N. (1992). Funds of knowledge for teaching: Using a qualitative approach to connect homes and classrooms. *Theory into Practice, 31*(2), 132–41. https://doi.org/10.1080/00405849209543534

Mu'awanah, N., Sumardi, S., & Suparno, S. (2021). Using Zoom to support English learning during Covid-19 pandemic: Strengths and challenges. *Jurnal Ilmiah Sekolah Dasar, 5*(2), 222–30. https://doi.org/10.23887/jisd.v5i2.35006

Munyua, H. M., & Stilwell, C. (2012). The applicability of the major social science paradigms to the study of the agricultural knowledge and information systems of small-scale farmers. *Innovation: Journal of Appropriate Librarianship and Information Work in Southern Africa, 2012*(44), 10–43. https://doi.org/10.10520/EJC127163

Mwombe, S. O., Mugivane, F. I., Adolwa, I. S., & Nderitu, J. H. (2014). Evaluation of information and communication technology utilization by small holder banana farmers in Gatanga District, Kenya. *The Journal of Agricultural Education and Extension, 20*(2), 247–61. https://doi.org/10.1080/1389224X.2013.788454

Naghdipour, B., & Manca, S. (2023). Teaching presence in students' WhatsApp groups: Affordances for language learning. *E-learning and Digital Media, 20*(3), 282–299. https://doi.org/10.1177/20427530221107968

Nchanji, E. B., & Lutomia, C. K. (2021). COVID-19 challenges to sustainable food production and consumption: Future lessons for food systems in eastern and southern Africa from a gender lens. *Sustainable Production and Consumption, 27*, 2208–20. https://doi.org/10.1016/j.spc.2021.05.016

Ndung'u, N., Thugge, K., & Otieno, O. (2011). Unlocking the future potential for Kenya: The Vision 2030 (Chapter 4, Preliminary Draft). https://citeseerx.ist.psu.edu/viewdoc/download?doi=10.1.1.410.6020&rep=rep1&type=pdf

Ngulube, P. (2015). Qualitative data analysis and interpretation: Systematic search for meaning. In E. Mathipa & M. Gumbo (Eds.), *Addressing research challenges: Making headway for developing researchers* (pp. 131–56). Mosala-MASEDI Publishers & Booksellers CC.

Njagi, L., & Kombo, H. (2014). Effect of strategy implementation on performance of commercial banks in Kenya. *European Journal of Business and Management, 6*(13), 62–67.

Olsen, W. (2004). Triangulation in social research: Qualitative and quantitative methods can really be mixed. *Developments in Sociology, 20*, 103–18.

Omotayo, A., & Isiaka, B. (2005). Video self training methods: It's effectiveness in disseminating agricultural information to rural farmers in southwest Nigeria. *Agricultural Research and Extension Network, Newsletter, 52*, 7. https://cdn.odi.org/media/documents/5253.pdf

Parker, R., Morris, K., & Hofmeyr, J. (2020). *Education, inequality and innovation in the time of COVID-19*. Johannesburg, SA: JET Education Services.

Pollett, S., & Rivers, C. (2020). Social media and the new world of scientific communication during the COVID-19 pandemic. *Clinical Infectious Diseases, 71*(16), 2184–86. https://doi.org/10.1093/cid/ciaa553

Reimers, F. M. (Ed.) (2022). *Primary and secondary education during COVID-19: Disruptions to educational opportunity during a pandemic*. Springer Nature.

Ribot, J. C., & Peluso, N. L. (2003). A theory of access. *Rural Sociology, 68*(2), 153–81. https://doi.org/10.1111/j.1549-0831.2003.tb00133.x

Schmidhuber, J., Pound, J., & Qiao, B. (2020). *COVID-19: Channels of transmission to food and agriculture*. FAO.

Soliman, A. A., & Salem, M. S. (2014). Investigating intention to use mobile instant messenger: The influence of sociability, self-expressiveness, and enjoyment. *The Journal of American Academy of Business, 19*(2), 286–93.

Tata, J. S., & McNamara, P. E. (2016). Social factors that influence use of ICT in agricultural extension in southern Africa. *Agriculture, 6*(2), 15. https://doi.org/10.3390/agriculture6020015

UN Migration Agency (2020). Displacement tracking matrix (DTM) COVID-19 regional overview on mobility restrictions. https://dtm.iom.int/reports/eastand-horn-africa-%E2%80%94-covid-19-regional-overview-mobility-restrictions-16-july-2020

Vision 2030. (2007). Vision 2030 (The Popular Version). http://www.vision2030.go.ke/wp-content/uploads/2018/05/Vision-2030-Popular-Version.pdf

Vygotsky, L. S. (1978). *Mind in society: The development of higher psychological processes.* Harvard University Press.

Wenger, E. (2001). Supporting communities of practice: How to make sense of this emerging market understand the potential of technology and set up a community platform [PDF]. https://guard.canberra.edu.au/opus/copyright_register/repository/53/153/01_03_CP_technology_survey_v3.pdf

Wyche, S., & Olson, J. (2018). Kenyan women's rural realities, mobile internet access, and "Africa Rising". *Information Technologies & International Development, 14*, 33–47.

Yeboah, J., & Ewur, G. D. (2014). The impact of WhatsApp messenger usage on students performance in tertiary institutions in Ghana. *Journal of Education and Practice, 5*(6), 157–64.

# 16 Conclusion

*Ataharul Chowdhury and Gordon A. Gow*

The concept of "participation" and "participatory development" has historically been promoted as a means of empowering marginalized communities (such as rural, agricultural, women, and other disadvantaged segments) in initiatives related to communication for development and community-based social change. These ideas have gained widespread acceptance in development and academic literature as essential mechanisms for enabling underprivileged individuals to challenge existing power structures, assert their rights, voice their concerns, and shape their own development objectives (Manyozo, 2012; Melkote & Steeves, 2015; Quarry & Ramírez, 2009; Servaes, 2018).

On the other hand, the development problem in agriculture and rural development has long been attributed to the information deficit model. This model suggests that there is a lack of science-based evidence, and public participation is viewed as a consultation and decision-making process informed by scientific and reliable information. While we have witnessed some progress by incorporating social, political, and cultural factors to shape public participation and communication, agricultural scientists and scientific communities need to make a more serious effort to understand the concerns and questions of the public on political, economic, and environmental governance (Broad & Biltekoff, 2022; Leeuwis & Aarts, 2021). Without progress beyond the deficit models, participation of agri-food and rural community actors will be limited, leading to growing incidences of misperceptions and mistrust among agri-food stakeholders (Broad & Biltekoff, 2022). This is particularly important in an age where agricultural development is embracing digital tools and platforms for networking, dialogues, knowledge sharing, and learning among stakeholders (Djanibekov et al., 2023; Fielke et al., 2020; James, 2023; Kabir et al., 2023; Kaushik et al., 2018; Klerkx, 2021; Rijswijk et al., 2019; Riley & Robertson, 2022; Sugihono et al., 2022) and emerging technologies (e.g., precision farming, artificial intelligence, robotics) for boosting production while promising to contribute information and knowledge to support environmental and other domains of sustainability (Ingram & Maye, 2023; Rijswijk et al., 2021).

This edited book tackles the issue of participation in the digital age by reflecting on, analyzing, and sharing experiences from practitioners and researchers from around the world. A central thread running through the chapters is that digitalization presents both opportunities and challenges for participation, practically and

DOI: 10.4324/9781003282075-19

theoretically. Following the COVID-19 pandemic and subsequent lockdowns, re-mote participation has become increasingly taken up for many development and livelihood activities, making it even more crucial to engage in this debate. The book's authors examine the innovative use of digital tools and methods before and during the pandemic, with the aim of providing new insights into how rural and agricultural communities can participate in the digital age. They focus on defining the significance of participation and identifying strategies for development and environmental change initiatives.

The book's chapters showcase a range of perspectives and experiences that of-fer a variety of viewpoints. They highlight how digital communications present challenges to some of the assumptions that have informed the long-held principles of participatory practices. While some chapters emphasize the positive impact that digital media and tools have on facilitating connections between agricultural and rural communities, their peers, organizations, and constituents, other chapters pre-sent examples that illustrate how digital media and platforms can undermine the 'non-negotiable' components of participation.

Although the critiques, experiences, and examples do not provide straightfor-ward solutions or recommendations for meaning and tactics, they offer intriguing viewpoints and lessons on how the very meaning of participation in the digital realm can be reconsidered. In this concluding chapter, the key themes and insights are synthesized from the previous chapters into a set of lessons and future direc-tions for research and action. These are set out in Figure 16.1, in a 'participation

*Figure 16.1* Participation triangle for the digital age.

triangle for the digital age.' The triangle represents three major lessons learned from this edited volume that provide a direction of research and understanding of the concept of digital participation and harness its benefits for marginalized communities. The lessons can be used to facilitate and support capacity development initiatives for agricultural and rural communities to participate in the digital age.

## Lesson 1: The 'non-negotiable' principles of participation are often at odds with prevailing business models of digital platforms

The initial two chapters of the book provide insights into participation and urge us to reconsider its definition, particularly in light of contradictions that have emerged in the digital era. Ramírez lays out a set of 'non-negotiable' principles that must be taken into account when defining participation in the digital age. One of the main principles is to explicitly and honestly state the type of participation that users can have in various phases of a project's initiative. However, these non-negotiables are often at odds with the business models that digital platforms usually adhere to. Reilly's chapter draws attention to this contradiction by pointing out, for example, that the general public doesn't have the required understanding of the algorithms that shape their interactions on commercial social media platforms. This chapter also brings up an important point about the need for participants to provide consent when using digital platforms for information search, idea sharing, and networking. When individuals participate online without knowing how their data is being used, they become part of a computer-mediated participatory process with limited opportunities to make informed decisions about it. Recent studies (e.g., Bronson, 2022; Bronson & Sengers, 2022; Ingram & Maye, 2023) have highlighted similar concerns, particularly with regard to the use of digital tools for knowledge sharing, participation, and decision-making by agri-food stakeholders, which include smallholder farmers and development agencies that provide them with agricultural and environmental services. The studies raised concerns that the data collected through the participation of the users on these platforms are utilized to fulfill the goals of investors rather than the community themselves.

Reilly argues that participation in digital platforms aims to regulate and discipline societal actors to achieve economic growth while ignoring the need for care and social development thereby undermining the 'non-negotiable' principles of participatory development. The digital age has also created new roles as 'data intermediaries' or 'digital moderators,' which engineer power sharing that aligns with the goal of accruing benefits in uneven or biased ways. Reilly argues that our rich human relations are reduced to the signaling of needs and receipt of services as we participate as nodes in these digital networks. The goal of fulfillment is defined as service delivery, rather than, say, producing a benefit for society or forming sustained relationships. Using the concept of logistics brokerage, Reilly argued that it generates efficiencies and conveniences. Additionally, there was also the expectation of immediacy. While producers and consumers coincide with each other in time and space for the brief time during which a service is rendered, they do not actually coincide with each other at the level of affective relationships.

This undermines the potential for the production of relations of care. Momentary connections are prioritized over solidarity, social support, or civil society.

It can be inferred that the values informing the design and use of commercial digital platforms do not fully align with the 'non-negotiable' elements of participation. For instance, Ramírez states that it is crucial to engage with various stakeholders and to comprehend their objectives as an essential element. However, Reilly argues that participation in a digitally networked society (Castells, 2015) implies that participants are involved in a layered network with varying goals, and the quantity of engagement becomes more significant than the quality of interactions. Leeuwis and Aarts (2021) explain that the early models of communication and participation in agricultural development were primarily focused on providing information about problems, potential solutions, and alternative options to change individual beliefs and attitudes. However, with the current trend of using digital tools and platforms in agricultural development, the goal is still to provide information about 'best practices' in a quantifiable manner, such as through digital repositories or portals, voice messages, text messages, video clips, and alerts that are directed toward individuals. The authors argue that digital tools should aim for interactive and sociologically enhanced communication and participation, rather than just influencing individual behaviors. These models should cater to the formation of collective identity, the design of institutions, the maintenance of trust, and the effective use of power and sanctioning systems in a community of actors.

How can we overcome the challenges of ensuring non-negotiable principles of participation? This is especially important as corporate and commercial platforms continue to enter the agri-food digital ecosystem (Birner et al., 2021; Wolfert, 2023). Other chapters in the book provide a glimpse into the ways digital tools were used in participatory development prior to and during the COVID-19 pandemic, suggesting that the answer to this question is complicated insofar as rural and agricultural communities have different ways of relating to digital media and commercial social media platforms.

## Lesson 2: Rural and agricultural communities have different ways of relating to digital media and technologies when participating in development initiatives

This book's chapters highlight that rural and agricultural communities have distinct social, cultural, and political structures. They necessitate a unique perspective to understand how they engage with and utilize digital media and technologies for participation in development initiatives. This also entails comprehending the social and cultural context of technology usage, devising leadership strategies, matching technology with livelihood needs, and creating strategies for integrating technology as a process of mediated and face-to-face interactions.

In their chapter, Lutomia et al. share their experience of using WhatsApp to support learning and capacity building of smallholder farmers in Kenya during the COVID-19 pandemic. They successfully facilitated access to technology, including video content, and helped farmers develop skills that helped them learn about agricultural practices. However, utilizing local leadership was critical to their success as it enabled them to maintain trust and monitor feedback among various stakeholders, such as international teams, local farmers, and agencies. They identified existing WhatsApp networks (local and international networks spanning Kenya, Nigeria, Ghana, and Bangladesh) and tapped into them, rather than creating new ones, and developed a virtual pull strategy to understand the problems and offer learning content that was responsive to local needs. The approach relied on trust-building in an intrinsic rather than coerced or leveraged manner, utilizing local language and culturally responsive means.

Their case study emphasizes the significance of having distributed leadership in the digital age. It is imperative for organizations to cultivate leadership that is spread out and not centralized in order to effectively support the participation of communities in the digital landscape. According to Larson and DeChurch (2020), digital teams require a shared and emergent leadership structure, which can take informal forms. Leaders in the digital age should consider how technology and behavior interact to create opportunities for leading and collaborating with teams. Lutomia et al. in this edited volume, found that maintaining prior relationships with project stakeholders, managing time differences, and providing regular communication and feedback were crucial in managing communication and learning in WhatsApp-mediated stakeholder networks. Giving feedback can enable capacity development and foster a culture of team building and recognition of good work, which can be propagated through these networks.

In their chapter, McMahon et al. examine the adoption of digital tools, such as Facebook, YouTube, and Instagram, as part of their community-engaged scholarship project called DigitalNWT. The project team had to adopt digital tools during COVID-19 to continue working with the Northern and remote Indigenous communities in Canada. The authors note findings similar to those of Lutomia et al. and found that effective digital engagement required a combination of expertise from partners based in participating communities and universities. They observe that remote, digitally mediated engagements were most effective when accompanied by opportunities for team members to meet and interact in person. The authors suggest that a successful digitally mediated community engagement requires a combination of intensive in-person interaction during training and initial contact, along with regular periods of remote engagement through email, text messaging, telephone calls, and videoconferencing. The chapter shows that digitally mediated research, education, and communication processes can be adapted effectively to engage with remotely located community participants using digital tools, even in unforeseen and difficult circumstances, such as the COVID-19 pandemic, if relations of trust and reciprocity are established.

Bakardjieva et al., in this edited volume, discuss three studies on the use of the Internet by rural people for livelihoods communication and participation in Canada, Chile, and Vietnam. Although the studies were conducted before the COVID-19 pandemic, they offer an interesting lens into how to analyze and understand when rural people participate and communicate using digital tools. The authors categorized the use of the Internet by rural communities into three main 'use genres,' namely Internet-powered business, fostering new-Internet lifestyles, and Internet-augmented education. The studies found that rural people have different ways of relating to technology, and effective use of digital tools will depend on the perceived usefulness of those tools to their lifeworld and livelihood needs. The authors argue that rural areas do not necessarily follow the same path as urban communities when it comes to adopting digital tools for participation and communication. This is in line with other recent studies, such as Dilleen et al. (2023), Higgins et al. (2023), and Rijswijk et al. (2021), that evidenced agricultural stakeholders, especially farmers, deploy digital tools to support their livelihoods and communication needs in a heterogeneous manner, involving peers, advisors, and technology providers, across diverse socio-material relationships with varying levels of trust across different platforms in relation to development goals as well as their own knowledge, practices, and priorities.

A key takeaway from these chapters is that digital participation in rural areas will require a dialogue with the inhabitants and attention to their social practices as they encounter and explore the possibilities of digital technologies in different situations. It is important to help rural inhabitants understand and make sense of the benefits of new technologies for education, better job opportunities, and business growth, etc. However, the adoption of these technologies is not straightforward when it comes to being respectful of local resistance from community members, such as parents and elders who may fear that young people will lose traditional values and knowledge when they are exposed to modern culture and values through digital platforms. Thus, it is essential to be aware of and to be prepared to negotiate a complex set of responses to digital technology in participatory development. Above all, the chapters point to the enduring importance of fostering local leadership and maintaining relationships of trust as a foundation for any type of participatory project.

### Lesson 3: Monitor and assess the emergent and unexpected outcomes of digital technology in participatory development

In the opening section of the book, Ramírez suggested that we need to appreciate the interactions between new media and different stakeholders as a complex adaptive system that requires ongoing monitoring and course correction. The case studies of three Internet genres, the WhatsApp network and DigitalNWT demonstrate that adapting digital media to the specific context of local realities is essential for successful appropriation of the platform for rural people's participation and communication. Similarly, Pavez Andonaegui analyzed how mobile devices fulfilled educational needs and empowered children during pandemic-imposed school closures in rural Chile. The case suggests that mothers and students had different perceptions about the affordances of smartphones. While mothers considered the device as a means for social connections and entertainment, their perceptions reversed after using smartphones for continuing children's education during COVID-19. Students felt empowered by using mobile phones but viewed such devices as useful mainly for entertainment and social media. The development of students' digital skills was supported by other family members, such as older siblings or cousins.

Ramírez also reminds us that participating in digital realms can lead to emergent and unexpected outcomes. In Chapter 5, Gow criticizes the use of corporate social media in agricultural extension since they often embrace the logic of corporate goals and profits. Gow argues that these practices create lock-in or path dependencies, meaning that they reproduce existing institutions and reinforce long-term path dependencies for digital technology use by agricultural stakeholders. To overcome these challenges, Gow suggests diversifying the existing digital landscape is necessary to create spaces for inclusive participation. Similarly, Abdulai's chapter highlights the growing dependency of digital advisory services in the Global South on commercial platforms. While solving some immediate problems, this situation also underscores a number of enduring concerns that continue to hinder farmers' participation and engagement in innovations. These include the affordability cost of digital services, limited access to the Internet, cultural barriers, and limited literacy capabilities.

In Chapter 4, Kabir et al. discuss the current state of agricultural advisory services and how they have been transformed by the introduction of digital tools and platforms. While they recognize the benefits of social and collaborative media in supporting communication and innovation processes, they also offer several critical perspectives. They point out that post-COVID-19 trends in digitalized advisory services, farmers and the public will face major challenges of exposure to misleading information. They highlight the growing problem of 'information disorder,' which refers to the abundance of credible and false information on digital platforms, often driven by economic interests, propaganda, and the power agenda of various stakeholders. While acknowledging that the challenges of maintaining information integrity are historical, the rapid advancement of agricultural technologies and media has led to ever-growing controversies and polarization on topics related to crops, fish, and livestock.

The authors discuss the impact of social media platforms and search engines on the accessibility and visibility of agricultural information. These platforms use algorithms to determine the ranking of information, but the lack of transparency and understanding of how these algorithms can lead to misinformation about farming practices, food safety, and environmental impacts. This can distort users' understanding of these topics. Additionally, the authors identify socio-cultural factors, such as confirmation bias, lack of scientific literacy, and political ideologies, as confounding factors that can affect the accuracy of information in the digital sphere.

In their chapter, Ramjattan et al. analyze the perception of 165 members of virtual agri-food communities of practices in Trinidad and Tobago. The study aims to understand how misinformation influenced their participation online and whether it affected their decision-making processes. The findings indicate that the majority of respondents reported the circulation of misinformation through Facebook and WhatsApp. The most common topics of agri-food misinformation were related to GMOs, the use of chemicals and environmental problems, food and nutrition, and climate change. Respondents reported that misinformation can potentially erode trust, and willingness to share information and incite conflicts among members. They were concerned about the amount of misinformation they were exposed to due to their membership in social media groups, individuals sharing materials that support their ideology or opinion, or following what family members posted, which can be misleading. These concerns reflect the difficulties in promoting inclusive participation, as people join groups of like-minded individuals to avoid information that contradicts their socio-technical and cultural repertoires.

Hauck provides a contrasting political perspective, highlighting the experience of Heyang villagers in China, where collective development and social goals are determined by the concept of mass line. The mechanisms entailing the concept enable the Communist Party and various classes of people to learn from each other by sharing experiences and developing goals together to address shared issues. Hauck explains how the local residents felt marginalized due to the market-oriented development approach in the village. They missed the mass line, which allowed them to define projects and realize developmental efforts. However, during the COVID-19 pandemic, the mass line was revived in response by the Chinese Communist Party. This included using technology to stay informed about the situation and mobilizing the masses to tackle the issue collectively. Hauck argues that technology has enabled rural people in China to fulfill the mass line's communicative function of learning "from the people" and propagating knowledge "to the people." Reflecting on the contemporary development experiences in Heyang and the possibilities of digital technologies to support participation, Hauck provides a cautious approach, highlighting that technology must enable people to achieve their goals through mutually dependent participatory action that may require reviving older practices in response to emergent or unexpected situations.

### Capacity development for digital participation: future directions for facilitation and harnessing the benefits in a post-COVID-19 age

The findings discussed in this edited volume indicate that respecting the 'non-negotiable' principles of participation requires us to move away from the current trend of applications of digital tools for participation as a mechanical process of transferring data and information. Instead, there is a need to adopt a multi-faceted approach that embraces socio-economic, cultural, and livelihood-oriented approaches. This is particularly important for commercial platforms, as highlighted by several chapters.

The book's chapters focus on different areas of skills and capacities that are necessary to ensure safe and inclusive participation, particularly in supporting essential principles. Digital literacy skills remain a fundamental requirement for digital participation. For example, Abdulai highlights that in countries like Ghana, farmers still lack the basic digital skills required to harness the benefits of growing digital advisories. Masambuka-Kanchewa and Rodriguez highlighted that the design of digital tools and services often neglects the context of marginalized users, especially when it comes to developing content and strengthening the relevance and reach of digitally mediated communication. The success of DigitNWT, described by McMahon in this book, however, illustrates how a thoughtfully managed process of digital skills development can be integrated into participatory projects with Northern communities in Canada.

Lawry and Manyozo look to the next generation of community-engaged professionals and point out that the curriculum of higher education should be geared toward imparting capacities among the next generation of C4D and CfSc scholars to develop a critical understanding of digitally mediated community engagement. This will be essential if they are to help people influence decision-making on issues that affect their lives. The curriculum of this order will entail the integration of theory-informed critical thinking abilities (e.g., critical analysis and reflection, power and participation, community-engaged governance, diversity, and social movement) and practical skills in responsible digital engagement (e.g., skills to understand and utilize emerging digital tools responsively).

Kabir et al. and Ramjattan et al. emphasize the need to move beyond basic digital skills and focus on enhancing the critical information literacy of farmers and community members. This is particularly important due to the growing risks of misinformation and emerging information disorders affecting participation in the agricultural and rural development sector. This is important from the perspective of policy formulation (Wijerathna-Yapa et al., 2023), professional practice (Diekman et al., 2023), advising farmers and rural communities (Chowdhury et al., 2023, 2024; Klerkx, 2021), and rural education systems (e.g., Seto, 2022). Critical digital literacy involves assessing communication content, examining the intentions behind the use of digital tools to generate and spread this information, and evaluating how this information can benefit—or harm—the livelihoods of farmers and rural community members (Cunliffe-Jones et al., 2021; Gow et al., 2023; Radovanović, 2023). The suggestions complement the idea that rural and agricultural communities also need informational capabilities to manage inherent tensions between data-driven knowledge and information facilitated by digital tools, and their tacit and intuitive understanding of agro-ecological context (Tzachor et al., 2023; van der Velden et al., 2023). Other chapters offer forward-thinking approaches to consider. For example, Gow et al. recount their experience of implementing the technology stewardship approach in the Global South, which focuses on community-level leadership that incorporates micro-level innovation with digital technology. The approach embraces an action learning methodology to promote participation in technology decisions at the community level.

To overcome the challenges of misinformation, Kabir et al., in this edited volume, suggest a power, profit, and propaganda (PPP) framework—an innovative means to help scholars and practitioners in the fields of Communication for Development (C4D) and Communication for Social Change (CfSC) to understand how power dynamics, economic incentives, and deliberate propaganda within the industry influence the flow of information, the narratives that emerge, and the decision-making processes. The framework takes into account various contextual factors such as social and online platforms, limited scientific literacy, confirmation bias, political interests, media landscape, and globalization. All of these elements contribute to information disorders. The authors emphasize the need to prioritize scientific literacy, critical thinking, and diverse perspectives.

Masambuka-Kanchewa and Rodriguez propose a series of steps that combine Rogers' Diffusion of Innovation theory and Media Richness Theory to co-create and co-design ICT messaging. The aim of these steps is to enhance the potential of messaging in digital platforms to reach and fulfill the needs of the most vulnerable individuals. The authors also recognize the importance of other enabling factors, such as infrastructure access to digital services and institutional capacities to respond to the needs of farmers, as the determining factors for the success of their proposed approach.

Lawry and Manyozo propose that there are differences between digital engagement (DE) and community engagement (CE). While DE is about reaching out, CE in the digital context needs to be nuanced as it entails thinking of how to execute CE using various digital tools. According to the authors, certain CE strategies are not transferable to the digital realm but can be negotiated to suit it, which might involve adjusting goals such as inclusion, interaction, and participation.

In conclusion, it is hoped that the aforementioned insights shared regarding facilitation and capacity-development strategies and techniques will be helpful for those who are utilizing the advantages of digital development as the reader can witness how digital technology played a crucial role in supporting participatory development during the global pandemic. It also revealed the limits of this technology in relation to the non-negotiable principles of projects that aspire to authentic and inclusive participation with agricultural and rural communities. It is the hope of the editors that the collection of contributions gathered in this volume shed light on the inherent challenges of integrating digital platforms and technologies into participatory practices, while also pointing to shedding some light on a pathway forward that involves a combination of capacity development and carefully considered deployment of technology in the service of inclusive development.

Looking ahead, it is expected that post-pandemic agricultural advisory, C4D and CfsC services, both in low and high-income regions, will see increased adoption of artificial intelligence (AI) powered, text-based, and decision-support systems, as well as various precision farming technologies (Dilleen et al., 2023; Ingram & Maye, 2023; James, 2023; Rijswijk et al., 2023; van der Velden et al., 2023; Wolfert et al., 2023). As the use of AI-powered digital platforms in rural and agricultural contexts is explored, the lessons learned in this edited volume and in recent years will guide our efforts going forward.

## References

Birner, R., Daum, T., & Pray, C. (2021). Who drives the digital revolution in agriculture? A review of supply-side trends, players and challenges. *Applied Economic Perspectives and Policy*, *43*(4), 1260–85. https://doi.org/10.1002/aepp.13145

Broad, G. M., & Biltekoff, C. (2022). Food system innovations, science communication, and deficit model 2.0: Implications for cellular agriculture. *Environmental Communication*, *17*(8), 868–74. https://doi.org/10.1080/17524032.2022.2067205

Bronson, K. (2022). *The Immaculate Conception of data: Agribusiness, activists, and their shared politics of the future.* McGill-Queen's University Press.

Bronson, K., & Sengers, P. (2022). Big tech meets big ag: Diversifying epistemologies of data and power. *Science as Culture, 31*(1), 15–28. https://doi.org/10.1080/09505431.2021.1986692

Castells, M. (2015). *Networks of outrage and hope: Social movements in the Internet age.* Wiley.

Chowdhury, A., Kabir, K. H., Abdulai, A.-R., & Alam, M. F. (2023). Systematic review of misinformation in social and online media for the development of an analytical framework for agri-food sector. *Sustainability, 15*(6), 4753. https://doi.org/10.3390/su15064753

Chowdhury, A., Kabir, K. H., Asafo-Agyei, E., & Abdulai, A.-R. (2024). Participatory and community-based approach in combating agri-food misinformation: A scoping review. *Advancement in Agricultural Development. 5*(2), 81–104. https://doi.org/10.37433/aad.v5i2.349

Cunliffe-Jones, P., Diagne, A., Finlay, A., Gaye, S., Gichunge, W., Onumah, C., Pretorius, C., & Schiffrin, A. (2021). *Misinformation policy in Sub-Saharan Africa: From laws and regulations to media literacy.* University of Westminster Press. https://doi.org/https://doi.org/10.16997/book53

Diekman, C., Ryan, C. D., & Oliver, T. L. (2023). Misinformation and disinformation in food science and nutrition: Impact on practice. *The Journal of Nutrition, 153*(1), 3–9. https://doi.org/10.1016/j.tjnut.2022.10.001

Dilleen, G., Claffey, E., Foley, A., & Doolin, K. (2023). Investigating knowledge dissemination and social media use in the farming network to build trust in smart farming technology adoption. *Journal of Business & Industrial Marketing, 38*(8), 1754–65. https://doi.org/10.1108/jbim-01-2022-0060

Djanibekov, N., Kurbanov, Z., Tadjiev, A., Govind, A., & Akramkhanov, A. (2023). *Farmers' social media groups for better extension and advisory services.* [Policy brief, no. 46, June 2023.] Leibniz Institute of Agricultural Development in Transition Economies.

Fielke, S., Taylor, B., & Jakku, E. (2020). Digitalisation of agricultural knowledge and advice networks: A state-of-the-art review. *Agricultural Systems, 180*(August 2019), 102763. https://doi.org/10.1016/j.agsy.2019.102763

Gow, G. A., Dissanayeke, U., Chowdhury, A., & Ramjattan, J. (2023). Digital literacy and agricultural extension in the Global South. In D. Radovanović (Ed.), *Digital literacy and inclusion: Stories, platforms, communities* (pp. 129–44). Springer International Publishing. https://doi.org/10.1007/978-3-031-30808-6_9

Higgins, V., van der Velden, D., Bechtet, N., Bryant, M., Battersby, J., Belle, M., & Klerkx, L. (2023). Deliberative assembling: Tinkering and farmer agency in precision agriculture implementation. *Journal of Rural Studies, 100,* 103023. https://doi.org/10.1016/j.jrurstud.2023.103023

Ingram, J., & Maye, D. (2023). "How can we?" the need to direct research in digital agriculture towards capacities. *Journal of Rural Studies, 100,* 103003. https://doi.org/10.1016/j.jrurstud.2023.03.011

James, J. (2023). *Recent trends in contemporary digital rural advisory services, Global Rural Advisory Services.* https://www.g-fras.org/en/gfras/1126-recent-trends-in-contemporary-digital-rural-advisory-services.html

Kabir, K. H., Rahman, S., Hasan, M. M., Chowdhury, A., & Gow, G. (2023). Facebook for digital agricultural extension services: The case of rooftop gardeners in Bangladesh. *Smart Agricultural Technology, 6,* 100338. https://doi.org/10.1016/j.atech.2023.100338

Kaushik, P., Chowdhury, A., Hambly Odame, H., & van Passen, A. (2018). Social media for enhancing stakeholders' innovation networks in Ontario, Canada. *Journal of Agricultural & Food Information, 19*(4), 331–53. https://doi.org/10.1080/10496505.2018.1430579

Klerkx, L. (2021). Digital and virtual spaces as sites of extension and advisory services research: Social media, gaming, and digitally integrated and augmented advice. *Journal of Agricultural Education and Extension, 27*(3), 277–86. https://doi.org/10.1080/13892 24X.2021.1934998

Larson, L., & DeChurch, L. (2020). Leading teams in the digital age: Four perspectives on technology and what they mean for leading teams. *The Leadership Quarterly, 31*(1), 101377. https://doi.org/10.1016/j.leaqua.2019.101377

Leeuwis, C., & Aarts, N. (2021). Rethinking adoption and diffusion as a collective social process: Towards an interactional perspective. In H. Campos (Ed.), *The innovation revolution in agriculture: A roadmap to value creation* (pp. 95–16). Springer International Publishing. https://doi.org/10.1007/978-3-030-50991-0_4

Manyozo, L. (2012). *Media, communication and development: Three approaches.* Sage Publications.

Melkote, S. R., & Steeves, H. L. (2015). *Communication for development: Theory and practice for empowerment and social justice.* Sage Publications.

Quarry, W., & Ramírez, R. (2009). *Communication for another development: Listening before telling.* Zed Books.

Radovanović, D. (Ed.) (2023). *Digital literacy and inclusion: Stories, platforms and communities.* Springer.

Rijswijk, K., de Vries, J. R., Klerkx, L., & Turner, J. A. (2023). The enabling and constraining connections between trust and digitalisation in incumbent value chains. *Technological Forecasting and Social Change, 186*, Part A, 122175. https://doi.org/10.1016/j. techfore.2022.122175

Rijswijk, K., Klerkx, L., Bacco, M., Bartolini, F., Bulten, E., Debruyne, L., Dessein, J., Scotti, I., & Brunori, G. (2021). Digital transformation of agriculture and rural areas: A socio-cyber-physical system framework to support responsibilisation. *Journal of Rural Studies, 85*, 79–90. https://doi.org/10.1016/J.JRURSTUD.2021.05.003

Rijswijk, K., Klerkx, L., & Turner, J. A. (2019). Digitalisation in the New Zealand agricultural knowledge and innovation system: Initial understandings and emerging organisational responses to digital agriculture. *NJAS - Wageningen Journal of Life Sciences, 90–91*, 100313. https://doi.org/10.1016/j.njas.2019.100313

Riley, M., & Robertson, B. (2022). The virtual good farmer: Farmers' use of social media and the (re)presentation of "good farming". *Sociologia Ruralis, 62*(3), 437–58. https:// doi.org/10.1111/soru.12390

Servaes, J. (2018). *Handbook of communication for development and social change.* Springer.

Seto, A. (2022). Hallway pedagogy and resources loss: Countering fake news in rural Canadian schools. In K. R. Foster & J. Jarman (Eds.), *The right to be rural* (pp. 51–68). University of Alberta Press.

Sugihono, C., Juniarti, H. A., & Nugroho, N. C. (2022). Digital transformation in the agriculture sector: Exploring the shifting role of extension workers. *STI Policy and Management Journal, 7*(2). https://doi.org/10.14203/stipm.2022.350

Tzachor, A., Devare, M., Richards, C., Pypers, P., Ghosh, A., Koo, J., Johal, S., & King, B. (2023). Large language models and agricultural extension services. *Nature Food, 4*, 941–48. https://doi.org/10.1038/s43016-023-00867-x

van der Velden, D., Klerkx, L., Dessein, J., & Debruyne, L. (2023). Cyborg farmers: Embodied understandings of precision agriculture. *Sociologia Ruralis*. https://doi.org/10.1111/soru.12456

Wijerathna-Yapa, A., Henry, R. J., Dunn, M., & Beveridge, C. A. (2023). Science and opinion in decision making: A case study of the food security collapse in Sri Lanka. *Modern Agriculture*. https://doi.org/10.1002/moda.18

Wolfert, S., Verdouw, C., van Wassenaer, L., Dolfsma, W., & Klerkx, L. (2023). Digital innovation ecosystems in agri-food: Design principles and organizational framework. *Agricultural Systems*, *204*. https://doi.org/10.1016/j.agsy.2022.103558

# Index

For Product Safety Concerns and Information please contact our EU
representative GPSR@taylorandfrancis.com
Taylor & Francis Verlag GmbH, Kaufingerstraße 24, 80331 München, Germany

www.ingramcontent.com/pod-product-compliance
Lightning Source LLC
Chambersburg PA
CBHW052119230326
41598CB00080B/3892

9 781032 252094